T0139068

Underground Infrastructure of Urban Areas 2

Editors

Cezary Madryas, Beata Nienartowicz & Arkadiusz Szot
Wrocław University of Technology, Wrocław, Poland

CRC Press
Taylor & Francis Group
Boca Raton London New York Leiden

CRC Press is an imprint of the
Taylor & Francis Group, an **informa** business

A BALKEMA BOOK

CRC Press/Balkema is an imprint of the Taylor & Francis Group, an informa business

© 2012 Taylor & Francis Group, London, UK

Paper: Top of rock investigations for secant piles at the Bronx shaft,
© J. Jakubowski & J.B. Stypułkowski

Typeset by MPS Limited, a Macmillan Company, Chennai, India

Printed and bound in Great Britain by TJ International Ltd, Padstow, Cornwall

Published by: CRC Press/Balkema
                P.O. Box 447, 2300 AK Leiden, The Netherlands
                e-mail: Pub.NL@taylorandfrancis.com
                www.crcpress.com – www.taylorandfrancis.co.uk – www.balkema.nl

ISBN: 978-0-415-68394-4 (Hbk+CD-ROM)
ISBN: 978-0-203-12820-6 (ebook)

Underground Infrastructure of Urban Areas 2 – Madryas, Nienartowicz & Szot (eds)
© 2012 Taylor & Francis Group, London, ISBN 978-0-415-68394-4

Table of Contents

Underground Infrastructure of Urban Areas 2 – Madryas, Nienartowicz & Szot (eds)
© 2012 Taylor & Francis Group, London, ISBN 978-0-415-68394-4

Preface

Research in the underground infrastructure of cities should be primarily devoted to studies aimed at developing integration models of underground transportation systems with network systems for media transportation in the newly developed underground space. This requires primarily the identification and description of the technical condition of the underground infrastructure in cities. Failure to undertake or delay in research work – both theoretical and in laboratories and on existing structures – will prevent improvement of the standard of living and sustainable city development which is closely related to improved quality of freight and passenger transportation.

In order to meet contemporary expectations, it is necessary to build transportation tunnels, underground parking lots, underground rail in cities which results in a need to construct, reconstruct and modernise many elements of network infrastructure. The scale of the problem concerning adjustment of the network infrastructure to the needs specified above is confirmed with the statistics concerning the construction of the first underground line in Warsaw of 21 km which was commissioned on 25 October 2008 after almost 25 years of construction work. The completion of the task required reconstruction of about 17 km of sewage and water supply networks, 16 km of gas pipelines and 6.3 km of heat distribution networks plus lots of power supply and telecommunications cables. The situation is similar in other large cities, not only in Poland. As the discussed example shows, the scale of the issue is gigantic and adjustment of cities to the expectations of their inhabitants with respect to living standards requires enormous intensification of efforts not only in terms of science (theoretical) but also practical, concerning modernised operation programs of underground infrastructure facilities and modern designs with respect to modernisation and expansion thereof.

Special attention should be focused on the strategy to adjust the sewage network to municipal needs which is due to the fact that combined sewers in the structure – contrary to other pipelines – very often have very large diameters, often similar to the diameter of underground transportation tunnels. Further, it is a system that occupies the largest part of underground structures in cities (without a developed underground rail system) if we include inspection chambers, pumping stations, holding tanks and other facilities. Therefore, in order to adjust cities to intensified underground developments it is necessary first to undertake research work to improve the processes of operation, modernisation and potential reconstruction of sewage systems. Without such efforts, sustainable development of urban agglomerations will not be possible.

The monography is a compilation of work by many authors who presented the results of their analytical work of many years concerning designing, construction and operation of underground facilities. Many authors also identified the development directions required for an optimum use of underground space in the studied cities. Thanking the authors for their intellectual contribution to the monography, I trust that it will be of interest to people interested in underground infrastructure in urban agglomerations and will be helpful to their work.

Main editor

Cezary Madryas

Underground Infrastructure of Urban Areas 2 – Madryas, Nienartowicz & Szot (eds)
© 2012 Taylor & Francis Group, London, ISBN 978-0-415-68394-4

Reviewers

Han ADMIRAAL, President of Dutch Group ITA-AITES (NL)
Claude BERENGUIER, Executive Council ITA-AITES (CH)
Rolf BIELECKI, President EFUC (D)
Bert BOSSELER, Wissenschaftlicher Leiter des IKT (D)
Sorin CALINESCU, President of Romanian Group ITA (RO)
Józef DZIOPAK, Rzeszów University of Technology (PL)
Bernhard FALTER, University of Applied Science-Münster (D)
Piergiorgio GRASSO, Vice-President of ITA-AITES, President of GEODATA (I)
Alfred HAACK, Past President of STUVA (D)
Jens HÖLTERHOFF, President of GSTT (D)
Ivan HRADINA, President of Czech Group ITA (CZ)
Pal KOSCONAYA, Executive Council ITA-AITES (H)
Andrzej KULICZKOWSKI, President of PFTB (PL)
Marian KWIETNIEWSKI, Warsaw University of Technology (PL)
In-Mo LEE, President of ITA-AITES (Korea)
Dariusz ŁYDŻBA, Wrocław University of Technology (PL)
Cezary MADRYAS, Wrocław University of Technology, President of PSTB (PL)
Dietmar MÖLLER, Universität Hamburg (D)
Harvey PARKER, ITA-AITES (USA)
Anna POLAK, University of Waterloo (CAN)
Chris ROGERS, University of Birmingham (UK)
Anna SIEMIŃSKA-LEWANDOWSKA, President of Polish Group ITA-AITES (PL)
Ray STERLING, Louisiana Tech. University (USA)
Roland W. WANIEK, President of IKT (D)
Adam WYSOKOWSKI, University of Zielona Góra (PL)
Jian ZHAO, Ecole Polytechnique Federalne de Lausanne (CH)

Underground Infrastructure of Urban Areas 2 – Madryas, Nienartowicz & Szot (eds)
© 2012 Taylor & Francis Group, London, ISBN 978-0-415-68394-4

Sponsors

PLATINIUM SPONSORS

HERRENKNECHT AG

HOBAS System
Polska Sp. z o.o

GOLD SPONSORS

INFRA S.A.

SILVER SPONSORS

BETONSTAL

Grupa Górażdże

ZPB Kaczmarek

PGE Górnictwo i
Energetyka
Konwencjonalna SA
Oddział Elektrownia
Turów

ViaCon Polska
Sp. z o. o.

OTHER SPONSORS

Dolnośląska
Okręgowa Izba
Inżynierów
Budownictwa

KWH Pipe
(Poland) Sp. z o.o.

METRO WARSZAWSKIE
Sp. z o.o.

Underground Infrastructure of Urban Areas 2 – Madryas, Nienartowicz & Szot (eds)
© 2012 Taylor & Francis Group, London, ISBN 978-0-415-68394-4

Selected aspects of maintaining and repairing ferroconcrete wells and sewerage chambers

T. Abel

Wroclaw University of Technology, Wroclaw, Poland

ABSTRACT: The article concerns methods used for assessing technical condition of wells and chambers located in sewerage systems, as well as testing the quality of material and its usability for further use. The paper will include descriptions of computer analyses carried out by means of finite element method. Results of the analyses have been used to determine actual values associated with the loss of load capacity and increase in deflections of structures. Possible applicable solutions will be presented, depending on condition of an object, with particular consideration given to instances that require reconstruction of structural layer. Finally work stages will be described, consisting in the strengthening of structure, following examples of completed renovations.

1 INTRODUCTION

1.1 *Elements of underground networks*

Line objects, such as tunnels, collectors and transmission networks transporting various fluids are basic elements of underground transport infrastructure enabling the functioning of large arteries and industrial areas. In order to ensure continuous operation of these objects, they must be maintained in good technical condition protecting them against sudden breakdowns. Due to location of all types of pipelines their repairs involve a lot of technical complications and social impediments. In majority of cases networks are located under traffic routes and on intensely urbanized and industrialized areas, which prevents or considerably limits free access to those objects by using conventional opencast techniques. Both in case of repairs and replacements of pipelines the use of trenchless technologies is necessary, which is a positive phenomenon due to the investment costs and social costs associated with hindrances caused by earth works, which are an essential element of each construction using conventional repair methods.

Integral parts of underground networks are objects used in running maintenance work and ensuring access to individual pipeline sections. Such elements are all types of inspection chambers, decompression chambers and wells. These objects are essential elements of network and should also be subject to systematic inspections and maintenance, which, depending on their condition, can be carried out as a surface repair, or in case of weakened load bearing capacity – as a reinforcement and reconstruction of structural layer. Structures such as chambers and wells in case of pipelines with small diameters are made of prefabricated elements, which is a very common solution. Other widely used materials are plastics, used in production of system wells mounted at the stage of system construction. In case of large-diameter pipelines their chambers and wells most often are made as ferroconcrete elements made according to individual designs. Quite often brick chambers can be found. Ferroconcrete structures are a very good solution, due to their freedom of geometry shaping and very good strength-related parameters of the material. These objects, like the entire underground network, require repair works, because, due to very unfavourable environmental conditions, they undergo fast destructive processes. The study will present factors influencing the worsening technical condition of structures forming underground networks. Most frequently conducted field and laboratory tests will be characterized, which enable obtaining information on the actual technical condition of a technical object and selecting proper repair technology. The subject area will concern ferroconcrete chambers and wells, due to loads acting on these structures and lack of possibility of replacing completely their elements as it is in case of system wells built of other materials.

1.2 *Structure degradation process*

Operating environment of a/m objects, due to transported fluids such as domestic or industrial sewage, is very aggressive and causes many damages and creeping degradation of a structure. Corrosion is the basic threat; it is a destructive influence of the environment on the material, which causes decrease in its parameters and usability. The most frequent phenomena adversely affecting the structure are (Czarnecki, 2002):

– frost corrosion,
– concrete pollution,
– loss of protective capabilities of cladding.

Environments with which the elements of wells and chambers are in contact with on the outside are ground, underground water and atmosphere, and inside they are in contact with sewage and polluted atmosphere (aggressive gases). Frost corrosion causes cracks in concrete, as a result of changes in water volume during freezing and resultant strong tensions. Concrete pollution occurs as a result of chemically aggressive substances penetrating its structure. These substances may react with ingredients of concrete, leading to its corrosion, which in turn causes reduction in its protective properties towards reinforcement.

The first stage of chemical corrosion comprises dissoluting and washing out dissolved components. Then chemical reactions occur between the concrete, steel and aggressive environment that lead to material degradation. The course of this process depends on the degree of sewage aggressiveness, ambient temperature as well as other factors and their mutual relations.

A layer of concrete which is a protection layer for the reinforcement, with time undergoes carbonatization because of adverse influence of carbon dioxide; a negative result of this process is a reduction in alkalinity of concrete, which leads to decrease in protective capabilities of concrete cladding up to total loss of protection for the reinforcement.

Other phenomenon adversely affecting ferroconcrete structures is cracking. This process occurs as a result of volumetric changes in hardening concrete, as well as environmental influences. In case of objects situated in networks transporting aggressive media it is particularly dangerous because it causes further cracks and their propagation. Primary reasons for cracks can be errors at the designing stage, incorrect selection of reinforcements, technological errors including improper curing of concrete and overloads during use exceeding admissible stresses in the concrete (Czarnecki, 2002).

1.3 *Synergy*

The discussed group of objects is exposed to simultaneous occurrence of phenomena described above, which causes significant increase in the pace of degradation processes. Results of joint influence are considerably greater than effects of single phenomena. In many cases isolated influences might be insufficient to cause corrosion and other damages, however simultaneous occurrence causes quick destruction. The pace of destruction processes depends on many factors, inter alia temperature, humidity and porosity of the concrete. The speed of degradation of structure depends on the slowest process. Example of degraded structure is shown in Figure No. 1. Stages of corrosion process are shown in Figure No. 2.

2 ASSESSMENT OF TECHNICAL CONDITION

2.1 *Tests carried out in objects*

The type and scope of tests to be carried out depend on technical condition of examined object. The methods used should enable obtaining sufficient knowledge for applying suitable repair technology. On-site visit carried out on the object along with visual assessment and inventory of damages enables determining necessary tests to be carried out in situ, as well as laboratory tests. In case of discussed

Figure 1. View of corroded inner surface of sewerage well (KAN-REM, 2010).

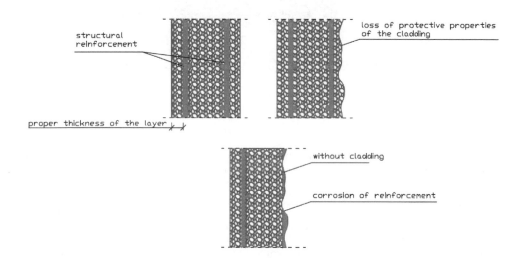

Figure 2. Corrosion progress stages in ferroconcrete structure.

group of objects, basic degrading factors described above include aggressive environment, which has indirect influence on boundary value of load capacity and usability.

An essential element of tests carried out on the object is the strength test and homogeneity test of the concrete. These parameters can be determined by means of non-destructive tests. Sclerometric method and ultrasonic method are used most frequently. Another group of tests includes semi-destructive tests, which consist in extracting blocks or anchors embedded in hardened concrete. The pull-out method is most frequently used. Also steel rings torn off from the concrete surface are

Figure 3.   Checking the neutralization degree of concrete (Grosel, 2009).

used – this is a pull-off method. Another semi-destructive method is the break-off method. The last group of methods that can be carried out are destructive methods; these are conducted on samples cut in the structure. In order to make an inventory of rods – their diameters should be measured and their spacing should be determined. This can be carried out on the basis of non-destructive electromagnetic tests, radar tests, radiographic and ultrasonic tests. Additional tests should be conducted, consisting in visual determination of the grade of steel in the openings made, on the basis of shape and ribbing of reinforcement bars, as well as conducting strength tests on samples collected from the structure and chemical tests of the steel (Drabiec, 2010). A very important group of tests that should be carried out in order to determine the condition of structure are chemical tests. These are aimed at determining the degree of carbonatization, concentration of chlorides and sulfates as well as alcalic reactivity of the aggregate. Field tests should include an assessment of neutralization process in sub-surface layer of the concrete. The essence of this method consists in spraying a solution of specially matched compound of chemical reagents identifying individual pH values ranging from 5 to 13.

   The pH reaction equal to 11, commonly deemed a limit value below which the natural ability of concrete to passivate the reinforcement is reduced, and corresponds to concrete colouring to violet. Colour transition from violet to green (pH = 9) indicates pH decrease below the value deemed a limit and potential corrosion hazard to the reinforcement (Grosel, 2009). Tests of neutralization degree of the concrete are illustrated in Figure No. 3.

Figure 4.   Device measuring the pull-off force (Grosel, 2009).

Tests of the tensile strength of concrete should be carried out in accordance with binding guide-lines (PN-EN 1542, 2000). Measurement may be made by means of a device that uses metal rings with a diameter of 50 mm. The essence of this method, generally speaking, is measurement of the force required to pull off a metal disc with a known surface area, glued to the tested surface. A centric notch 10–15 mm deep is made around the disc. The recorded value of the pull-off force, divided by the surface area onto which the load is transferred, gives as a result the value of the tensile strength of concrete, also called the peel strength. The tests should be carried out after precise removal of the weakened concrete layer. According to guidelines on the requirements for the concrete surface which condition the possibility of conducting modern surface repairs on it, e.g. in the form of PCC materials, the following conditions should be met (Grosel, 2009):

– average compressive strength of concrete should not be less than 25 MPa,
– average value of peel strength, determined in a given measurement location for all measurements, should not be less than 1.5 MPa,
– minimum value of a single measurement should not be less than 1.0 MPa.

The layer of concrete that meets the above parameters is classified as the proper one to make surface repairs. The course of the tests is shown in Figure No. 4.

Semidestructive tests carried out on objects are a very important source of information on the concrete. The pull-out method is one of research methods based on measuring the value of force needed for extracting a steel anchor from the concrete, the anchor varies according to adopted technical solution. The anchor used in the pull-out test may earlier be embedded in concrete or seated in a drilled opening. Specific destruction of the concrete as a result of anchor extraction enables defining strict correlation between the recorded pull-out force and strength of the concrete. This method, with few exceptions, is independent of the influence of the cement type and value of water/cement ratio, setting conditions, content of ingredients. The pull-out force is applied to the anchor through a weighting element, resting on the surface of the element to be tested. Most often such element is a steel ring with inner diameter D, located centrally above the anchor. As a result of the force action, a cement fragment is extracted with a shape of truncated cone and height "h", with the diameter of the greater base corresponding to resistance frame D and smaller diameter

Figure 5.   Example of semidestructive tests – capo test.

of a mandrel wedging the anchor in the concrete – "d" (Drabiec, 2009). The course of the test is shown in Figure No. 5.

The hardness test, which is a sclerometry, belongs to commonly used concrete quality control methods. According to a definition, hardness means material resistance to deformations caused by concentrated forces, therefore the methods are divided depending on the way of applying the pressure into static and dynamic methods, as well as depending on the method of measuring plastic strains – into imprint method and rebound method. In the imprint method the most frequently devices used are called Poldi improved hammers and HPS and in sclerometric methods using the magnitude of rebound – Schmidt sclerometers are most frequently applied (Drabiec, 2009).

2.2   *Numerical analyses of structures*

In order to properly select the repair method and determine the strength parameters of the layer to be repaired, as well as to establish the percentage increase in the value of load capacity of the structure – strength calculations are necessary. Analyses can be carried out by means of the finite element method, by means of specialist computer software (Abaqus, 2008). In case of objects qualified for repair a series of computer simulations were conducted, reflecting real operating conditions for a/m structures. In the first case a chamber was modelled, as an object without damages, subject to external loads. In the second case a model was adopted in the form of an object with narrowed thickness of ferroconcrete wall, which was to simulate cavities caused by corrosion, additionally an assumption was adopted of no internal reinforcement as an effect of degradation processes. Such state can be deemed an emergency condition that might lead to a construction disaster. The computer analyses carried out show that changes in the values of stresses as well as deformations occurring as a result of changes in their strength parameters may considerably influence the safety of the structure. Both the boundary state of load capacity and boundary state of usability are exceeded in case of the model with damages. Values obtained have shown an increase in the stresses in the structure within the range 80–100% depending on the diagram of the structure and average increase of 150% in deflections in relation to admissible values, and thus considerable excess of boundary usability. Results of the analyses carried out are illustrated in example charts of stresses and deflections shown in Figure No. 6 and 7. Calculations show the necessity of reconstructing the structural layer that will protect the object against the loss of stability and disaster.

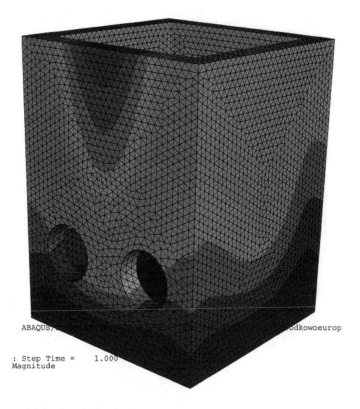

ABAQUS/   dkowoeurop

: Step Time =   1.000
Magnitude

Figure 6.   Charts of deflections in the chamber structure.

3   REPAIRS TO A STRUCTURE

3.1   *Technology*

The way of repairing a structure will be discussed quoting the example of two objects representing elements of gravitational underground networks. The first of them is a chamber in a sanitary sewerage system at the passage under a river. It is a ferroconcrete structure made in classic technology. Tests carried out have shown that the sub-surface layer of concrete in all examined fragments of the structure is characterized by acid reaction. The pH values found range from 5 to 9 (after removal of corroded concrete), on average pH value below 9.0 reaches the depth of approximately 10 to 15 mm, neutralization reaching 50 mm has been found locally. As a result of a very high aggressiveness of sewage the chamber surface has undergone corrosion. The layer of concrete cladding has been destructed and consequently the internal layer of structural reinforcement has undergone total degradation. The object condition is shown in Figure No. 8.

Preparation of the concrete surface is the first stage of each repair. Surfaces to be repaired are always polluted, damaged or neutralized. Surface preparation is necessary due to the need of ensuring further protection for reinforcements and obtaining a surface that can be appropriately connected with the corrective layer. Activities associated with preparation of the surface are very important, because they largely determine the quality of repair. Very important is the depth to which the concrete is removed, as well as replacement or remounting of reinforcement bars and protection of steel surface. Hydrodynamic cleaning removing a loose layer of corroded concrete is an initial stage of surface preparation. The next stage consists in mechanical loosening of concrete layer that has lost its strength and chemical properties. Depending on the reach of corrosion, it is very often necessary to remove internal reinforcement of the structure, which has also lost its

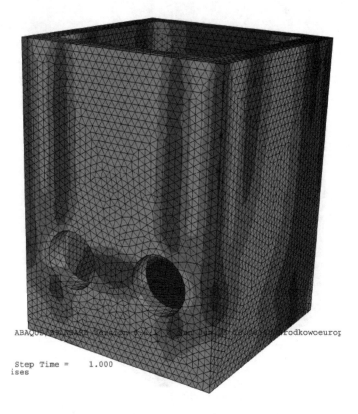

Figure 7.   Charts of deflections in the chamber structure.

Figure 8.   View of internal surface of the chamber after initial cleaning (KAN-REM, 2010).

Figure 9.   View of a new internal layer of structural reinforcement (KAN-REM, 2010).

Figure 10.   Mounting of polyethylene layer creating internal surface of the chamber (KAN-REM, 2010).

carrying properties. In case of good quality of reinforcing steel, its surface should be cleaned so as to obtain appropriate adhesion between the corrective material and material to be repaired. Due to total degradation of internal reinforcing layer, the next stage of work was the assembly of newly made reinforcing fabric, shown in Figure No. 9. In order to protect the newly made layer of concrete

Figure 11. The object after renovation, before putting into operation (KAN-REM, 2010).

Figure 12. Interior of the chamber after renovation, before putting into operation (KAN-REM, 2010).

against corrosion processes, additionally an insulating layer was made by mounting polyethylene plates. Polyethylene as a material with a very high chemical resistance guarantees full protection against aggressive media. Polyethylene plates were used, thus permanent adhesion of insulating layer and concrete was obtained, which is shown in Figure No. 10.

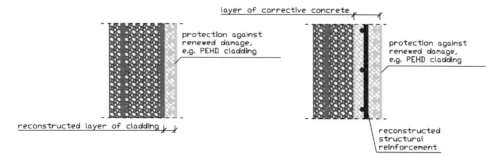

Figure 13.   Repair of ferroconcrete structure depending on the range of damages.

As a result of the repair – reconstruction of structural layer and raising its strength parameters to initial values have been obtained. Additionally the object is protected against renewed corrosion through mounting of PEHD cladding. The object after the repair is shown in Figure No. 11.

The second object subjected to structural repair was a decompression chamber on a sanitary sewerage system. In this case the stage of initial hydrodynamic cleaning also revealed the damage of concrete cladding and corrosion of internal structural reinforcement. Work stages proceeded similarly as in a/m case. The object is shown in Figure No. 12; Figure No. 13 illustrates the repair diagram.

3.2   Conclusions

As can be seen from the material presented above, objects such as chambers and wells located on sewerage systems are largely exposed to degradation processes occurring inside. Such situation may lead to very serious breakdowns and even damages to the structures and construction disasters. Correct diagnostics of a/m objects, as well as taking preventive actions protect network users against occurrence of sudden breakdowns. Appropriate diagnosis of technical condition of the structure enables making computer calculations and simulations providing information on boundary load capacity and usability, which is a necessary condition for ensuring correct operation of entire underground system. Vast variety of systems and corrective means available on the market enable very good selecting strength parameters dependent on technical condition of the structure. Thanks to such wide range of available products it is possible to optimize costs and apply materials with properties corresponding to conditions prevailing in the object.

REFERENCES

Abaqus – Workshop preliminary, 2008. Simulia. USA. Dassault Systemes.
Czarnecki L., Emmons P.H. 2002. Repairs and protection of concrete structures. Kraków. Polish Cement.
Drabiec L., Jasiński R., Piekarczyk A., 2010. Methodology, field- and laboratory tests of concrete and steel. Warsaw. Polish Scientific Publishers PWN.
Grosel J., Madryas C., Moczko A., Przybyła B., Wysocki L., Abel T. 2009. Technical expert evaluation about a sewer – The U series report no. 20/2008. Wrocław, Wrocław Univeristy of Technology.
KAN – REM – Technical & informative materials. 2010. Wrocław. KAN-REM.
PN-EN 1542. 2000. Products and systems for protecting and repairing concrete structures. Testing methods. Measuring adhesion by pull-off method.

Underground Infrastructure of Urban Areas 2 – Madryas, Nienartowicz & Szot (eds)
© 2012 Taylor & Francis Group, London, ISBN 978-0-415-68394-4

Concrete resistant to aggressive media using a composite cement CEM V/A (S-V) 32.5 R – LH

D. Dziuk & Z. Giergiczny
BETOTECH Technology Center, Dabrowa Gornicza, Poland
Silesian University of Technology, Faculty of Civil Engineering, Gliwice, Poland

M. Sokołowski & T. Pużak
BETOTECH Technology Center, Dabrowa Gornicza, Poland

ABSTRACT: The production of cements with mineral additives has lately acquired greater significance due to the implementation of CO_2 emission fees into atmosphere. The volume of cement clinker production ought to be executed considering granted limits. Thus, the increase of mineral additives share in cement content and the decrease of clinker ratio is the fundamental solution creating new ways of diminishing carbon dioxide emission by cement industry. Domestic industrial experience prove wider used of cements with mineral additives (CEM II; CEM III; CEM IV; CEM V). The production of such cements, despite significant reduction of production costs owing to the substitution of some Portland clinker part with mineral additives, gives binders with higher resistance to aggressive media actions.

Hereby article presents the properties of concrete made with composite cement CEM V/A (S-V) 32.5R – LH. Selected examples of applications complete the whole paper.

1 INTRODUCTION

Recently, one can observe the increase of the production of composite cements containing one or more mineral additives as essential ingredients of cement (CEM II CEM ÷ V). In European Union countries, such cement production reaches a level close to 70%. This trend, to a large extent, creates the introduction of fees for CO_2 emissions from production processes. Increasing share of non-clinker components in the cement composition is an effective solution providing opportunities to significant reduction of carbon dioxide emissions from cement industry. Standard EN 197-1:2002 lists eight non-clinker components of common use cements: granulated blast furnace slag (S), a natural pozzolana (P), a natural burnt pozzolana (Q), silica fly ash (V), calcareous fly ash (W) burnt shale (T), limestone (L, LL) and silica dust (D).

Mineral additives impart to cements (concretes) the properties, never achieved by the use of Portland cement CEM I (Ballim & Graham 2009, Chłądzyński & Garbacik 2008, Dinkar et al. 2007, Lothenbach et al. 2011).

The widespread accessibility of cement with additives (CEM II CEM ÷ V) and their properties lead to their increasing applications in construction and building works such as bridge elements, the construction of sewage treatment plants, foundations, road construction, precast production and the production of a new generation concretes (self compactive concretes SCC, high performance concretes BWW, fibre concretes). Cements with a high content of mineral additives (CEM II CEM ÷ V) are characterized by low heat of hydration (LH), free-to-moderate dynamics of strength increase, high compressive strength in longer terms and high resistance to chemical attack (Giergiczny et al. 2002, Temiz et al. 2007, Li et al. 2006). Fly ash (V) and granulated blast furnace slag (S) are the most common additives in Poland applied in cement composition. They equally may be a mixture of components, both in the composition of Portland composite cement, as well as in the composite cement CEM V/A, B (S-V).

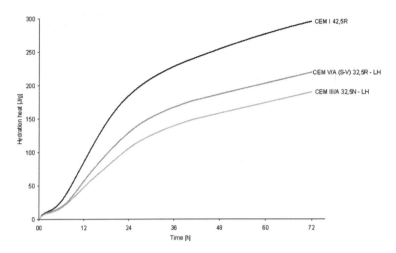

Figure 1. Hydration heat of cements.

Hereby paper presents the results of resistance to sulfate aggression of pastes made of composite cement CEM V/A (SV) 32.5 R – LH, and determines the rate of migration of chloride ions through concrete produced with this type of cement.

2 CHARACTERISTICS OF COMPOSITE CEMENT CEM V/A (S-V) 32.5R – LH

Analyzing the impact of the synergies of siliceous fly ash and blast furnace slag in the composition of cement composite CEM V, it was confirmed, that the mechanical and physical properties of cements containing two non-clinker components are better than cements containing one non-clinker component. Such cements providing the optimum proportions of the ingredients bring mortars and concrete with a great rheological mixture characteristics and good mechanical properties with also high resistance to aggressive media. The study and practice of manufacturing composite cements CEM V show, that the optimal performance are guaranteed by a similar share of mineral additives – slag S and fly ash V in cement composition. This ratio provides the maximum synergistic effect of fly ash and slag impact on cement properties (Giergiczny et al. 2002, Lothenbach et al. 2011, Temiz et al. 2007).

Composite cement CEM V/A (SV) 32.5 R – LH has a lower heat of hydration than Portland cements CEM I and composite Portland cements CEM II. (Wang & Lee 2010, Chen & Brouwers 2007, Ballim & Graham 2009.) It is classified in accordance with the standard EN 197-1: 2002/1:2005 as a low hydration heat cement (LH) (Puzak et al. 2011). The curve of heat release during the hydration of composite cement CEM V/A 32.5 R-LH, in relation to other cements is shown in Figure 1. This type of cement has low heat of hydration and is preferred to the production of massive concrete (hydraulic engineering, foundations, and containers for sewage treatment plants).

Initial setting time as well as the mechanical properties of cement containing the mixture of granulated blast furnace slag and fly ash depends on the mutual proportion of components (Tab. 1).

Out of these characteristics the one which should emphasized are the comparable properties of composite cement CEM V in relation to slag cement CEM III/A 32.5N – LH/HSR/NA and Portland composite cement CEM II/B-M (S-V) 32.5R, as shown in Figure 2. With this similar water demand, composite cement CEM V forms the consistency and workability of concrete mixture. Concrete does not show any tendencies to give the water back to the surface (very small bleeding) (Puzak et al. 2011).

A relatively low compressive strength in the initial period of hardening (after 2 days) is the characteristic feature of composite cement CEM V/A 32.5 R-LH. However, after a long period of

Table 1. Properties of composite cements made in laboratory

Cement name	Component content [%] Fly ash (V)	Slag (S)	Setting time – Initial [min]	Compressive strength [MPa] after 2 days	7 days	28 days	90 days	180 days
CEM II/B-M	20	15	215	15.0	28.8	46.1	62.7	68.8
	15	20	235	16.6	29.9	48.8	62.3	67.8
	15	15	205	19.0	33.9	49.8	61.3	67.5
CEMV/A	20	20	260	14.9	28.6	47.8	62.0	64.3
	20	30	280	11.1	24.1	45.8	63.3	70.4
	26	26	325	8.1	21.1	37.9	50.1	56.3
	30	30	375	7.6	18.2	40.3	58.8	63.8
CEMV/B	35	35	>420	5.1	12.6	36.4	53.4	61.0
	38	38	510	4.1	16.0	33.3	42.2	47.6
	35	45	>420	2.6	11.1	33.9	46.5	51.4

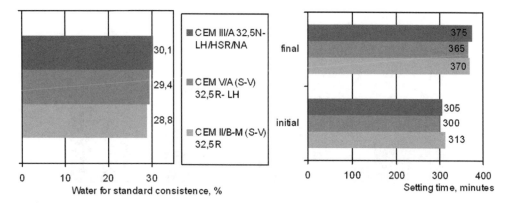

Figure 2. Physical properties of composite cement CEM V/A (S-V) 32.5R – LH (Puzak et al. 2011).

hardening (90, 180 days) the strength of composite cement CEM V/A (SV) 32.5 R-LH reaches the values close or higher to other cements containing fly ash and/or granulated blast furnace slag-composite Portland cement CEM II/B-M (SV) 32.5 R and slag cement CEM III/A 32.5 N – LH/HSR/NA (Fig. 3) (Puzak et al. 2011).

Designing the composition of concrete with composite Portland cement CEM V/A one should pay an attention to water-cement ratio (w/c). The lower the w/c ratio means a higher strength (Fig. 4), which is not irrelevant to concrete durability, especially subjected to corrosive actions of aggressive environments.

3 RESISTANCE OF COMPOSITE CEMENT CEM V/A (S-V) 32.5R – LH TO CHEMICAL ATTACK

One of the key issues for examination when introducing new components to concrete is to assess its durability, especially resistance to aggressive environmental corrosion. The increased resistance to chemical corrosion of cement-containing mineral additives with pozzolana-hydraulic character (fly ash, granulated blast furnace slag) is influenced by the following factors (Al-Dulaijan 2007, Kurdowski 1991, Neville 2000, Bonakdar & Mobasher 2010, Dinkar et al. 2007):

- limiting the content of clinker phases susceptible to corrosion, mainly tricalcium aluminate in cement composition, by reducing the clinker share in the binder for fly ash and granulated blast furnace slag,

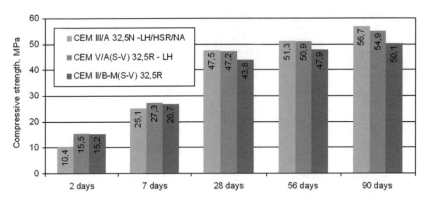

Figure 3.   Mechanical properties of composite cement CEM V/A (S-V) 32.5R – LH (Puzak et al. 2011).

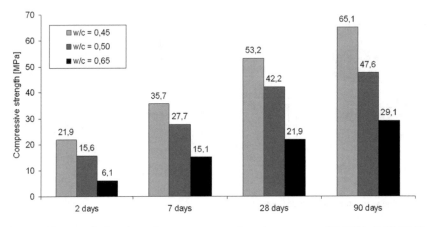

Figure 4.   The ifluence of w/c ratio on the concrete strength made on cement CEM V/A (S-V) 32.5R – LH (350 kg of cement in 1 m^3).

- reduction of Ca (OH)$_2$ in the hardened binder matrix; due to binding of calcium ions in the pozzolanic reaction – calcium hydroxide readily undergoing the chemical corrosion under the influence of different aggressive agents, especially when it creates a concentration of crystals of significant size or coating on the aggregate grains, which can occur in concretes with Portland cement,
- a change in the microstructure of hardened cement paste due the course of pozzolanic reaction of ash and hydration of slag, to form a gel product (so-called CSH phase), tightly filling the available space in paste or concrete, with proper adhesion to the aggregate/reinforcement, characterized by low own porosity (so-called micropores with nanometric size and capillary pores, with the micrometric size); such structure with the progress of conversion thickens transforms gradually without damaging the initially created a rigid matrix; the access to water or aggressive media inside concrete is thus effectively blocked.

Many research results prove that the use of mineral additives in cement or concrete composition allows reduction or elimination of the adverse impact of the alkali reaction of cement composition with reactive components of aggregates on concrete properties. It is due to the low availability of alkalis contained in the slag and fly ash. They are usually embedded in a glassy phase and are very poorly soluble, resulting in reduced concentration of OH-ions in the pore solution of concrete (Temiz et al. 2007).

The most frequent type of corrosion is the sulphate and chloride corrosion. Resistance of concrete to sulfate aggression depends largely on the mineral composition of cement. A good resistance is

Table 2. Properties of composite cement CEM V/A (S-V) 32.5R – LH.

Property	Unit	Test results
Setting time – initial	Minutes	290
Compressive strength after 2 days	MPa	15.9
Compressive strength after 28 days	MPa	46.5
Sulfates content (SO^3)	%	2.4
Chloride content (Cl^-)	%	0.06

Table 3. Properties of composite cement CEM V/A (S-V) 32.5R – LH.

Component	Component content [%]	
	Siliceous fly ash	Ground granulated blast furnace slag
L O I	1.5	0.0
SiO_2	51.0	37.6
Al_2O_3	28.3	6.9
Fe_2O_3	6.7	2.0
CaO	3.8	45.4
MgO	2.7	5.8
SO_3	0.6	0.3
Na_2O	1.5	0.6
K_2O	3.2	0.6
Cl^-	0.01	0.04

determined by a low content of tricalcium aluminate (C_3A) and a moderate content of alite (C_3S), which leads to the appearance of a significant amount of calcium hydroxide – Ca $(OH)_2$. Increasing resistance of concrete to sulfate ions attack can be achieved by the use of hydraulic or pozzolanic additives reducing the content of $Ca(OH)_2$ and bringing on the appearance of more C-S-H phases. As a result of these reactions the permeability and porosity of the paste considerably reduces (Al-Dulaijan 2007, Kurdowski 1991, Li et al. 2006).

The permeability (diffusivity) of the cement paste is crucial for the course of the chloride corrosion, as the corrosion process involves the migration of Cl^- ions deep inside and OH-ions transport to the concrete surface. Cl-ions are the fastest diffusing ions through the cement paste (Neville 2000). The negative effect of chloride ions action consists in the destruction of the concrete structure by the reaction with cement hydration products to form expansive compounds, and by the reaction of the present in the reinforced concrete reinforcement contributing to the corrosion of reinforcing steel due to a significant reduction in pH of the paste (Kurdowski 1991, Temiz et al. 2007, Yildirim et al. 2011).

A properly made concrete with mineral additives (slag, fly ash) usually shows a higher corrosion resistance to chloride attack than concretes without additives. Chlorides diffusion coefficients in the slag cement CEM III are 10-fold less, while in the cement pastes with 30% ash, silica – 3 times smaller than in the paste on Portland cement CEM I (Short & Page 1982).

3.1 Resistance to sulfate attack

Cement CEM V/A (SV) 32.5 R – LH was subjected to the test of resistance to sulfate attack according to the method described in PN-B 19707, "Cement. Special cement. Composition, requirements and conformity criteria", Appendix C. The test based on this procedure consists in observing the changes in hardened standard cement samples length (beams size $20 \times 20 \times 160$ mm), made on analyzed cement and maturing for 28 days in standard conditions (in water at 20°C), immersed in Na_2SO_4 solution at ions concentration $SO_4^2 = 16$ g/dm^3 for a period of one year. During this time,

17

Table 4. Expansion of the composite cement CEM V/A (S-V) 32.5R – LH in comparison to sulfate resistance cements CEM III/A

Test time [weeks]	Expansion of the cement sample [%]		
	CEM V/A (S-V) 32.5R – LH	CEM III/A 32.5N – LH/HSR/NA	CEM III/A 42.5N – HSR/NA
4	0.00	0.00	0.01
8	0.01	0.02	0.02
12	0.01	0.02	0.03
16	0.01	0.02	0.05
20	0.02	0.03	0.08
24	0.02	0.03	0.09
28	0.02	0.05	0.10
32	0.03	0.06	0.11
36	0.04	0.07	0.11
40	0.04	0.08	0.12
44	0.04	0.08	0.13
48	0.05	0.09	0.13
52	0.07	0.11	0.15

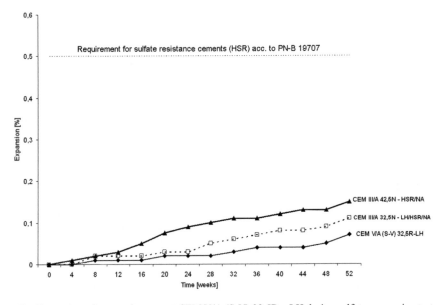

Figure 5. Expansion of composite cement CEM V/A (S-V) 32.5R – LH during sulfate aggression test.

the length of the samples is examined in every four weeks (detailed results are shown in Tab. 4). After each measurement the solution of sulfate is exchanged for the new one. Parallel to the samples stored in sulfate solution the changes of the length of comparative samples stored in distilled water are defined. The result of the indication will be the volume change which is the difference between the length of samples (beams) stored in a solution of sodium sulfate in relation to changes in the length of the samples (beams) stored in distilled water.

The research was carried on cement CEM V/A (SV) 32.5 R – LH produced on an industrial scale, with properties shown in Table 2. Table 3 shows the chemical composition of fly ash and slag used in production process of composite cement CEM V/A. Table 4 and in Figure 5 present

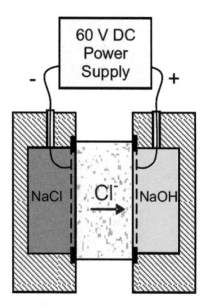

Figure 6.   Sample during chloride ion penetration test.

Table 5.   Chloride Ion penetrability based on charged passed.

Charged passed [Coulombs]	Chloride Ion penetrability
>4000	High
2000–4000	Moderate
1000–2000	Low
100–1000	Very low
<100	Negligible

the test results of the expansion of composite cement CEM V/A (S-V) 32.5R – LH stored in the solution of Na_2SO_4 for 52 weeks.

The results of sulfate resistance test shown in Figure 5 indicate, that the composite cement CEM V/A (SV) 32.5 R – LH, taking into account the criterion of the Polish standard for special cements PN-B19707, can be classified as a sulfate resistant cement HSR.

Nowadays, composite cements CEM V, due to the fact that they were not included in the Polish standard PN-B 19707 as high sulfate resistant cements (HSR) can not be labeled as sulfate resistant cements. However, this situation will change, once new European standard for common use cements EN 197-1:2009 has been implemented, in which the CEM V cements will be classified as sulfate resistant, in some European countries including, inter alia, Poland.

3.2   *Resistance to chloride corrosion*

The test of the chloride ion penetration was carried out in accordance with the American standard ASTM C 1202-05 *Standard Test Method for Electrical Indication of Concrete's Ability to Resist Chloride Ion Penetration*. The measurement was performed after 28 and 90 days of hardening in water on cylindrical concrete specimens with a diameter of 100 mm and height of 50 mm. Before the test the cylindrical specimens had been saturated with water and placed between NaCl and NaOH solutions. The constant voltage of 60 V for 6 h (Fig. 6) was applied between the electrodes.

The ability of chloride ions to migrate in the examined composite is measured by the amount of charge in Coulombs [C], which went through the sample during the six-hour test. The evaluation is done according to Table 5 (ASTM C 1202-05).

Table 6.   Composition of concrete.

Component	Unit	Concrete CEM I 42.5R	Concrete CEM V/A
Cement	kg/m³	350	350
Water		158	
Sand 0–2 mm		675	
Gravel 2–8 mm		510	
Gravel 8–16 mm		695	

Table 7.   Properties of hardened concrete.

Component	Unit	Concrete CC I	Concrete CC V
Compressive strength after 2 days	MPa	33.1	18.5
Compressive strength after 7 days		50.4	33.1
Compressive strength after 28 days		59.0	47.2
Compressive strength after 90 days		67.6	73.0
Water permeability after 28 days	mm	9	21
Water permeability after 90 days		9	9
Chloride ion penetration after 28 days	C	2840	1570
Chloride ion penetration after 90 days		1590	970

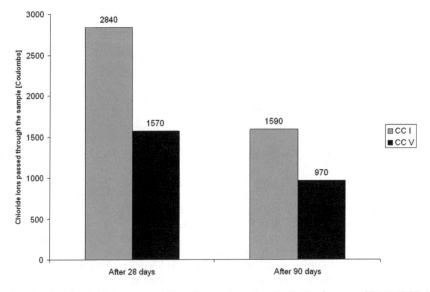

Figure 7.   Results of chloride ion permeability of concrete made with Portland cement CEM I 42.5R (CC I) and composite cement CEM V/A (S-V) 32.5R – LH (CC V).

The test of chloride ion migration was performed in two concretes, which composition is shown in Table 6. For the production of concrete the two following cement types had been applied: CEM I 42.5 R and CEM V/A (SV) 32.5 R – LH.

The properties of hardened concrete are presented in Table 7.

Verifying the results of the tests of chloride ion permeability through concrete, shown in Figure 7, one can observe that concrete made with composite cement CEM V/A (SV) 32.5 R – LH has a much lower permeability of chloride ions comparing to concrete made with Portland cement CEM I 42.5 R.

The obtained results (Fig. 7) prove thesis that concrete made of cements containing additives with hydraulic and/or pozzolanic properties are characterized by a lower permeability against the corrosive media in relation to concrete on Portland cement CEM I (Page et al. 1981, Short & Page 1982, Song & Saraswathy 2006, Yildirim et al. 2011).

Comparing the test results of hardened concrete included in Table 7 and Figure 7, it can be seen, that after 28 days of hardening, concrete with composite cement CEM V/A (SV) 32.5 R – LH reached lower compressive strength and greater depth of water penetration under pressure, it though, was characterized by greater resistance to penetration of chloride ions in relation to concrete on Portland cement CEM I 42.5 R. It proves a different microstructure of cement paste with the participation of composite cement CEM V/A.

Comparing the results of chloride ion penetration with the classification of the standard ASTM C 1202-05 (Tab. 5) it can be stated that concrete made of composite cement CEM V/A (SV) 32.5 R – LH after 28 days of hardening can be classified as a low chloride ions permeability, though, after 90 days of hardening as concrete with very low permeability. After 90 days of hardening concrete made of composite cement CEM V/A (SV) 32.5 R – LH is characterized by a higher compressive strength and lower ability to penetrate chloride ions (Tab. 7, Fig. 7).

4 CONCLUSIONS

The obtained results show that the cement composites (mortar, concrete) made from composite cement CEM V/A (SV) 32.5 R – LH characterize with high sulfate and chloride resistance. Concrete produced appropriately with composite cement CEM V/A 32.5 R – LH provides structural protection against aggressive actions of solutions containing sulfate and chloride ions. Additional positive factors influencing the durability of concrete with the composite cement CEM V/A 32.5 R – LH are the significant increases in compressive strength in longer periods of setting (90 and 180 days). It can be associated with the change of porosity (improvement of density) resulting from the pozzolanic (fly ash) and pozzolana – hydraulic (granulated blast furnace slag) activity of the components of cement composition. In concrete, after a long period harden-ing (90 days and longer), the migration of chloride ions (and others) impedes.

REFERENCES

Al-Dulaijan, S.U. 2007. Sulfate resistance of plain and blended cements exposed to magnesium sulfate solutions. *Construction and Building Materials* (21): 1792–1802

Ballim, Y. & Graham, P.C. 2009. The effects of supplementary cementing materials in modifying the heat of hydration of concrete. *Materials and Structures* (42): 803–811

Bonakdar, A. & Mobasher, B. 2010. Multi-parameter study of external sulfate attack in blended cement materials. *Construction and Building Materials* (24): 61–70

Chen, W. & Brouwers, H.J.H. 2007. The hydration of slag, part 2: reaction models for blended cement. *J Mater Sci* (42): 444–464

Chładzyński, S. & Garbacik, A. 2008. *Composites cements in civil engineering*. Cracow: Polish Cement Association.

Dinkar, P., Babu, K.G. & Sanhanam, M. 2007. Corrosion behavior of blended cements in low and medium strength concretes. *Cement and Concrete Composites* (29): 136–145

Giergiczny, Z., Małolepszy, J., Szwabowski, J. & Śliwiński, J. 2002. *Cements with mineral additives as a component of new generation concretes*. Opole: Silesian Institute

Giergiczny, Z. 2006. *Role of calcareous and siliceous fly ashes in properties creation of present binders and cement compisites*. Cracow: Cracow Technical University Editor

Kurdowski, W. 1991. *Chemistry of cement*. Warsaw: Polish Science Editor

Li, Y-X., Chen, Y-M., Wei, J-X., He, X-Y., Zhang, H-T. & Zhang, W-S. 2006. A study on the relationship between porosity of the cement paste with mineral additives and compressive strength of mortar based on this paste. *Cement and Concrete Research* (36): 1740–1743

Lothenbach, B., Scrivener, K. & Hooton, R.D. 2011. Supplementary cementitious materials. *Cement and Concrete Research* (41): 217–229

Neville, A.M. 2000. *Properties of concrete*. Cracow: Polish Cement Association

Page, C.L., Short, N.R. & EL Tarras, A. 1981. Diffusion of chloride ions in hardened cement pastes. *Cement and Concrete Research* (11): 395–406

Puzak, T., Sokolowski, M. & Dziuk, D. 2011. Composite cement CEM V/A (S-V) 32.5R – properties and possibilities of application in civil engineering. *MATBUD Conference Proceedings*. Cracow: Cracow Technical University Editor

Short, N.R. & Page, C.L. 1982. Diffusion of chloride ions through Portland and blended cement pastes. *Silicates Industriels* (47): 237–240

Song, H-W. & Saraswathy, V. 2006. Studies on the corrosion resistance of reinforced steel in concrete with ground granulated blast-furnace slag—An overview. *Journal of Hazardous Materials* (B 138): 226–233

Temiz, H., Kose, M.M. & Koksal, S. 2007. Effects of portland composite and composite cements on durability of mortar and permeability of concrete. *Construction and Building Materials* (21): 1170–1176

Wang, X-Y. & Lee, H-S. 2010. Modeling the hydration of concrete incorporating fly ash or slag. *Cement and Concrete Research* (40): 984–996

Yildirim, H., Ilica, T. & Sengul, O. 2011. Effect of cement type on the resistance of concrete against chloride penetration. *Construction and Building Materials* (25): 1282–1288

Underground Infrastructure of Urban Areas 2 – Madryas, Nienartowicz & Szot (eds)
© 2012 Taylor & Francis Group, London, ISBN 978-0-415-68394-4

What is new in German liner design code?

B. Falter
University of Applied Sciences Münster, Department of Civil Engineering, Germany

ABSTRACT: In the last years the second edition of code M 127-2 for the structural design of linings has been finally discussed by the Work Group DWA-ES 8.16 and is now prepared for the yellow print. The paper describes the new items and additional outlines based on a ten years experience in application of the first edition in Germany and in many countries abroad.

An important change is the introduction of the partial safety factor principle well-known from the Euro codes. The format of the safety proof is now $S_d R_d \leq 1$ (formerly $\gamma \geq$ req γ) describing more precisely the statistical nature of the problem by different values of γ_F (forces) and γ_M (material properties). Further topics are the imperfections and tables to calculate the liner stresses for GRP linings and egg shaped structures by hand.

The influence of the new code on the wall thickness required for linings is demonstrated by a few examples.

1 INTRODUCTION

The leaflet ATV-M 127-2 appeared in 2000. It has been applied and tested in practical use even for difficult lining projects with large diameters and different shapes. This is suitable not only for Germany but for many projects abroad where the leaflet had been prescribed for structural design (cf. the English version of M 127-2).

Damage cases that came up were explained by site faults, insufficient curing or wrong application of the code. Reasons of actual failures were normally detected with the help of M 127-2. By special lectures in No Dig techniques many engineers from practice have been skilled for the principles of liner design since more than ten years.

The leaflet is based on the application of the equation of Glock (1977) for the ideal circular "pipe-in-pipe-problem" and the enhancement to three imperfections to describe the real situation in the deteriorated sewer. Additionally the section forces factors have been developed allowing manual stress proves. The factors were evaluated by a nonlinear theory.

To perform a stress calculation has priority over the stability proof for different liner material (e.g. for non circular liners, thick walled CIPP, mortar liner or in case of big imperfections).

For the work on the leaflet many experimental results have been evaluated to identify each single influence on the buckling behaviour e.g. the state of the old pipe, the shape of the liner, the imperfections, the material behaviour (creep effects, fracture modes), loadings like cover height, traffic loads and water table. The approaches of the code show good coincidence with the tests (Guice et al., 1994).

For the 2nd edition further research was necessary e.g. on the residual safety of deteriorated masonry sewers (Steffens et al. 2002), the creep buckling of polyethylene liners with egg shape (Falter et al. 2008a) and sewers with shallow cover made of different material (Falter & Wolters, 2008b).

In the 2nd edition names of variables were changed. Some of them are listed in Table 1.

2 DIMENSIONING CONCEPT WITH PARTIAL SAFETY FACTORS

In the code a new dimensioning concept using partial safety factors has been adopted. These factors have been defined to keep the formerly safety niveau (req $\gamma = 2$). For a CIPP subjected to water pressure yields $\gamma_F \cdot \gamma_M = 1.5 \cdot 1.35 = 2.025 \cong 2.0 =$ req γ.

Table 1. Variables in the 1st and 2nd edition of M 127-2

Description	1st edition	2nd edition
Host pipe wall thickness	s	t
Liner wall thickness	s_L	t_L
Local imperfection related to radius	$w_v/r_L \cdot 100\%$	ω_v
Global imperfection related to radius	$w_{GR,v}/r_L \cdot 100\%$	$\omega_{GR,v}$
Annular gap related to radius	$w_s/r_L \cdot 100\%$	ω_s
Shrink measure of mortar	–	ω_M
Partial safety factor for forces	–	γ_F
Partial safety factor for resistance	–	γ_M
Characteristic value	–	index k
Design value	–	index d

Table 2. Imperfections for circular and egg shaped linings

		Circular cross section			Egg shape
Imperfection	Host Pipe State (load case)	Cured in Place Pipe (CIPP)	Deformed and Redeformed Liner	Spiral Wound Liner	Cured in Place Pipe (CIPP)
ω_v	I	$\geq 2\%$	$\geq 2\%$	$\geq 1\%$	$\geq 0.5\%$ of $r_S^{2)3)}$
	II				$0.5\% + \omega_{GR,v}/10$ [2]
$\omega_{GR,v}$ cracked four times	II + III	$\geq 3\%$	$\geq 3\%$	$\geq 3\%$	$\geq 3\%$
ω_s	I – III (p_a)	$\geq 0.5\%$	$\geq 2\%$	$\omega_M \geq 0.1\%$ 0.25% [1]	$\geq 0.5\%$ of $r_C^{4)}$
	III (q_v)	0	0	0	0

[1] If no tests are performed. [2] To be doubled for masonry sewers. [3] S: Springlines [4] C: Crown

The concept allows defining smaller values of γ_M for prefabricated linings (1.25) than for CIPP that are cured under sewer conditions (1.35).

3 IMPERFECTIONS

The imperfections to be regarded in the structural model presented in Table 2 depend on the Host Pipe State (definition cf. DWA-M 127-2), the liner type and the cross section.

The minimum values of ω_v and ω_s are fixed in the code while $\omega_{GR,v}$ is evaluated by the video prints or by in-situ measurements (for more see DWA-M 127-2).

For standard imperfections $\omega_v = 2\%$ and $\omega_s = 0.5\%$ a new diagram with reduction factors $\kappa_{v,s}$ was developed for the 2nd edition which allows a more economic dimensioning (Figure 1 and Falter

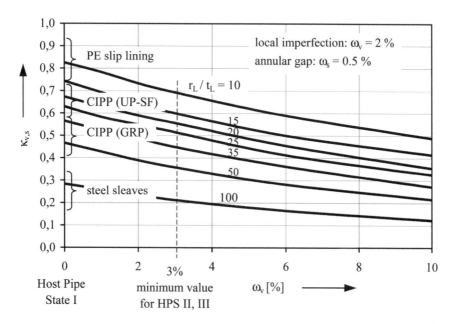

Figure 1. Reduction factors $\kappa_{v,s}$ for different values of the ovalisation (M 127-2, Diagram D2).

et al., 2003). The critical water pressure is calculated by the equation.

$$\text{crit } p_{a,d} = \kappa_{v,s} \cdot 2.62 \cdot \left(\frac{r_L}{t_L}\right)^{0,8} \cdot S_{L,d} \tag{1}$$

where $S_{L,d} = (1/\gamma_M) \cdot [E_{L,k}/12 \cdot (1 - \mu^2)] \cdot (t_L/r_L)^3$ is the long-term ring stiffness of the liner.

The equation is generally valid but a check of the liner stresses must be added regularly, see the following chapters.

4 STRUCTURAL ANALYSIS USING STRESSES

4.1 *Factors for bending moments and normal forces*

The appendices A4 and A5 of code M 127-2 with factors m, n and δ_{el} have been updated completely for the following reasons:

– For GRP-liners factors for the bending moments and normal forces were asked for from practice. They are based on the characteristic value of the Young's modulus $E_{L,k} = 6000$ N/mm².
– As the rehabilitation of egg shaped pipes became more important factors for this cross section should be added.
– For Host Pipe State II curves with 6% and for State III 9% ovalisation were added.
– The following parameters had to be actualised: $\omega_s = 0.5\%$ (formerly 1%), $E_{L,k} = 1400$ N/mm² (formerly 1800 N/mm²), $\mu = 0.35$ (instead of 0).

Factors for the newly defined Host Pipe State IIIa (old pipe without compression strength) are presented.

The effort for these amendments could be reduced by restriction to one nominal diameter each: ND 300 for circular and B/H = 600/900 mm for egg shaped profiles. An adaption to smaller or bigger cross sections is possible in an easy way.

In the following some new diagrams for ND 300 and B/H = 600/900 mm from the appendices A4 and A5 are presented and commented, cf. overview in Table 3.

Table 3. Diagrams for UP-SF liners from appendix A4 and A5 in M 127-2 with explanations

Figure	Diagram	Geometry	Host Pipe State	E_2 N/mm^2	Content
2	A4/1	circular	I + II	–	m_{pa}[1]
3, 4	–	circular	I + II	–	accuracy of m_{pa}
5	A4/3a	egg	I + II	–	m_{pa}
6	–	egg	I	–	function of m_{pa}
8	–	egg	I	–	accuracy of m_{pa}
9	A4/3b	egg	I + II	–	n_{pa}
10	A5.1/2a	circular	III	8	m_q
11	A5.1/2b	circular	III	8	n_q
12	D6	circular	III	3 + 8	crit $q_{v,d}$
13	A5.3/2a	circular	IIIa	8	m_q[1]

[1]$n_{pa} \cong -1.0$

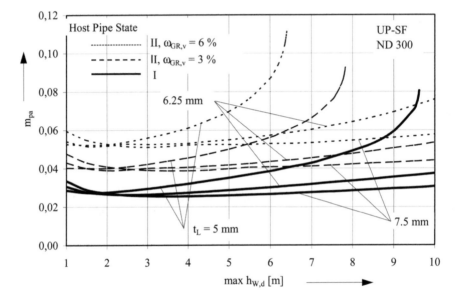

Figure 2. Factor m_{pa} for circular UP-SF liners (M 127-2, Diagram A4/1).

4.2 Diagrams for circular profiles, Host Pipe State I and II

To apply Figure 2 to other diameters the following calculation for the parameter t_L must be performed:

$$t_{L,ND} \cong t_{L,ND300} \cdot ND / 300 \tag{2}$$

E.g., if the factor m_{pa} for the diameter ND 600 is required the curve for $t_L = 5$ mm is approximately valid for the wall thickness

$$t_{L,600} \cong 5 \cdot 600 / 300 = 10 \text{ mm.}$$

Figure 2 has been developed for the design E-modulus $E_{L,d} = 1400/\gamma_M \cong 1000 \text{ N/mm}^2$, comparable diagrams are available for GRP with $E_{L,d} = 6000/\gamma_M \cong 4400 \text{ N/mm}^2$. Furtermore curves for an ovalisation of $\omega_{GR,v} = 6\%$ have been added.

26

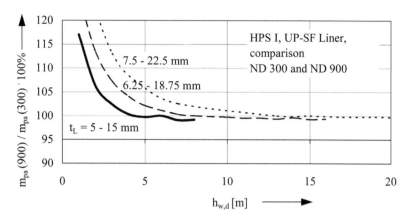

Figure 3.   Accuracy of m_{pa} for the diameter ND 300 if applied to ND 900 (HPS I).

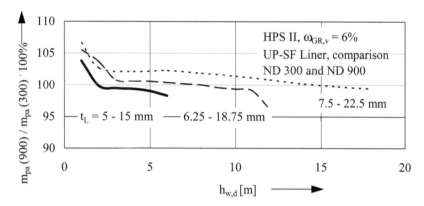

Figure 4.   Accuracy of m_{pa} for the diameter ND 300 if applied to ND 900 (HPS II).

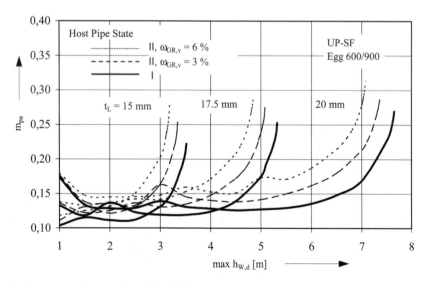

Figure 5.   Factors m_{pa} for egg shaped UP-SF liners (M 127-2, Diagram A4/3a).

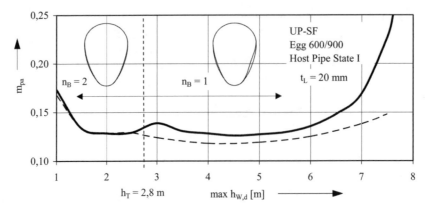

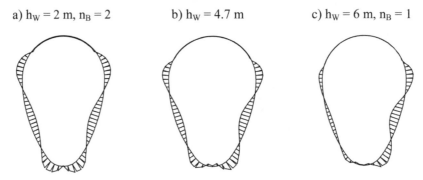

Figure 6. Function of the factors m_{pa} for egg shaped liners, deflections, $n_B =$ Number of buckles.

a) $h_W = 2$ m, $n_B = 2$ b) $h_W = 4.7$ m   c) $h_W = 6$ m, $n_B = 1$

Figure 7. Bending moments of egg shaped liners subjected to increasing water table.

The curves m_{pa} tend to infinity in the near of the buckling load (e.g. for $t_L = 5$ mm at $h_{W,d} = 6.3$ m) and the stresses grow excessively. Therefore the buckling test is done by the stress test as well for the following conditions:

1. Calculation with γ_F fold loads and resistance divided by γ_M (regularly E_L is divided by γ_M),
2. application of imperfections that are similar to the buckling mode (this condition must be valid only approximately) and
3. calculation of the equilibrium at the deformed system.

The accuracy of the factors m_{pa} for differing diameters is presented in the Figures 3 and 4. The differences to the unsafe side are less than 2% (Figure 3) for Host Pipe State I and utmost 4% (Figure 4) für Host Pipe State II with 6% ovalisation.

4.3   Diagrams for egg shaped profiles, Host Pipe State I and II

In Figure 5 the factors m_{pa} for needle felt liners with egg shape B/H = 600/900 mm are presented. Again, each curve tends to infinity for a certain water table, the buckling load of the profile.

Comparing the Figures 2 and 5 non regular curves for egg profiles are obvious. The reason is the transition of the system from *two mode* and approximately symmetric deflections to *one mode* deflections for increasing loads up to the buckling load, cf. $h_T \cong 2.8$ m in Figure 6.

The transition can be seen more clearly from the bending moments than from the deflections. Figure 7 shows the gradual transition from a symmetric moment distribution ($h_W = 2$ m) to a one side stress distribution on the right side with the local imperfection: For $h_W = 6$ m the liner fits closely to the host pipe at the left side. This effect has a mechanical basis; however, in case of symmetric solutions the program Linerb (Falter, 2011a) shows a warning, that *buckling* of an egg shaped liner must always be non symmetric ($n_B = 1$).

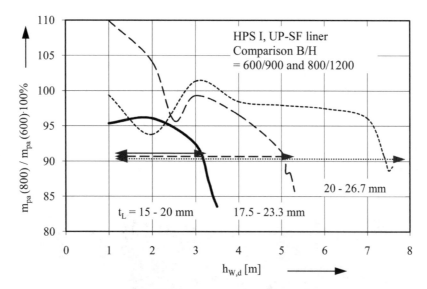

Figure 8. Accuracy of the factors m_{pa} for egg shaped profiles with B > 600 mm.

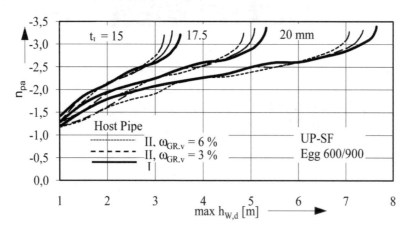

Figure 9. Factors n_{pa} for egg shaped UP-SF liners (M 127-2, Diagram A4/3b).

The examination of accuracy for egg shaped profiles B = 800 mm shows a relatively good coincidence of the factors up to the buckling load (steep part of the curves in Figure 5). The deviation to the unsafe side is normally less than 7.5%, cf. Figure 8.

When judging the accuracy of the diagrams additional reading and interpolation faults must be accepted. The diagrams are well suitable to estimate liner wall thicknesses. In cases of bigger diameters and/or extensive loading a computer analysis of the necessary wall thickness is unavoidable.

In circular profiles the liner normal forces for Host Pipe States I and II can be approximately calculated by $n_{pa} = -1.0$ (=vessel formula for outside pressure). For egg and circular profiles in Host Pipe State III this is not possible any longer, cf. Figures 9 and 11.

4.4 Diagrams for Host Pipe State III

Beside the soil and traffic loads for Host Pipe State III soil properties (deformation modulus E_2, factor of the horizontal soil pressure K_2, angle of inner friction φ') and properties of the old pipe compression zones (compression strength and hinge excentricity e_G, cf. Table 4) must be introduced into the analysis.

29

Table 4. Hinge excentricity e_G, depending on the state of the old pipe contact zones

	Good state (no spalling, high compression strength, no corrosion, new pipe)	Normal (no or small spalling, acceptable compression strength, little corrosion) = regular case	Severe damages (obvious spalling, poor compression strength, severe corrosion)
Compression zone in the springlines			
Hinge excentricity e_G/t	≤ 0.45	~ 0.35	~ 0.25
Width of the compression zone b_D/t	≥ 0.13	~ 0.40	~ 0.67
Example			

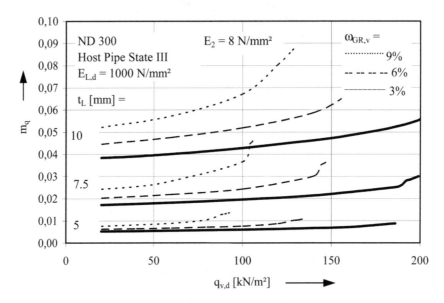

Figure 10. Factors m_q for circular UP-SF liners, Host Pipe State III (M 127-2, Diagram A5.1/2a).

In case of high compression strength the excentricity can be supposed as $e_G/t = 0.45$, while corrosion or spalling in the region of the springline compression area reduce the excentricity e.g. to $e_G/t = 0.25$, cf. Table 4. The parameter e_G/t has an important influence on the section forces factors and the overall stability of the system. Careful inspection and in cases of doubt the test of the compression strength of the old sewer material are condition for a safe design.

In Figure 10 the bending moment factors m_q in case of soil and traffic loads are presented for UP-SF liners ND 300. The evaluation for other diameters is possible with Eq. (2) again. For standardised egg profiles no factors are necessary as the kinematic of the two crown quarter parts in

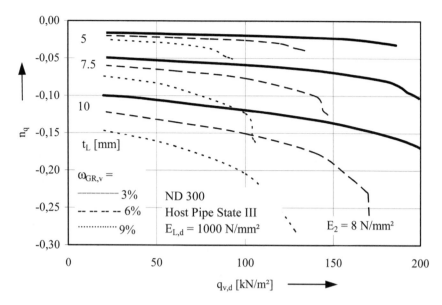

Figure 11.   Factors n_q for circular UP-SF liners, Host Pipe State III (M 127-2, Diagram A5.1/2b).

Host Pipe State III coincidises with the movement of a broken circular profile. Thus the calculation as a circular profile with the crown radius of the egg shape is sufficiently accurate.

The factors for the normal forces n_q in Figure 11 are significantly smaller than for Host Pipe States I and II ($n_{pa} \cong -1.0$) as the old pipe has to take a major amount of the compression forces.

The compression forces in the old pipe are transferred in the contact areas of the longitudinal springline cracks:

$$N_S \cong -q_v \cdot (1 + \delta_h) \cdot OD / 2 \qquad (3)$$

where δ_h ($\cong \delta_v$) is the horizontal deformation of the system related to OD.

Assuming a parabolic distribution of the stresses in the pressure zone b_D (Table 4) in the old pipe compression stresses result that must be compared with the material strength.

4.5   Critical total vertical load crit $q_{v,d}$

Like the curves for m_{pa} the factors m_q show increasing functions which indicate the critical load crit $q_{v,d}$ of the host pipe-soil system. The values crit $q_{v,d}$ for UP-SF liners ND 300 are drawn in Figure 12 with the assumption of an unfavourable hinge excentricity $e_G/t = 0.25$ (cf. Table 4).

The dotted lines (curve paramter $t_L = 0$ in Figure 12) are valid for the pipe-soil system alone. The curves for an increasing wall thicknesses t_L show the reinforcement of the pipe-soil system by the liner. To archieve a remarkable fortification a wall thickness of ca. 7.5 mm or more is necessary.

By enlarging the wall thickness the critical load and the load bearing capacity of the system is increased. On the other side the factors m_q and the bending moments of the liner are amplified (cf. Figure 10) as the flexible liner undergoes the same deformations like the old pipe.

4.6   Diagrams for Host Pipe State IIIa

It cannot be excluded that the old pipe is deteriorated severely e.g. by corrosion. In this case the compression forces cannot be transferred by the pipe. It must be regarded as gravel which defines the liner's bedding together with the surrounding soil. The liner must now be calculated as an elastically bedded construction which is the well-known system for newly layed pipes.

The factors m_q for this case are presented in Figure 13. If compared with Host Pipe State III (Figure 10) now significantly bigger wall thicknesses are required to get sufficient buckling loads.

31

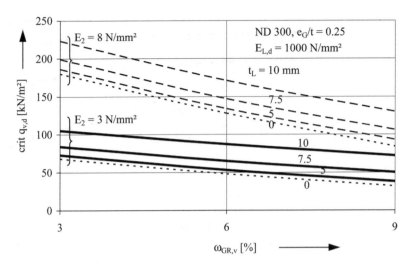

Figure 12.   Critical vertical loading crit $q_{v,d}$ for UP-SF liners, ND 300, $K'_2 = 0.2$ (M 127-2, Diagram D7).

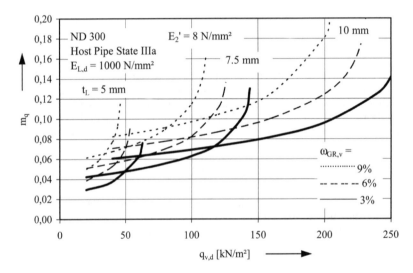

Figure 13.   Factors m_q for circular UP-SF liners, HPS IIIa (M 127-2, Diagram A5.3/2a).

E.g., if a cover height of 5 m is assumed the loading is $q_{v,d} = 1.5 \cdot 20 \cdot 5 = 150\,kN/m^2$ and a wall thickness $t_L = 9$ or $10\,mm$ will be required, cf. Figure 13. For HPS III and a small ovalisation 5 mm would be sufficient, cf. Figure 10.

5   EXAMPLES

5.1   *Calculations of circular and egg profiles for Host Pipe State II*

The structural analysis of liners can be performed manually in very few steps with help of the diagrams in the appendices A4 (Host Pipe State I, II) and A5 (Host Pipe State III, IIIa) of code DWA-M 127-2. As the factors have been developed by non linear calculations the stability proof is included in Table 5. As a control the results of a computer program (Falter, 2011a) are added – the deviation is in the order of the reading accuracy.

32

Table 5. Calculation steps for a circular and an egg shaped UP-SF liner, HPS II, $\omega_{GR,v} = 6\%$

	Variable	Unit	Equation/Source	Circle ND 600 manual	Circle ND 600 EDV	Egg B/H = 600/900 manual	Egg B/H = 600/900 EDV
1	$h_{W,k}$	m	Specification	3		4	
2	$h_{W,d}$	m	$h_{W,k} \cdot \gamma_F = h_{W,k} \cdot 1.5$	4.5		6	
3	$p_{a,d}$	N/mm^2	$\gamma_W \cdot h_{W,d} = 10 \cdot h_{W,d}/10^3$	0.045		0.06	
4	$E_{L,d}$	N/mm^2	$E_{L,k}/\gamma_M = 1400/1.35$	$\sim$1000		$\sim$1000	
5	$\sigma_{fb,d}$	N/mm^2	$\sigma_{fb,k}/\gamma_M = 20/1.35$	14.8		14.8	
6	$\sigma_{D,d}$	N/mm^2	$\sigma_{D,k}/\gamma_M = 25/1.35$	18.5		18.5	
7	t_L	mm	chosen	10		20	
8	r_L	mm	$(ND - t_L)/2$	295		290	
9	n_{pa}	–	Figure 9	-1.0	-1.16	-2.6	-2.51
10	$N_{pa,d}$	N/mm	$n_{pa} \cdot p_{a,d} \cdot r_L$	-13.3	-15.4	-45.2	-43.6
11	m_{pa}	–	Figure 2/Figure 5	0.066	0.0647	0.19	0.176
12	$M_{pa,d}$	Nmm/mm	$m_{pa} \cdot p_{a,d} \cdot r_L^2$	258	253	959	890
13	A	mm^2/mm	$1 \cdot t_L$	10		20	
14	W	mm^3/mm	$1 \cdot t_L^2/6$	16.7		66.7	
15	$\sigma_{i,d}$	N/mm^2	$\cong N_{pa,d}/A + M_{pa,d}/W$	$+14.3$	13.8	$+12.4$	$+11.5$
16	$\alpha_{ki,a}$	–	$1 \pm t_L/3/r_L$	1.011		1.023	
17	$\sigma_{a,d}$	N/mm^2	$\cong N_{pa,d}/A - M_{pa,d}/W$	-16.6	-16.6	-16.3	-15.2
18	Proof	–	$\max \sigma_d/\sigma_{fb,d} \leq 1$	0.97 < 1	0.93 < 1	0.84 < 1	0.78 < 1
19	Proof	–	$|\min \sigma_d|/\sigma_{D,d} \leq 1$	0.90 < 1	0.90 < 1	0.88 < 1	0.82 < 1

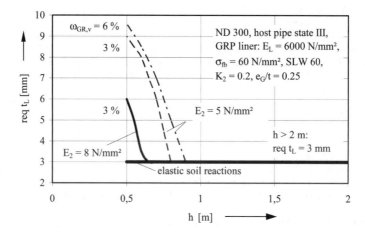

Figure 14. Required wall thickness depending on ovality $\omega_{GR,v}$ and deformation modulus E_2, cover height $h \geq 0.5$ m, no water table (Falter & Fingerhut 2011b).

For Host Pipe State III the number of parameters to be regarded is bigger than for State I or II. The dependence of the liner wall thickness from the parameters e_G, E_2 and $\omega_{GR,v}$ for shallow covers and heavy truck loads is published by Falter & Fingerhut (2011b).

5.2 Calculations for Host Pipe State III and shallow covers

For longitudinally cracked old pipes buried under streets with shallow covers a calculation due to Host Pipe State III must be performed independent of the size of ovalisation $\omega_{GR,v}$. This analysis sometimes results in remarkably bigger wall thicknesses than a buckling analysis by Eq. (1) and Host Pipe State II assuming the minimum water table of $h_W = 1.5$ m, cf. Figure 14.

The steep slope of the curves for required t_L at decreasing covers has mainly the following reason: The bedding reactions at the sides the pipe fragments are restricted by the passive soil pressure.

It is a big safety and economic feature to introduce realistic values for the dominant parameters of the problem, e.g. the soil group, the deformation modulus E_2 of the surrounding soil, the factor K_2 of the horizontal soil pressure, the old pipe compression strength with its influence on e_G and the ovalisation of the cross section $\omega_{GR,v}$.

6   SUMMARY

Some news of the 2nd edition of the German code DWA-M 127-2 (draft 2011) for the structural analysis of linings and assemly procedures are presented. The new basis is the concept of partial safety factors. The imperfections to be applied and the hinge excentricities in the old pipes contact zones are discussed.

The validity of the appendices with factors for section forces to evaluate stresses has been extended remarkably allowing manual dimensioning for GRP linings and egg profiles. The application of the diagrams is explained by two examples (circular and egg shaped profiles, Host Pipe State II).

The author thanks the Liner Design Working Group DWA-ES 8.16 for the constructive collaboration on the new edition of the liner code.

REFERENCES

ATV-M 127-2:2000 and DWA-M 127-2:2011, draft 2nd ed. Static calculations for the rehabilitation of sewers with lining and assembly procedures. *DWA*, Hennef.

Falter, B., Hoch, A. & Wagner, V. 2003. Hinweise und Kommentare zur Anwendung des Merkblattes ATV-M 127-2 für die statische Berechnung von Linern. *Korrespondenz Abwasser* 50, 451–463.

Falter, B., Eilers, J., Müller-Rochholz, J. & Gutermann, M. 2008a. Buckling experiments on polyethylene liners with egg-shaped cross-sections. *Geosynthetics International* Vol. 15 No. 2 152–164.

Falter, B. & Wolters, M. 2008b. Mindestüberdeckung und Belastungsansätze für flach überdeckte Abwasserkanäle (MIBAK). Co-operative Research Project IV-9-042 3E1, IV-7-042 3E1 0010, funded by MUNLV. *Final Report*.

Falter, B. 2011a. Computer Program Linerb 7.20. Structural Analysis of Linings. Münster, Germany.

Falter, B. & Fingerhut, S. 2011b. Host Pipe State and Necessary Wall Thickness of Sewer Linings. Contribution to the 29th *International NoDig Conference* May 2–5 2011 in Berlin, Germany.

Glock, D. 1977. Überkritisches Verhalten eines starr ummantelten Kreisrohres bei Wasserdruck von außen und Temperaturdehnung. *Stahlbau* 46, pp. 212–217.

Guice, L. K., Straughan, T., Norris, C. R. & Bennett, R. D. 1994. Long-Term Structural Behaviour of Pipeline Rehabilitation Systems. *TTC Technical Report #302*, Louisiana Tech University Ruston, Louisiana, USA.

Steffens, K. (ed.), Falter, B., Grunwald, G. & Harder, H. 2002. Abwasserkanäle und -leitungen, Statik bei der Substanzerhaltung und Renovierung (ASSUR). Co-operative Research Project 01RA 9803/8, funded by BMB+F. *Final Report*, Insitute for Experimental Statics, Univ. of Appl. Sc. Bremen.

Underground Infrastructure of Urban Areas 2 – Madryas, Nienartowicz & Szot (eds)
© 2012 Taylor & Francis Group, London, ISBN 978-0-415-68394-4

High Performance Concrete (HPC) in concrete pipe production

Z. Giergiczny
BETOTECH Technology Center, Dabrowa Gornicza, Poland
Silesian University of Technology, Faculty of Civil Engineering, Gliwice, Poland

T. Pużak & M. Sokołowski
BETOTECH Technology Center, Dabrowa Gornicza, Poland

H. Skalec
P.V. Prefabet, Kluczbork, Poland

ABSTRACT: Concrete ought to comply with the specific requirements due to its repeatedly hard exploitation environment and as a product for the production of prefabricated concrete pipes. Gaining the concrete with desired properties requires proper selection of components as well as adequate design duration considering either the character of the precast plant and the mixture parameters – appropriate mixture liquidity, transport and concreting duration, etc.

Hereby paper describes the results of laboratory tests and industrial attempt of High Performance Concrete (HPC) designed for concrete pipe production. The study presents also concrete durability test results such as: frost resistance, water absorbability and permeability. The entire paper is completed with practical remarks and observations made during the production process of concrete pipes.

1 INTRODUCTION

Densely inhabited urban areas, increased traffic and an increasing number of installations, impedes the open development of channels. In this situation the only solution is to build a system of underground installations with trenchless method. The advantage of this method is that the construction of the pipeline is made without obstructing traffic and disrupting pedestrians and vehicles (Fig. 1). If there is a need to carry out works in the area of highways, railways and major roads and thoroughfares has no other reasonable alternative. The main advantages of trenchless technology are:

– reducing the earthworks by about 85% compared to the open pits at the same pipe diameter and the same channel length,
– no need to lower the groundwater level,
– minimize the amount of output,
– elimination of the trench line,
– stable ground level increasing safety of the pipeline (Hakenberg 1999 & Madryas et al. 2002)

2 REQUIREMENTS FOR CONCRETE PIPES

Application of micro-tunneling is increasingly important in water and sewage economy. For effective jacking, in addition to soil parameters and the length of time and jacking, properties of concrete used for the manufacture of reinforced concrete pipe are also important (Hakenberg 1999).

High-strength pipes "High Performance – Pipe" present a perfect combination of strength requirements by optimizing the volume of excavated soil during microtunnelling. Reinforced concrete jacking pipes are made in accordance with PN EN 1916, *"Concrete pipes and fittings, unreinforced, steel fiber and reinforced"*.

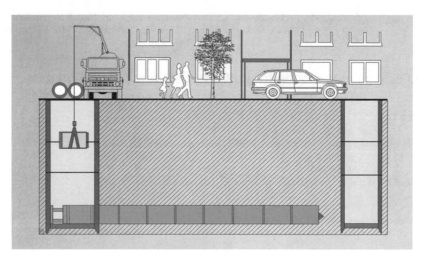

Figure 1.   Trenchless technology assumption.

Table 1.   Relation of total output and workload for standard pipes and "High Performance Pipe" pipes.

| Diameter [mm] | Output [m^3] | | Additional workload of ground works |
	"High Performance Pipe"	Standard pipe	
500	34.25	47.9	40%
600	46.01	60.9	32%
800	73.22	95.10	30%

Concrete for the constructions, given the often extremely difficult operating conditions, must meet specific requirements, including: resistance to aggressive environmental attack due to the very different pollution load in water and soil, high water-resistance, frost resistance and reduced shrinkage.

Concrete Pipes "High Performance – Pipe" compared to the standard pipe outside diameters have a smaller thickness of the wall, making it possible to reduce the output by about 40%, reduction of a work time and the costs related (Tab. 1). Pipes are manufactured by casting a wet concrete and the process of concrete maturing takes place in steel forms, in which reinforcement is placed by polymeric spacers. The use of such distances largely eliminates the cracks on the pipe surface. With this technology we get a very smooth and uniform outer surface of the pipe. The smoothness of the surface of the pipe in conjunction with the reduced external surface, due to reduced wall thickness, strongly reduces the amount of frictional resistance occurring during jacking. Moreover, reduced friction allows the extension of jacked sections. Owing to this fact, "High Performance Pipe" pipes are suitable for jacking distances up to 200 m.

3   PROJECT ASSUMPTIONS

Obtaining high-strength concrete with the desired properties requires an appropriate selection of components and proper design process considering the density and dispersion of reinforcement, good texture of concrete, as well as transport duration and development of concrete.

The requirements regarding the properties of concrete and hardened concrete used in the production of "High Performance Pipe" pipes are shown in Table 2.

Table 2. Project assumptions for reinforced pipes "High – Performance Pipe".

Requirements for concrete mixture	Requirements for hardened concrete
Concrete mixture consistency: 50–55 cm measured by slump test	Minimum concrete class C 70/85 after 28 days of setting
	Compressive strength after 7 hours of infusion cycles (3 h of heating element, 4 h of cooling, temp 55°C) – minimum 20 MPa
Consistency kept in time by min. 30 minutes	Compressive strength of concrete after 1 day of setting – minimum 25–30 MPa
No segregation of components	Absorbability – no more than 5%
No "bleeding" effect	Water penetration depth – maximum 30 mm
	Frost resistance degree F 150

Table 3. Properties of slag Portland cement CEM II/B-S 52.5N.

Property	Requirements of PN-EN 197-1	Plant laboratory test result
Le Chatelier consistency change	≤ 10 mm	0.7 mm
Beginning of setting time	≥ 45 minutes	264 minutes
Compressive strength after 2 days	≥ 20 MPa	26.8 MPa
Compressive strength after 28 days	≥ 52.5 MPa	59.0 MPa

4 CHARACTERISTICS OF MATERIALS APPLIED IN RESEARCH

4.1 *Cement*

Given the desired level of early strength, providing unmoulding and transport of the finished element on the site of prefabrication plant, slag Portland cement CEM II/BS 52.5 N with the properties shown in Table 3 was applied in hereby study. This cement presents high early strength (1, 2 days) and during standard time (28 days), also the high heat of hydration. The binder is commonly used in the production of large and fine precast, prestressed elements, HPC and SCC concretes.

4.2 *Aggregate*

For the composition of the concrete mixture a high-quality local gravel aggregates were used. Particular attention was paid to the grain size of sand. Aggregates used allowed to achieve the proper flow of concrete mixture and correct filling of the space in the constructed element (densely arranged reinforcement).

4.3 *Chemical admixture*

The chemical admixture based on ether poly-carboxylates was used for the design and construction of concrete. The admixture is intended for the production of precast, reo-dynamic concretes with high early strengths and prestressed concretes. It enables the production of concrete with a low w/c ration, which leads to concrete with high strength, both in initial and standard period (28 days). Concretes with applied polymer additives present the ability to maintain consistency, even at high temperatures. This kind of an additive accelerates the process of cement phase's hydration, thus, while hardening, especially in the initial period of time, more heat is being precipitated. This results in a relatively high early strength of concrete.

4.4 *Fly ash*

In order to improve the rheological properties of concrete mixture high-quality fly ash, meeting the requirements of BS EN 450-1: 2006 "Fly ash for concrete", was used in the composition of concrete. Specially selected mineral additive, along with cement, provides adequate liquidity of the concrete mixture, and allows the extension of concrete workability time (convertibility). Concrete

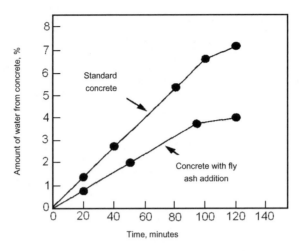

Figure 2. Amount of water from concrete without and with the addition of fly ash (Malhotra & Ramezanianpour 1994).

Table 4. Physical properties of fly ash.

Loss on ignition [%]	SO_3 [%]	CaO_{free} [%]	Cl^- [%]	Activity [%] after 28 days	after 90 days	Sieve residue 0.045 mm [%]	Density [g/cm^3]
2.24	0.67	0.07	0.007	78.4	89.7	34.0	2.13

Table 5. Chemical composition of fly ash.

SiO_2	Al_2O_3	Fe_2O_3	CaO	MgO	Na_2O	K_2O
51.50	27.83	7.50	3.68	2.51	1.07	2.97

mix containing fly ash is cohesive and has fewer tendencies to release cement wash, as shown in Figure 2. Addition of fly ahs, while reducing the cement content in concrete, decreases the shrinkage of concrete.

Concrete with fly ash is characterized by increased resistance to corrosion attack of chemically aggressive environments (Neville 2000, Malhotra & Ramezanianpour 1994, Giergiczny & Giergiczny 2010).

Siliceous fly ah with the composition and properties presented in table 4 and 5 was applied in hereby study

4.5 *Silica fume*

Silica fume is a by-product in the manufacture of metallic silicon or its alloys. It shows an amorphous nature and occurs in the form of empty beads with diameters of less than 10^{-6} m. Used in the production of concrete shall meet the requirements specified in PN-EN 13263-1:2006 "Silica fume concrete. Part 1: Definitions, requirements and conformity criteria."

The introduction of silica fume into the composition of concrete mixture changes its rheological properties of projecting the way its laying and compaction. Very fine particles of this additive affect the increasing density and reduction of concrete mix plasticity, resulting in an increase of water demand. This requires the application of an appropriate quality and quantity of plasticizers and the extension of their compaction (vibration) time (Neville 2000 & Nocuń-Wczelik 2005).

Table 6. Composition of concrete mixture.

Component	Amount [kg/m³]
Cement CEM II/B-S 52.5N	410
Sand 0/2	646
Gravel 2/8	360
Gravel 8/16	792
Fly ash	30
Mikrosilica	41
Superplasticizer	6.97
Water	136

Table 7. Properties of concrete mixture.

Property	Received result
Consistency of concrete mixture tested with slump method	54 cm
Air content	1.9%
Segregation of components	no
Consistency kept in time	30 minutes
Temperature of concrete mixture	18.5°C
Water "bleeding" effect	no

Silica fume very positively affects the strength of concrete, generally, by increasing it. The increase in compressive strength of concrete is accompanied by increase of Young modulus. It should be noted, that silica fume positively influences durability of concrete. Concrete with this addition is characterized by a greater tightness, lower water demand, high resistance to chemical attack and increased resistance to chemical attack. The silica fume used during the design of concrete was in the form of a liquid suspension.

5 COMPOSITION AND THE PROPERTIES OF CONCRETE MIXTURE

The composition of designed and tested concrete mixture is shown in Table 6 and its properties in Table 7.

6 PROPERTIES OF HARDENED CONCRETE

Concrete samples were stored after molding for 4 hours at 55°C, followed by another 3 hours matured in the laboratory conditions (temperature $20 \pm 1°C$ and a relative humidity above 50%). After 7 hours samples were unmolded. Such simulation was to reflect the technological conditions at the production plant. Technological scheme of heat treatment in prefabrication plant is shown in Figure 3. Having obtained the desired strength and durability parameters the industrial laboratory trial was conducted (Fig. 4).

The following tests were conducted on hardened concrete:

• Compressive strength after 7 hours; 1, 2 and 28 days,
• Depth of water penetration under pressure,
• Absorbability,
• Resistance to chemical attack,
• Frost resistance,
• Size of chloride ions penetration according to ASTM C 1202-05.

Table 8 shows the average compressive strength of concrete both from laboratory and industrial trial. Uzyskane wyniki badań potwierdzają osiągnięcie założeń projektowych (uzyskanie klasy C70/85)

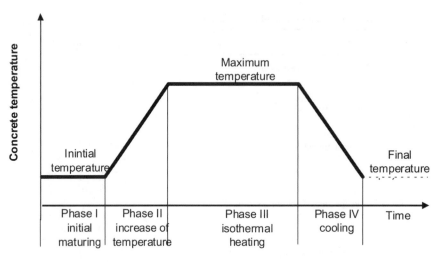

Figure 3. Technological scheme of concrete heat treatment.

Figure 4. Production process of "High Performance Pipe" pipes.

Table 8. The average compressive strength of concrete.

	Average strength $f_{ck,cube}$, [MPa], after				
	7 hours	1 day	2 days	28 days	Received strength class
Laboratory trial	40.3	56.1	67.1	101.9	C80/95
Industrial trial	32.9	45.1	56.2	93.9	C70/85

 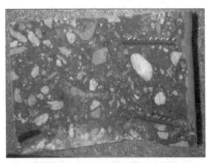

Figure 5. Depth of water penetration under pressure (visible lack of water penetration).

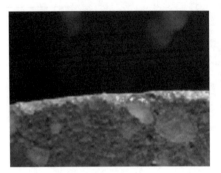 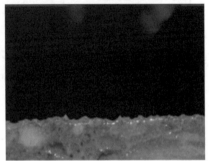

Figure 6. Microscopic magnification of pipe's outer layer.

The tested concrete was characterized by a high tightness. The test carried out according to the procedures comprised by the PN-EN 12390-8, "Tests of concrete. Part 9: The depth of water penetration under pressure", shows the lack of water penetration under pressure both in the sample derived from the laboratory paste as well as in the bore of industrial production, as shown in Figure 5. Following the German (DIN) recommendations, such concrete can be classified as watertight concrete (maximum depth of water penetration of 50 mm, or in case of concrete exposed to aggressive environments 30 mm). The average absorption for the analyzed concretes was 3.1% for laboratory tests, and 3.35% for the industrial sample. At the same time, the microscopic enlargement of the surface of pipe with high strength (High Pipe) in comparison with the standard concrete pipe shows a smoother and tighter structure of the subsurface layer of high-strength pipes (High Pipe), as shown in Figure 6.

Frost resistance tests were performed for F150 frost resistance degree according to PN-88/B-06250 "Standard concrete". The test procedure started after 28 days. The test results are presented on Figures 7 and 8. Analyzed concrete complied with all the requirements for frost resistance degree F150. Concrete was characterized with slight mass loss (0.6%) and slight compressive strength loss (9.0%).

Tests were also carried out on samples of hardened concrete (28 days) to assess resistance to aggressive environments (sea water and the environment sulfates). After 180 days of storage of

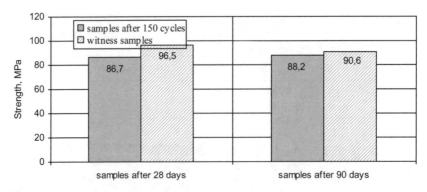

Figure 7.   Frost resistance concrete test – decrease of strength.

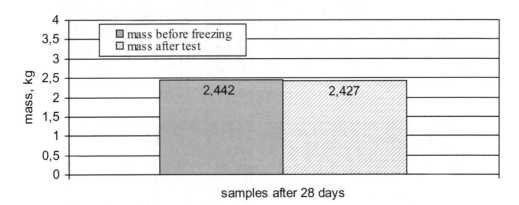

Figure 8.   Frost resistance concrete test – mass loss.

Table 9.   Compressive strength of samples stored in the aggressive solutions.

Sample	Type of environment	Time of storage [days]	Average compressive strength $f_{ck,cube}$ [MPa]	Strength decrease [%]
C 70/85	Artificial seawater consisting of: – 1000 g of water – 30 g of sodium chloride – 6 g of magnesium chloride – 5 g of magnesium sulfate – 1,5 g of calcium sulfate – 0,2 g of potassium carbonate	180	90.5	9.5
C 70/85	Sulfate solution (content of SO_3 16 g/dm^3)	180	96.6	3.4
C 70/85	Water (witness sample)	28	99.9	–

samples in the solutions consistent with Pr ENV 196-X "Methods of testing cement – Determination of the resistance of Cements to attack by sulfate solution or by seawater" samples were subjected to the compressive strength test according to PN-EN 12390-3. The results are shown in Table 9. Samples of concrete after the test maintained the expected level of strength and fit in the proposed strength class (C 70/85).

Chloride ion penetration test was performed after 28 and 90 days of concrete setting in accordance with the procedure contained in ASTM C 1202 to 2005 "Standard Test Method for Electrical Indication of Concrete's Ability to Resist Chloride Ion Penetration". The results compared to the

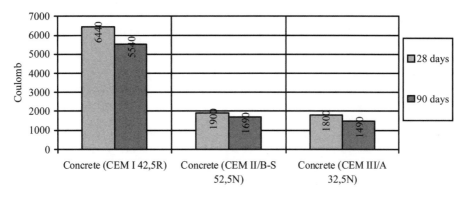

Figure 9.   Chloride ions permeability in hardened concrete.

Table 10.   Chloride Ion Penetrability Based on Charge Passed.

Charge Passed (Coulombs)	Chloride Ion Penetrability
Above 4000	High
2000–4000	Moderate
1000–2000	Low
100–1000	Very Low
Under 100	Negligible

concretes of similar composition, but on other cements are shown in Figure 9. In accordance with the standard requirements (Tab. 10) analyzed samples can be classified as samples with low permeability of chloride ions. This confirms the tight structure of concrete, preferably affecting its durability.

7   CONCLUSION

Hereby study demonstrates the possibility to obtain and use of high-strength concrete (C80/95, C70/85) in the production of reinforced concrete pipes in accordance with the requirements of BS EN 1916 standard. The concrete was characterized by high early and standard strength (28 days), high tightness and resistance to low temperatures (analyzed frost resistance degree of F 150) and aggressive agents. Based on laboratory studies and their afterward industrial verification, we can conclude, that it is reasonable to apply high-class concrete in the manufacture of reinforced concrete pipes. A prerequisite is the use of quality components and strict compliance with the technological regime both at the concrete plant, as well as precast production plant.

REFERENCES

Giergiczny, E., Giergiczny, Z. 2010. The influence of variable fly ash quality on the properties of fly ash-cement composites, *Cement-Lime-Concrete, no 3.*
Hakenberg, J. 1999. Mikrotunelling. *Trenchless Technologies no 3*: 56–59.
Madryas, C. Kolonko, A. Wysocki, L. 2002. *Sewer constructions.* Wrocław.
Malhotra, V.M. Ramezanianpour, A.A. 1994. *Fly ash in concrete.* CANMET. Canada.
Neville, A.M. 2000. Properties of concrete. *Polish Cement.* Cracow.
Nocuń-Wczelik, W. 2005. Silica dust. Properties and application in concrete. *Polish Cement.* Cracow.

Underground Infrastructure of Urban Areas 2 – Madryas, Nienartowicz & Szot (eds)
© 2012 Taylor & Francis Group, London, ISBN 978-0-415-68394-4

Rheological properties of fresh fibre reinforced self compacting concrete

J. Gołaszewski & T. Ponikiewski
The Silesian University of Technology, Poland

ABSTRACT: In the paper the methodology and test results of the investigation are presented and discussed of the influence of steel fibres on rheological and mechanical properties of Steel Fibre Reinforced Self-Compacting Concrete (SFRSCC). The rheological parameters of SFRSCC – behaves as Bingham body, their rheological parameters yield value g and plastic viscosity h were determined to using new kind of rheometer BT2 to mortar and concrete mix research. The mechanical parameter of SFRSCC – the cube compressive strength fc were presented as well. In the research, an experimental verification of and significance of an influence: volume fraction of fibres, fibres factor, lengths and shape of fibres on rheological properties of SFRSCC was investigated. In the paper the results obtained for mixes with 3 kind of steel fibres shapes are presented. Concrete mixtures are proportioned to provide the workability needed during construction and the required properties in the hardened concrete. The length of fibres does not have the significant influence on yield value g and plastic viscosity h of SFRSCC. The significant influence of the length of fibres on plastic viscosity h of tested hooked steel SFRSCC was observed only. The rheological properties of SFRSCC from workability point of view are better than for SCC with other types of fibres.

1 INTRODUCTION

Technology of self-compacting concrete allows shaping structure of engineering objects in the quicker and safer way than in case of concrete with traditional properties. Technological operations of concrete elements forming are in case of self-compacting concrete considerably simplified and end results allow to expose hardened concrete structures in more extended way (Kaszyńska 2003, Martinie 2009). One modification of considered concrete is to add to its volume various kinds of fibres as diffused reinforcement (Brandt 2000, Felekoğlu 2009). This is not a new issue in the technology of concrete, however in case of concrete with self-compacting properties it provides current area of research. Problems resulting from using modified in such way concrete mixes were determined based on carried out tests of workability of fresh self-compacting concrete mix modified with steel fibres in rheological context (Ghanbari 2009). Technological problems in applying self-compacting concrete modified with steel fibres as diffused reinforcement is the subject of the present article.

Analysing of influence of fibres on workability and durability parameters of concrete is one of new tendencies in research of self-compacting concrete (Barragán et al. 2004, Ding Y et al. 2004, Ponikiewski 2005, Ozyurt 2007). Research of steel fibres of various geometric parameters influence were presented to determine the impact of its volume fraction, the length and the shape on rheological and mechanical properties of self-compacting concrete.

Essence of applying steel, polypropylene and other fibres to cement mix has been already discussed in earlier publications (Ponikiewski et al. 2003–2005). General tendency of the improvement of hardened self-compacting concrete characteristics with the increase of contents of fibres in its volume, makes workability of these concrete mixes worse during forming (Brandt 2009, Felekoğlu 2009). The current problem, also in case of self-compacting concrete modified with steel fibres, is technological difficulty of its production and carrying out of technological processes in concrete works (Stroeven 2009). It compels to recognise the real nature of workability and to determine an impact of added fibres on phenomena taking place in fresh and hardened self-compacting concrete.

2 RHEOLOGICAL MODEL AND MEASUREMENTS OF RHEOLOGICAL PARAMETERS OF FRESH CONCRETE

It is well documented that fresh mortar and fresh concrete behaves as Bingham material, whose properties can be expressed by two rheological parameters, the yield stress and the plastic viscosity according the formula:

$$\tau = \tau_o + \gamma \eta_{pl} \tag{1}$$

where τ (Pa) is the shear stress at shear rate γ (1/s), τ_o (Pa) is the yield value and η_{pl} (Pa $\cdot$ s) is the plastic viscosity (Szwabowski 1999, Li 2007, Petit et al. 2007, Feys et al. 2008). The physical interpretation of yield value is that of the stress needed to be applied to a material in order to start flowing. When the shear stress is higher then yield value the mix flows and its flow resistance depends on plastic viscosity.

Rheological parameters of fresh mortar, like those of fresh concrete, can be measured using Two Point Workability Test (TPWT), by applying a given shear rate and measuring the resulting shear stress. Because of the nature of rheological behaviour of cement mixtures, the measurements should be taken at no less than two considerably different shear rates. The rheological parameters are determined by regression analysis according to the relation:

$$T = g + N h \tag{2}$$

where T is the shear resistance of a sample measured at rotation rate N and g (Nmm) and h (Nmms) are constants corresponding respectively to yield value τ_o and plastic viscosity η_{pl}. By suitable calibration of the rheometer, it is possible to express g and h in fundamental units. According to (Banfill 1991), in the apparatus like used in this work, $\tau_o = 7.9\,g$ and $\eta_{pl} = 0.78\,h$, but all results are given below in terms of parameters g and h. The principles of TPWT and rheological properties of fresh cement mortars and concretes are presented in existing literature (Tattersall & Banfill 1983).

It should be noted that rheological properties of cement pastes and cement binder mixtures (mortars and concretes) differ from each other. During the flow test cement paste reveals plastic characteristics with high degree of nonlinearity and with high ticsotropic effects (Szwabowski 1999). As far as Bingham model is adequate to characterize rheological properties of fresh mortar and concrete, characterization of rheological properties of cement paste demands more complex models. It was demonstrated in (Atenzi et al. 1985) that rheological properties of cement paste are best described by the following models: Herschel-Bulkley, Robertson-Stiff and Ellis model. In the same time it was stated that Bingham model may be used for characterization of properties of cement paste only in narrow range.

3 ASSUMPTIONS AND METHODOLOGY OF RESEARCH

Results of workability tests of self-compacting cement mixes modified with steel fibres in rheological context are presented in this paper. Testing carried out with method of rheometrical of workability test (RWT) were conducted with rheometer for mortars and concrete mixes – BT2 (Fig. 1). RWT method was discussed detailed in literature (Szwabowski 1999).

Approximation of measurement results conducted by two-parameter Bingham rheological model and three-parameter Hershell-Bulkey model was done (Fig. 2). It allowed determining two basic rheological parameters – yield value g and plastic viscosity h. The values were determined by two-parameter model.

Composition of the tested self-compacting mixture is presented in table 1. The concrete mix was modified – variable kinds and volume fraction of steel fibres were used. Steel fibres were selected out of a large number of fibres available on the market. Despite of their availability and the variety it is however difficult to purchase fibres of similar geometric parameters and shape. Results of testing of self-compacting mixtures modified with eleven kinds of steel fibres are presented in the article. Tests were carried out in two blocks for four levels of variability. In the first block tests were carried out for variable volume fraction of fibres in the matrix. In the second block a variable level

Figure 1.   Rheometer BT2 to determine rheological parameters of concrete mixes – general view of the apparatus during the measuring procedure.

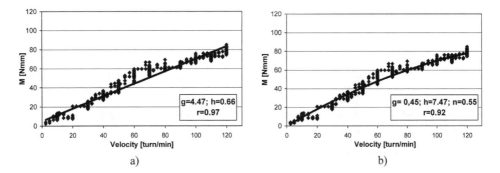

a)   b)

Figure 2.   The rheological behaviour of the fresh concrete; typical example of flow curve and determined rheological parameters as: a) Bingham model; b) Hershell-Bulkey model.

of the fibre reinforcement was examined (fibre factor – F_F), taking geometric parameters of fibres into consideration (length L and diameter d) as well as fibre volume fraction V_f in the mixture, according to the following pattern.

$$F_F = V_f \cdot \frac{L}{d} \tag{3}$$

Taking the level of the fibre reinforcement into consideration in testing (F_F) allows to determine the influence of each parameter that characterise the used diffused reinforcement on workability of self-compacting mixtures in rheological context in more reliable way.

In block I, tested fibre volume fraction in the concrete mixture was 0.5–1.0–1.5–2.0% what corresponds to 39.25–78.50–117.75–157.00 kg/m³ contents. In block II a level of variability (F_F) was considered 0.2–0.4–0.6–0.8, what corresponds to fibre mass that is subject to slenderness of fibres, as presented in Table 2.

Geometric characteristics of tested fibres and fibre volume fraction in concrete mixture according to level of fibre reinforcement were presented in Table 2. The shape of fibres due to variability of their geometry is an additional factor influencing test results but overlapping with considered remaining variable parameters of fibres.

Table 1. Composition of the self-compacting mixture

Component		For batch of concrete	For m³
CEM II B-S 42.5	[kg]	12.3	344
Fly ash	[kg]	4.9	138
Water	[kg]	5.9	164
SP Viscocrete 3	[1.5%]	0.19	5
Aggregate 2–8	[kg]	29.0	810
Sand 0–2	[kg]	27.8	776
Steel fibres	[%]	0.5 – 1.0 – 1.5 – 2.0	
W/(C+SP)		0.34	0.34

Table 2. Geometric characteristics of tested steel fibres and variability of fibres volume fraction level of fibres reinforcement (FF) in self-compacting concrete mixture.

Characteristics of fibres [mm]			Mass of fibres for variable (F_F) [kg]			
Shape	L	d	0.2	0.4	0.6	0.8
Straight	13	0.16	20.93	41.87	62.80	83.73
Straight	25	0.40	25.12	50.24	75.36	100.48
Straight	6	0.16	41.87	83.73	125.60	167.47
Wavy	50	1.00	31.40	62.80	94.20	125.60
Wavy	35	0.80	35.89	71.77	107.66	143.54
Wavy	30	0.70	36.63	73.27	109.90	146.53
Hooked	50	0.45	14.13	28.26	42.39	56.52
Hooked	60	0.65	17.01	34.02	51.03	68.03
Hooked	64	0.80	19.63	39.25	58.88	78.50
Hooked	60	0.80	20.93	41.87	62.80	83.73
Hooked	30	0.50	26.17	52.33	78.50	104.67

4 RESULTS OF TESTS AND DISCUSSION

Properties of self-compacting mixtures modified with steel fibres were tested to determine rheological parameters measured with RTU method. On the basis of pre-examinations, determining relationship between the time and the flow diameter measured with Abram's cone method, an estimated self-compacting limit was deter-mined for tested mixtures with steel fibres, according to the assumption: flow time T_{50} = max 9 seconds and the flow diameter R = min. 600 [mm]. The above mentioned assumptions of the self-compacting limit were obtained for maximal yield value g on the level 600 [Nmm]. Any plastic viscosity h value as a limit one for self-compacting mixtures with steel fibres was unambiguously determined. Figures 3 and 4 present influence of kind and volume fraction of straight steel fibres on rheological parameters of self-compacting mixtures – yield value g and plastic viscosity h value.

The increase of yield value g and plastic viscosity h value along with the increase of straight fibres volume fraction in the considered research area of modified self-compacting mixtures was shown. In this research group (straight fibres), addition of 13 × 0.16 fibres to the mixture resulted in the biggest increase of g parameter and, what follows, workability becomes wrong. Addition of 6 × 0.16 fibres to the mixture resulted however in the smallest increase in the g parameter. Thus the smallest worsening of the considered mixture workability was obtained. In case of plastic viscosity h, the biggest value of this parameter was also obtained for the self-compacting modified mixture with 13 × 0.16 fibres, what also makes workability of considered mixture worse.

Addition of 6 × 0.16 fibres to the mixture resulted in the smallest increase of the h parameter. Thus the smallest worsening of the considered mixture workability was obtained. Similar results

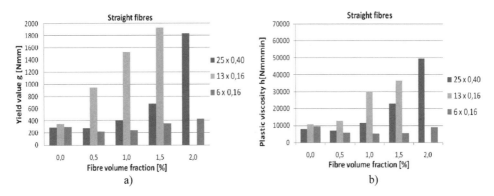

Figure 3. Influence of kind and volume fraction of straight steel fibres on: a) yield value g, b) plastic viscosity h value.

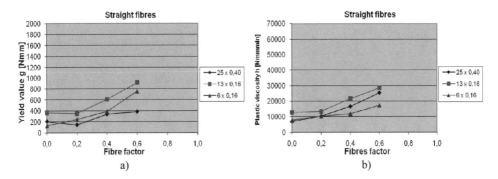

Figure 4. Influence of kind and fibre factor of straight steel fibres on: yield value g, b) plastic viscosity h value.

of examinations of self-compacting mixtures modified with straight fibres were obtained research blocks I and II. Mixtures with the addition of 13×0.16 fibres started not to fulfil conditions for self-compacting mixtures sooner.

Figures 5 and 6 present influence of kind and volume fraction of wavy steel fibres on yield value g and plastic viscosity h value. The increase of yield value g and plastic viscosity h value along with the increase of wavy fibres volume fraction in self-compacting mixtures was shown. In this research group (wavy fibres), addition of 50×1.0 fibres to the mixture resulted in the biggest increase of g parameter in both research blocks and, what follows, the biggest worsening of workability of modified self-compacting mixtures. Condition of self-compacting was obtained for all considered wavy fibres in the whole range of variability of volume fraction. In case of the factor (F_F), self-compacting limit for all wavy fibres was level 0.6. All tested hooked fibres except of the discussed above 64×0.80 fibres fulfilled self-compacting condition within the whole range of fibre reinforcement.

On the basis of carried out tests it is possible to feature estimated brackets of properties of self-compacting mixtures with steel fibres of various geometrical parameters and volume fraction. Hooked steel fibres were the next considered fibres. Influence of these fibres on rheological parameters of self-compacting mixtures was presented on Figures 7 and 8. The increase of yield value g and plastic viscosity h value along with the increase of hooked fibres volume fraction in self-compacting mixtures was shown. In this research group addition of 64×0.80 fibres to the mixture resulted in the biggest increase of g and h parameters and limitative fulfilling of self-compacting condition for volume fraction 0.5. Similar parameters were obtained for fibres 30×0.5. Addition of 60×0.65 fibres to the mixture resulted in the smallest increase in the g parameter. Condition of self-compacting was obtained for volume fraction close to 1.0%.

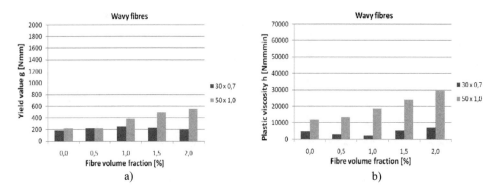

Figure 5.   Influence of kind and volume fraction of wavy steel fibres on: a) yield value g, b) plastic viscosity h value.

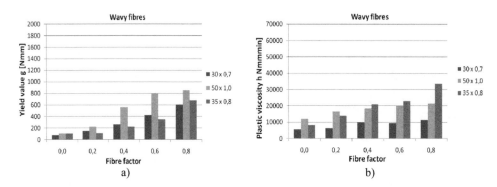

Figure 6.   Influence of kind and fibre factor of wavy steel fibres on: yield value g, b) plastic viscosity h value.

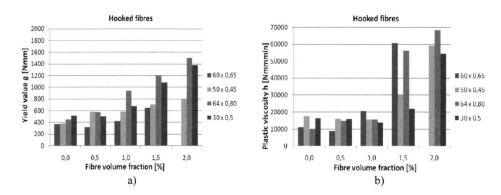

Figure 7.   Influence of kind and volume fraction of hooked steel fibres on: a) yield value g, b) plastic viscosity h value.

Table 3 presents brackets of properties of self-compacting for variable volume fraction together with fibres weight quantity. Table 4 presents brackets of properties of self-compacting for variable level of fibre reinforcement (F_F), together with fibres weight quantity. Lack of self-compacting effect of mixtures in the whole considered block I of added straight fibres 13×0.16 was shown. Total – in block I – self compacting effect of mixtures modified with steel fibres was stated in case of two types of wavy fibres $30 \times 0.7 - 50 \times 1.0$ and straight fibres 6×0.16. For two types of fibres the tests were not carried out in block II.

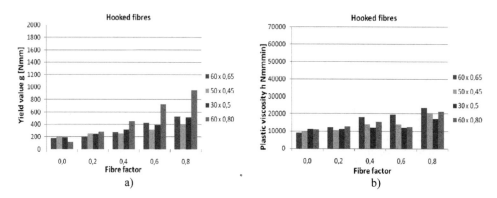

Figure 8.   Influence of kind and fibre factor of hooked steel fibres on: yield value *g*, b) plastic viscosity *h* value.

Table 3.   Brackets of properties of self-compacting for variable volume fraction (V_f) together with fibres weight quantity.

Characteristics of fibres [mm]			Mass of fibres for variable V_f [kg]			
Shape	L	d	0.5	1.0	1.5	2.0
Straight	13	0.16	–	–	–	–
Straight	25	0.40	39.2	78.5	–	–
Straight	6	0.16	39.2	78.5	117.7	157.0
Wavy	50	1.00	39.2	78.5	117.7	157.0
Wavy	35	0.80	nd	nd	nd	nd
Wavy	30	0.70	39.2	78.5	117.7	157.0
Hooked	50	0.45	39.2	78.5	–	–
Hooked	60	0.65	39.2	78.5	–	–
Hooked	64	0.80	39.2	–	–	–
Hooked	60	0.80	nd	nd	nd	nd
Hooked	30	0.50	39.2	–	–	–

Description to tables 3; (nd)-no data available, (–)-condition of self-compacting not fulfilled.

Table 4.   Brackets of properties of self-compacting for variable level of fibre reinforcement (F_F), together with fibres weight quantity.

Characteristics of fibres [mm]		Mass of fibres for variable V_f [kg]			
Shape	L/d	0.2	0.4	0.6	0.8
Straight	75.0	20.9	41.8	–	–
Straight	62.5	25.1	50.2	75.4	100.5
Straight	37.5	41.8	83.7	–	–
Wavy	50.0	31.4	62.8	94.2	–
Wavy	43.8	35.8	71.8	107.6	–
Wavy	42.9	36.6	73.3	109.9	–
Hooked	111.1	14.1	28.3	42.4	56.5
Hooked	92.3	17.0	34.0	51.0	68.0
Hooked	80.0	nd	nd	nd	nd
Hooked	75.0	20.9	41.8	–	–
Hooked	60.0	26.2	52.3	78.5	104.7

Description to tables 4; (nd)-no data available, (–)-condition of self-compacting not fulfilled.

There is a lack – insignificant though – of consequence in results of tests. Straight 13×0.16 fibres in research block I have not indicated any self-compacting effect within the total research area, however in block II these properties were kept up to F_F value 0.4 i.e. for 42 kg/m^3.

Wavy fibres 50×1.0 fibres in research block I have indicated self-compacting effect within the total research area i.e. maximum 157 kg/m^3, however in block II these properties were not kept for F_F value 0.4 i.e. for 125.6 kg/m^3.

Hooked fibres 30×0.5 were the last incorrect case. In block I they indicated self-compacting properties for $V_f = 1.0\%$ i.e. at most for 78.5 kg/m^3, however in block II self-compacting properties were indicated within whole considered research area i.e. even for 104.7 kg/m^3. Any impact of the length of fibres on changes of rheological parameters of the considered modified mixtures was unambiguously determined.

5 THE SUMMARY AND FINAL CONCLUSIONS

Analysis of mutually exclusive factors taking place as a result of adding steel fibres to self-compacting concrete: workability worsening or even loss of self-compacting properties and improvement of self-compacting concrete mechanical properties was the subject of the present article. Presented results of testing self-compacting concrete modified with steel fibres show influence of fibre addiction to worsen workability of fresh mixture and increase in compressive strength of hardened fibre concretes made out of self-compacting mixtures.

To keep self-compacting effect of mixtures modified with steel fibres, the volume fraction of 2.0% seems to be recommended to ensure its maintenance. This in not however the case with all fibres taken under consideration. The number of possible to apply steel fibres to ensure self-compacting effects increases along with the decrease of fibres volume fraction but simultaneously probability to improve mechanical properties drops down.

Problems occur with homogenous filling of concrete volume with the added fibres and the required technological processes for this type of concrete make keeping homogenous structure even more difficult. Pumped self-compacting fibre concrete should be delivered directly to forming place, with limiting of horizontal relocation of mixtures within formed concrete structure. The slenderness and volume fraction of steel fibres in the mixture worsens its workability but improves strength parameters though not for all fibres. Keeping the homogeneity of steel fibres during the process of self-compacting concrete forming is the current research problem.

It seems recommendable to carry out broader re-search to determine influence of steel fibres on properties of fresh and hardened self-compacting concrete based on variability of so called fibre factor. Taking workability under consideration it seems to be proper to add shorter fibres with higher volume fraction into concrete mixture. This should ensure homogeneity of formed concrete structure.

Influence of added fibres shape, important from fibres anchorage energy in self-compacting concrete matrix, has not been unambiguously determined in the research. Currently the author conducts research of relationship between energy to draw fibres out of the concrete matrix and fibres geometric parameters as well as research of the influence of real distribution of diffused reinforcement on concrete compressive strength parameters. It is necessary to remember about the diversified shape of tested fibres together with their diversified slenderness. It is recommendable to carry out additional re-search to eliminate overlapping of variable factors. The broad commercial offer of fibres imposes however some limitations.

REFERENCES

Atenzi, C, Massidda L, Sanna U. 1985 Comparison between rheological models for Portland cement pastes, *Cement and Concrete Research*, 15 (4): 511–519.
Banfill, P F G. 1991. The rheology of fresh mortar. *Magazine of Concrete Research*, 43 (154): 13–21.
Barragán B., Zerbino R., the R. ghetto, Soriano M., de la C. Cruz, Giaccio G., Bravo M. 2004. Development and application of steel fibre reinforced self-compacting concrete, *In 6th RILEM Symposium on Fibre-Reinforced Concretes (FRC) – BEFIB 2004*, Varenna, Italy, 457–466.

Boukendakdji O., Kenai S., Kadri E.H., Rouis F. 2009. Effect of slag on the rheology of fresh self-compacted concrete. *Construction and Building Materials*, Volume 23, Issue 7, 2593–2598.

Brandt A.M. 2000. Applying fibres as reinforcement in to concrete elements, In *Conference: Concrete on the threshold of the new Millennium*, Cracow, Poland, 9–10.11.2000, 433–444.

Brandt A.M. 2009. Cement-based composites, materials, mechanical properties and performance. Routledge, Taylor & Francis Group. London and New York. 526 p. ISBN10: 0-415-40909-8.

Ding Y., Thomaseth D., Niederegger Ch., Thomas A., Lukas W. 2004. The investigation on the workability and flexural toughness of fibre cocktail reinforced self-compacting high performance concrete, In *6th RILEM Symposium on Fibre-Reinforced Concretes (FRC) – BEFIB 2004*, Varenna, Italy, 467–478.

Felekoğlu B., Tosun K., Baraban B. 2009. Effects of fibre type and matrix structure on the mechanical performance of self-compacting micro-concrete composites. *Cement and Concrete Research*, Volume 39, Issue 11, 1023–1032.

Feys, D., Verhoeven, R., De Schutter, G. 2008. Fresh self compacting concrete, a shear thickening material, *Cement and Concrete Research*, 38 (7): 920–929.

Ghanbari A., Karihaloo B.L. 2009. Prediction of the plastic viscosity of self-compacting steel fibre reinforced concrete. *Cement and Concrete Research*, Volume 39, Issue 12, 1209–1216.

Jau W-Ch., Yang Ch-T. 2010. Development of a modified concrete rheometer to measure the rheological behavior of fresh concrete. *Cement and Concrete Composites*, Article in Press. Available online 13 January 2010.

Kaszyńska M. 2003. Mix design of the self-compacting concrete, In: *Proc. Int. Symp. 'Brittle Matrix Composites 7'*, A.M. Brandt, V.C. Li, I.H. Marshall, Warsaw, 13–15.10.2003, 331–338.

Li, Z. 2007. State of workability design technology for fresh concrete in Japan, *Cement and Concrete Research*, 37 (9): 1308–132

Martinie L., Rossi P., Roussel N., 2010. Rheology of fiber reinforced cementitious materials: classification and prediction. *Cement and Concrete Research*, Volume 40, Issue 2, 226–234.

Ozyurt N., Mason T.O., Shah S.P. 2007. Correlation of fiber dispersion, rheology and mechanical performance of FRCs. *Cement and Concrete Composites*, Volume 29, Issue 2, 70–79.

Petit, J-Y., Wirquin, E., Vanhove, Y., Khayat, K. 2007 Yield stress and viscosity equations for mortars and self-consolidating concrete, *Cement and Concrete Research*, 37, (5): 655–670.

Ponikiewski, T., Szwabowski J. 2003. The influence of selected composition factors on the rheological properties of fibre reinforced fresh mortar, In: *Proc. Int. Symp. 'Brittle Matrix Composites 7'*, A.M. Brandt, V.C. Li, I.H. Marshall, Warsaw, 13–15.10.2003, 321–329.

Ponikiewski T. 2004. Aspects of the selection of fibres from the point of view of the technology of the concrete mixture, In *6th Rheological Seminar*, Gliwice, Poland, 2004.

Ponikiewski T. 2005. Influence of steel fibres on rheological and mechanical properties of self-compacting concrete, In *7th Rheological Seminar*, Gliwice, Poland, 2005.

Stroeven P., He H. 2009. Patches in concrete: recent experimental discovery of a natural phenomenon – supporting evidence by dem In: *Proc. Int. Symp. 'Brittle Matrix Composites 9'*, A.M. Brandt, J. Olek, I.H. Marshall, Warsaw, 2009, 399–408.

Szwabowski J., 1999. *Rheology of mixtures on cement binders*, Printing House of Silesian Technical University, Gliwice, Poland.

Tattarsall, G. H., Banfill, P. F. G. 1983. *The Rheology of Fresh Concrete. Boston*: Pitman Books Limited. 356 p.

Underground Infrastructure of Urban Areas 2 – Madryas, Nienartowicz & Szot (eds)
© *J. Jakubowski and J.B. Stypułkowski, ISBN 978-0-415-68394-4*

Top of rock investigations for secant piles at the Bronx shaft

J. Jakubowski
AGH University of Science and Technology, Krakow, Poland

J.B. Stypułkowski
HNTB Corporation, New York, USA

ABSTRACT: This paper describes the results of phased geotechnical investigation on determination of depth to rock leading to installation of the secant pile wall on the shaft site. Top of rock (TOR) predictions from borings at subsequent stages of the project has been shown and compared with top of rock from pile drills and top of rock scanned after soil excavation inside the shaft. Commonly used linear interpolation method is compared with multiple linear least squares regression. Models have been evaluated with inaccuracy measures: confidence intervals and deviations from reference top of rock elevations. Further simulations are presented to examine TOR prediction inaccuracy for various shaft locations and diameters. Applied data analysis methods and general conclusions and recommendations have been presented.

1 INTRODUCTION

1.1 *Harlem River Tunnel project and Bronx shaft*

The Harlem River utilidor tunnel is a part of the new underground transmission feeder, and is housing a number of utility lines: one 345 kV Oil-Filled Feeder, three 345 kV Solid Dialectric Feeders, one 16" OD Gas Main, and sixteen 15 kV Distribution Feeders.

The major elements of the project include: a 700 feet long, 13 by 12 feet horseshoe-shaped tunnel bored 150 feet below ground surface in hard rock, two 24 feet ID shafts, concrete lined shallow vaults/pull chambers connected to the shafts, ventilation and cable chamber and the head house. The shaft is located in the parking lot adjacent to Metro-North railroad tracks.

The shaft wall consisted of 44 secant piles 1 m in diameter overlapping piles in closed circular alignment installed at the 35 ft diameter (Fig. 1). The piles were installed using Delmag RH28 drill rig: the primary piles first, then the secondary piles between the primary ones. The secondary piles hold a steel I-beam in their core as reinforcement for full depth. A grouting tube was attached to the I-beam to facilitate contact grouting (Mooney et al. 2008; Stypulkowski et al. 2009; Stypulkowski et al. 2010).

The site is geologically located in the New England Upland, a division of the Appalachian Highlands that includes all of New England and parts of New York, New Jersey and Pennsylvania. The area consists primarily of metamorphic and igneous rocks. The site lies on a subdivision of the Manhattan Prong called the Inwood Lowland, which consists of Inwood Marble. The units have been subjected to folding and faulting. The Inwood Marble consists of calcite and dolomite marbles and calcareous schist. Across the site, the rock is overlain by a relatively thin, but fairly continuous, layer of glacial basal till. In the Bronx, the varved soil and till was eroded away when a layer of clean outwash sand was later deposited in its place. The corresponding outwash is overlain by silts and fine sands typical of a low energy environment, such as a lake or abandoned stream meander. Above this lies a layer of soft organic silty clay and peat.

1.2 *Top of rock definitions and identification criteria*

Top of rock estimation is an important element of secant piles installation for technical and economic reasons, and it is a component of bid documentation. Its overestimation or underestimation may

Figure 1.   Shaft after partial soil excavation with cap beam and secant piles installed.

result in technical problems and additional costs. Both contractor and investor have interest in accurate estimation of the TOR.

The data for top of rock estimation prior to secant piles installation are usually provided by core drilling. First borings are usually done at the planning stage of the project, during initial field investigation of geological conditions or in order to confirm available data. TOR estimation is usually not the main purpose of these borings. They are not necessarily located in the vicinity of secant piles installation since it is rarely fixed at that stage. For technical and legal reasons, the location of secant piles installation is often not available during field investigations. The initial borings are often done in phases, by several contractors and often with different precision.

Defining the top of rock from drilling operations is difficult, especially where large boulders exist, below irregular residual soil profiles. Determination of the top of rock must be done with care, as an improper identification may lead to miscalculated rock excavation volume or erroneous pile length estimation. In the USA an ASTM D 2113 is commonly used, where TOR is simply defined as a depth when core drilling procedures are used when formations encountered is too hard to be sampled by soil sampling methods. In addition a penetration of 1 inch or less by a 2 inches diameter split-barrel sampler following 50 blows using standard penetration energy can be used by the field inspector. In NYC depth of weathering is limited and often less then 5 ft which is a standard interval of SPT testing. In selected instances, geophysical methods, such as seismic refraction, can be used to assist in evaluating the top of rock elevations.

NYC Building Code is very specific about type of rock which is acceptable for foundations. For the purpose of secant pile installation at the shaft site it was anticipated that "intermediate rock" would have to be found to provide adequate bearing for piles. The intermediate rock characteristics are: "the rock gives dull sound when struck with pick or bar; does not disintegrate after exposure to air or water; broken pieces may show weathered surfaces; may contain fracture and weathered zones up to 1 inch wide spaced as close as 1 foot; the RQD with a double tube, NX size diamond core barrel is generally 35 to 50% for each 5 feet run, or a core recovery with BX-size core of generally 35 to 50% for each 5 feet run". The designer opted for defined top of the rock in GBR as: "the level at which rock core recovery is 50% or greater for each 5 feet run". This approach follows older edition of NYC BC recommendations for intermediate rock.

1.3   Borings in the area of the shaft

Subsurface investigation performed in and before 2007 resulted in five borings in the vicinity of the shaft, designated M-5, M-23, M-33, M-35 and M-36. Those five borings were available to the bidding contractor to establish depth to rock and plan secant pile installation work. The second set of five borings numbered KB-60 and KB-75, LB-5, LB-6 and LB-8 was performed for contractor prior to the secant wall installation. The secant pile wall was installed between October 2008 and December 2008. More borings have been drilled after completion of the secant pile wall. Among them 16 were associated with secant piles, outside and inside the outer diameter of the secant pile wall Additional 4 borings were drilled for special investigation by the United States Geological Services (USGS).

2 APPLIED APPROXIMATION METHODS

2.1 *General remarks*

Top of rock is measured at borings as the level where soft drilling equipment meets refusal and these measurements are the base for top of rock surface estimation. The collected data for the site includes boring location coordinates and depth to top of rock from the borings. The series of these measurements is a base for the surface of TOR approximation. When having a model of surface approximated, one can predict TOR elevation at several secant piles locations, which is the final result of the procedure. It is important to understand, that TOR surface model is estimated based on data from borehole locations and then used for TOR prediction at secant piles locations. It is certainly desirable that boreholes locations are selected as close to secant piles as possible, but this is not always possible, especially at the early stage of a project. It is also desirable, that inaccuracy of this TOR prediction at secant piles is estimated.

For the purpose of TOR prediction at secant piles various surface models can be applied. The application of linear interpolation and ordinary least squares regression model is presented below. The general least squares regression presented here provides solution to both linear and nonlinear regression, but planar model with two independent variables (coordinates) is considered to be the optimal model here. Regression analysis allows to examine planar model residuals and reject the model, if data does not fit it. Even in such case, there is usually no evidence that any other, particular model better fits data, especially if few measurements are available. Rock surface in the Bronx shaft area is described in preexisting geological descriptions as horizontal and quite planar, with possible boulders in overlying soil that may lead to uncertain measurements. There is also a low number of borings at the stage of bidding documents preparation, so there is little alternative to planar model in practice.

2.2 *Linear interpolation*

The commonly used procedures of top of rock estimation are based on linear interpolation. Top of rock is measured in the available boring locations. Then linear interpolation and smoothing is used to produce a contour plot of the rock surface elevations with use of popular software tools such as Surfer or Statistica. Then elevations for all secant piles locations are estimated. Unlike in statistical approach, all measurements are treated as error-free. So the interpolation method is locally sensitive to measurement errors, so even single imprecise or erroneous TOR level measurement from boring may lead to possible large local errors. It can be particularly imprecise, when there are only few measurements available from borings located in some distance from the secant piles planed location. It also does not allow assessing the approximation inaccuracy based on data. The linear interpolation method inaccuracy is sometimes accepted as ±5 ft based on engineer's experience (MTA 2010).

2.3 *Least squares regression*

This model is a linear multiple regression model with two independent variables x_1, x_2 (coordinates) and TOR elevation from borings as dependent variable y. We consider a model, following the equation below for each case:

$$y = \beta_0 + x_1\beta_1 + x_2\beta_2 + \varepsilon \tag{1}$$

Let $\mathbf{X}$ be a $n \times 3$ matrix with ones in the first column and vertical coordinates of n borings in the second and third column. Each row $(1, x_1, x_2)$ of matrix $\mathbf{X}$ represents a single boring vertical coordinates. Vector $\mathbf{y}$ is a vector of measurements from n borings. The model can be presented in matrix notation as follows (Drapper & Smith 1966):

$$\mathbf{y} = \mathbf{X}\boldsymbol{\beta} + \boldsymbol{\varepsilon} \tag{2}$$

where $\boldsymbol{\beta}$ is a vector of 3 regression parameters $(\beta_0, \beta_1, \beta_2)$, $\boldsymbol{\varepsilon}$ is a vector of n independent residual variables. The expected value vector and covariance matrix are equal respectively:

$$E(\boldsymbol{\varepsilon}) = 0; \quad D^2(\boldsymbol{\varepsilon}) = \mathbf{I}\sigma^2 \tag{3}$$

The sum of $\boldsymbol{\varepsilon}$ element squares is equal to $\boldsymbol{\varepsilon}'\boldsymbol{\varepsilon}$. Minimization of $\boldsymbol{\varepsilon}'\boldsymbol{\varepsilon}$ gives:

$$(\mathbf{X}^T\mathbf{X})\mathbf{b} = \mathbf{X}^T\mathbf{y} \tag{4}$$

where $\mathbf{b}$ is the least squares estimator of $\boldsymbol{\beta}$. If the equations represented by (4) are independent, matrix $\mathbf{X}^T\mathbf{X}$ is not singular and:

$$\mathbf{b} = (\mathbf{X}^T\mathbf{X})^{-1}\mathbf{X}^T\mathbf{y} \tag{5}$$

Let $\mathbf{X_0}$ be the matrix of k secant piles locations, where we intend to predict TOR values with the regression model. The $\mathbf{X_0}$ matrix is a $k \times 3$ matrix with the first column of ones and the second and third column of vertical coordinated of secant piles. The predicted values vector $\hat{\mathbf{y}}_0$ is given by the model:

$$\hat{\mathbf{y}}_0 = \mathbf{X_0}\mathbf{b} \tag{6}$$

2.4 Outlier identification

The vector of residuals and the mean square estimate of variance are given by:

$$\mathbf{e} = \mathbf{y} - \hat{\mathbf{y}} \tag{7}$$

$$s^2 = \frac{(\mathbf{e}^T\mathbf{e})}{(n-3)} \tag{8}$$

Let $\mathbf{x}$ be a single observation vector and a row of $\mathbf{X}$ matrix. For this observation the 95% residual confidence interval about e can be calculated as:

$$e \pm t\{(n-3), 0.975\} \cdot s_{BT} \cdot \sqrt{1 - \mathbf{x}^T(\mathbf{X}^T\mathbf{X})^{-1}\mathbf{x}} \tag{9}$$

$$s_{BT}^2 = \frac{(n-3)s^2}{(n-3-1)} - \frac{e^2}{(n-3-1)\cdot(1 - \mathbf{x}^T(\mathbf{X}^T\mathbf{X})^{-1}\mathbf{x})} \tag{10}$$

where s_{BT} is the residual mean square for observation $\mathbf{x}$, when this observation is omitted, given by Beckman and Trussell (Chatterjee & Hadi 1986). These confidence intervals plotted with case order residual plots are used here for outliers identification.

2.5 Confidence intervals for least squares regression

Further we assume normal distribution of residuals $\boldsymbol{\varepsilon}$. Confidence interval at 0.05 confidence level, for expected average value of y_0 at point $\mathbf{x}_0(1, x_1, x_2)$ is described by the following expression:

$$\hat{y}_0 \pm t\{(n-3), 0.975\} \cdot s \cdot \sqrt{\mathbf{x}_0^T(\mathbf{X}^T\mathbf{X})^{-1}\mathbf{x}_0} \tag{11}$$

where $\hat{y}_0$ is the predicted value, t{} is a Student's t distribution with $n-3$ degrees of freedom and cumulative distribution value equal to 0.975, s^2 is the residual mean square estimate of σ^2.

Table 1. Regression summary.

	b_0	b_1	b_2	F statistics	F df	p-value
Regression 5	−63.3384	−0.0948	0.1056	4.091	2; 2	0.1964
Regression 4	−62.1147	−0.2000	0.1400	85.87	2; 1	0.0761
Regression 10	−61.5304	−0.0994	0.1122	3.116	2; 7	0.1077
Regression 9	−60.9210	−0.1854	0.1023	7.546	2; 6	0.0230
Regression 30	−61.2600	−0.1005	0.1006	17.81	2; 27	0.00001

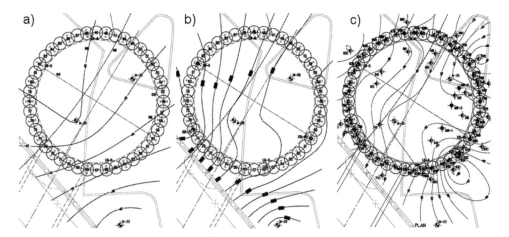

Figure 2. Top of rock surface projected with a linear interpolation based on (a) 5 borings prior to bidding documentation, (b) 10 borings prior to secant piles installation, (c) 30 borings.

3 TOP OF ROCK ESTIMATION AT THE SHAFT

3.1 TOR estimation at secant piles based on 5, 10 and 30 borings

Five borings were available for TOR estimation in bidding documents. Figure 2a shows result of linear interpolation between 5 borings and smoothing. Based on this map TOR surface elevation for secant piles locations is estimated.

Ordinary least squares regression based on 5 initial borings was estimated and evaluated. Residual case order plot (Fig. 3a) shows, that for boring M-23, the 95% residual confidence interval (see chapter 4.3) does not cross the zero reference line and this indicates that this measurement can be an outlier. M-23 is located a relatively long distance from secant piles and from the other borings so it influences the model to a large extend and therefore may easily distort it. Evaluation of the 4 points regression with M-23 excluded shows, that it is significant (at 0.1 level). Removing outlier in case of such small samples it is usually controversial, but we decide to remove M-23 and rely on 4 points instead of 5 (Tab. 1).

10 borings were available before installing secant piles. Following the steps in previous paragraph, linear interpolation (Fig. 2b) and linear least squares regression (Tab. 1) were used to estimate TOR at secant piles. Residual order plot indicated M-23 to be an outlier, so it has been removed from the sample (Fig. 3b). The remaining 9 points regression was significant.

20 subsequent borings were made before the shaft was excavated. For the full set of 30 borings linear interpolation (Fig. 2c) and ordinary regression (Tab. 1) were executed according to the same procedure to compare with previous models. Based on 30 measurements an ordinary regression was significant.

Table 1 presents an ordinary multiple regression summary. Figure 4 shows predicted values for secant piles locations estimated by linear interpolation and regression as described above.

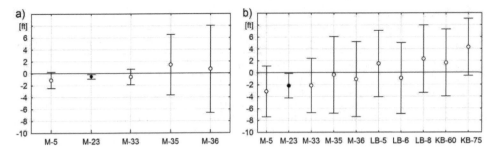

Figure 3.   Case order plots used for outliers identification. a) Case order plot for 5 points. b) Case order plot for 10 points.

4   INACCURACY OF TOP OF ROCK PREDICTION

4.1   *Prediction inaccuracy: uncertainty and error*

We consider approximation (interpolation and regression) procedure leading to TOR prediction at secant piles as an indirect measurement of TOR based on borings. Prediction inaccuracy, uncertainty and error is then consequently considered as referring to an indirect measurement.

Inaccuracy of prediction is a characteristic of the degree of deviation of a predicted value from the true value of the considered physical quantity. Quantitatively inaccuracy can be characterized either as a prediction uncertainty or a prediction error. Prediction error in absolute form is a difference between the predicted value and the known true value. Prediction uncertainty is an interval within which an unknown true value of a predicted physical quantity lies. Uncertainty is defined with its limits and corresponding confidence probability. The limits are statistics calculated as a confidence interval (11) or a prediction interval limits.

If there is no data available, quantitative estimation is difficult or not possible, prediction uncertainty can sometimes be suggested by engineer's experience.

4.2   *Model evaluation based on statistics*

The statistical approach allows to estimate the inaccuracy based on the sample – the same sample that is used for the model estimation. This approach is not applicable to interpolation procedure. For all regression models a confidence interval at given confidence level can be calculated for each secant pile. Figure 5 shows 95% confidence intervals for the Regression 9 model based on 9 borings. A confidence interval was computed for each secant pile location, giving confidence band. At each secant pile a confidence interval is different and depends on the measurement at borings, regression model features and the secant pile location. To compare and evaluate TOR regression models we need a single characteristic of confidence interval over all secant piles. We elect this confidence interval characteristic for a model to be a mean of 95% confidence half-intervals over all 44 secant piles (c_ω). The summary of mean confidence half-intervals for the considered regression models is presented in Table 2.

For the interpolation the statistical approach and uncertainty evaluation is inapplicable, so we must rely on the general engineer's experience. We assume 5 ft is the inaccuracy half-interval.

4.3   *Reference top of rock elevation*

It is impossible to compare TOR predictions achieved with the interpolation or regression with any real, directly measured value. Each TOR measurement procedure is determinating some different physical quantity or is based on different TOR definition. There are no known to the authors studies proving any clear relations between them. So there is no obvious alternative to TOR approximation procedure that could serve as a true reference for the values predicted with models. We propose one for the purpose of this study.

In case of Bronx shaft, in addition to TOR prediction from borings, two other procedures were applied for TOR measurement at secant piles locations: TOR from pile drills (T_{pd}) when pile drill

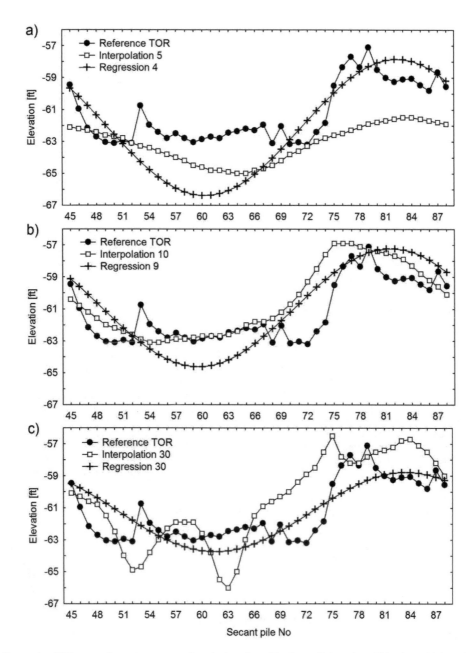

Figure 4. TOR approximated at secant piles: (a) based on 5 borings, (b) based on 10 borings, (c) based on 30 borings.

indicates rock surface, and from uncovered rock surface (T_{us}) after soil is excavated from the shaft and the rock surface is washed and cleaned. Outcome of both measurement methods application for each secant pile is presented on Figure 6.

It is clearly visible on this figure, that both above mentioned measurements T_{us} and T_{pd} are coupled but they measure different physical quantities, having shifted value (with the mean difference through all secant piles 4.27 ft). This is confirmed by Wilcoxon's signed rank test for dependent samples. The test indicates that there are significant differences between groups of

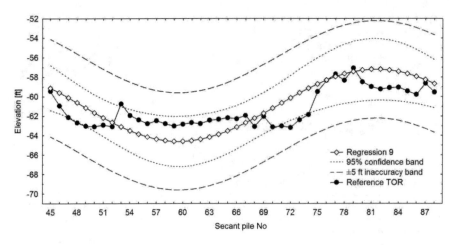

Figure 5.   Uncertainty bands at secant piles locations. 95% confidence band based on 9 borings regression, and ±5 ft inaccuracy band based on engineer's experience.

Table 2.   Model evaluation over all secant piles

		Uncertainty based model evaluation		Reference TOR based model evaluation			
Description	Model	Mean 95% confidence half-interval c_ω [ft]	Experience based [ft]	Mean difference d_ω [ft]	Mean absolute difference a_ω [ft]	Root mean sq. difference q_ω [ft]	Maximum absolute difference m_ω [ft]
Prior to bidding, 5 borings	Interpolation 5	–	5.00	1.86	1.92	2.26	4.80
	Regression 4	4.65	5.00	0.85	1.70	2.07	3.66
Prior piles installation, 10 borings	Interpolation 10	–	5.00	−0.73	1.08	1.51	4.25
	Regression 9	2.47	5.00	−0.34	1.43	1.60	3.05
After piles installation, 30 borings	Interpolation 30	–	5.00	−0.72	1.72	2.09	4.35
	Regression 30	1.09	5.00	0.00	1.02	1.19	2.31
Based on reference TOR at piles	Regression bls	1.11	–	0.00	0.89	1.09	2.32

T_{us} and T_{pd} measurements ($p < 0.000001$). When one compares two dependent samples: T_{us} and shifted $T_{pd} + 4.27$ ft, the same test does not indicate significant differences between two dependent samples, with high p-value of 0.879.

For the purpose of prediction evaluation we assume, that the real TOR is represented by the arithmetic mean of TOR from pile drill and TOR from uncovered rock surface. So for each secant pile location:

$$T_o = \frac{T_{pd} + T_{us}}{2} \qquad (12)$$

Let $\mathbf{T_0}$ be a vector of T_o values for all secant pile locations. We use $\mathbf{T_0}$ as a reference TOR for further model evaluation. It is presented on Figure 6 as line connected dots. The reference TOR changes rapidly between subsequent points, and the reference TOR difference between neighboring piles distanced 2.5 ft reaches 2.65 ft. This gives some hint for surface variation rank contributing to any model prediction inaccuracy.

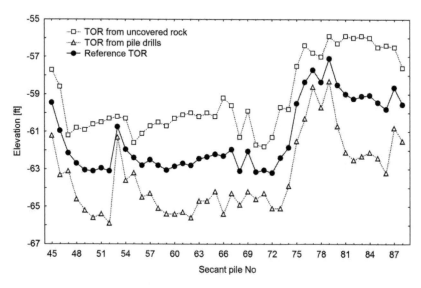

Figure 6.   Top of rock measured from pile drills (T_{pd}), from uncovered and washed rock surface (T_{us}) and the reference elevation (T_o) as a mean of the above.

4.4   Model evaluation based on the reference TOR

TOR prediction error changes between secant piles. For the purpose of model evaluation, we need an overall characteristics describing prediction error over all secant piles. For this purpose, we use the following characteristics of model ω:

1. Mean difference (d_ω), that is the arithmetic mean of difference between a model predicted result and a reference TOR, through all secant piles
2. Mean absolute difference (a_ω), through all secant piles
3. Root mean square difference (q_ω), through all secant piles
4. Maximum absolute difference (m_ω), through all secant piles

These characteristics were calculated for all considered interpolation and regression models. Results are presented in Table 2.

The best performing linear interpolation model is based on 10 (not on 30) borings. It reflects the major shortcoming of the interpolation method: it is sensitive to TOR measures from borings inaccuracies. Even a single inaccurate TOR measurement from borings can affect the interpolation model locally but very strongly. It is shown by interpolation traces through secant piles visible at Figure 4c. The increase of the number of borings from 10 to 30 does not result in better interpolation accuracy.

Regression models, due to statistical averaging effect, reduce the influence of a single erroneous measurement, and also improve performance with the number of observations (Tab. 2 and Fig. 4).

Based on 5 and 30 borings regression models gave significantly more accurate results, than respective interpolation models. For 10 borings, interpolation gives lower mean absolute difference but higher maximum absolute difference through secant piles, so in balance comparison is indistinct.

4.5   Regression based on the reference TOR at secant piles

All the previously considered interpolation and regression models were estimated from borings, and evaluated at secant piles locations. Now we consider a model estimated from reference TOR at secant piles and evaluate it with the same data. Since it is based on information available after the secant piles installation, it doesn't have any practical use for TOR prediction. But it can serve for other models evaluation.

The least squares regression model based on reference TOR at secant piles (Regression bls) is minimizing the sum of model residual squares. In this sense it is the "best possible" linear multiple regression model evaluated at secant piles.

The evaluations of the Regression bls model inaccuracies have been listed in Table 2. Inaccuracy measures c_ω and m_ω have little lower values for Regression 30 than Regression bls model because the later one is minimizing the sum of residual squares, not other inaccuracy measures. Still it can work as the "best possible" reference model for borings based models. It is limited by the given reference TOR distribution at secant piles. Regression 30 model evaluations fits very well, and one can't expect any notably better model (of linear multiple regression). Taking into account very low number of borings Regression 4 appeared to perform very well, either. Other models can also be evaluated, having Regression bls as a reference.

5 FURTHER INACCURACY CONSIDERATIONS

5.1 *The number of borings*

Model inaccuracy is a random variable and depends, among others, on the number of borings and on a particular borings selection. In other words, the n boring model result depends on the selection of the particular n borings.

The question is: how many borings do we have to drill to provide some level of TOR accuracy at secant piles? We would like to answer this question in general, for a model based on data available from all 30 borings. To solve this problem we apply MC simulation. This is the plan for the study:

1. Sample size n between 5 and 25 will be subsequently investigated.
2. Pick randomly n out of 30 available borings, with no repetition. For each sample run regression and evaluate it.
3. There are $30!/n!(30-n)!$ possible picks. We assume the sample of 1000 sufficient and representative for the investigated parameters for each n.
4. Repeat step 2 for 1000 times. Compute some summary statistics for each n.
5. The above 2–4 execute for $n = 5, 6, \ldots, 25$.
6. The result is a overall inaccuracy as a function of n.

The above procedure demands over 20000 regression analysis runs. The key issue is an automated model parameters estimation and evaluation, which is easy with Statistica or Matlab scripting languages. Results of the simulation procedure described above are presented on Figure 7. The following summary statistics are elected to characterize regression performance for each n:

$\bar{a}$ – mean a_ω through 1000 regression models,
$\bar{m}$ – mean m_ω through 1000 regression models,
$\bar{c}$ – mean c_ω through 1000 regression models,
$\bar{c}'$ – mean c_ω through 1000 significant regression models only.

Statistical significance at (0.1 level) is determined for each regression. If F test p-value is larger than 0.1 the regression is insignificant and another set of n borings is randomly sampled, until 1000 significant regressions are run. As it can be seen on Figure 7, the distinction between all regressions and significant regressions summary is observed up to $n = 10$, then there is no considerable distinction. This is because with growing n regression performance improves and the number of insignificant regressions is falling down rapidly.

The results of the above procedure are interesting but affected by the limited total number of 30 available borings. Locations of available borings (at least of some of them) is not random. If the goal for borings is the TOR surface identification at secant piles, the boring locations are spread around secant piles in their vicinity. Random sampling of n borings may result in selection of n borings on one side of the shaft, or all located long distance from the shaft. This is valid particularly for small n and is probably increasing the overall inaccuracy characteristics for small n. The limited number of 30 borings is lowering general variation between samples and within samples, that is might decrease the overall accuracy, especially for large n.

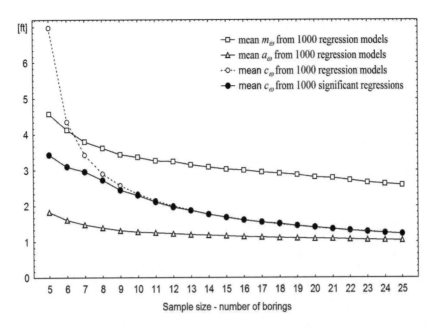

Figure 7. TOR at secant piles approximation inaccuracy as a function of the number of borings ($n = 5$ to 25). Summary of the MC simulation of 1000 random sets samples of size n. The inaccuracy is assessed by summary statistics for each n.

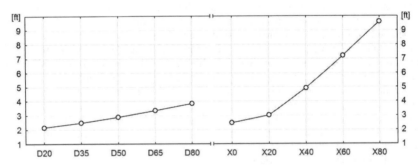

Figure 8. Mean 95% confidence interval through secant piles for various hypothetically considered shaft diameter and location.

5.2 Secant piles locations

It is interesting to investigate the inaccuracy of model for various possible secant pile set locations. The reference TOR elevation is known only for the given secant piles locations. So for considering approximation inaccuracy at other secant piles locations, only the statistical approach is available. As alternative locations of secant piles, we consider:

– shaft axis shifted 20, 40, 60 and 80 ft towards west, all shafts are 35 ft diameter,
– shaft diameter equal to 20, 35, 50, 65 and 80 ft, with no shaft axis shift.

For each case, we calculate 95% confidence interval at each secant piles. Then mean half-interval ( ) is calculated over all secant piles. The results are presented on Figure 8. D65 refers to shaft of diameter 65 ft (secant pile circle diameter in fact) and of original location. X40 refers to original shaft diameter of 35 ft and location shifted 40 ft towards west. Other cases are named respectively. D35 and X0 is the case referring to the original location of the shaft and its (secant piles circle) diameter is 35 ft.

65

Shaft diameter increase to 80 ft slightly increases the model inaccuracy measured by mean 95% confidence interval through secant piles. Shifting the shaft west more than 20 ft results in rapid increase of inaccuracy.

6 SUMMARY

Phased top of rock investigation for secant piles installation was presented. The study shows, that neither TOR from uncovered rock surface nor TOR from pile drills are projecting TOR from borings. These are clearly measurements of 3 different physical quantities. But the arithmetic mean of TOR from uncovered rock surface and TOR from pile drill, fits very well models based on borings, and seems to provide a good representation of the TOR from borings. It was not expected, but became an interesting observation.

Commonly used linear interpolation method was compared with multiple linear least squares regression. Models have been evaluated by inaccuracy measures: confidence intervals and errors based on a reference TOR.

In this study we consider 5+ borings sample, as it was available at Bronx shaft. Approximations from smaller samples are doubtful in general and not recommended. Some other models and inaccuracy estimation methods should be applied for such problem consideration.

The results showed the drawback of linear interpolation method that is its sensitivity to measurement error, resulting in large local inaccuracies of interpolation model. Unlike interpolation, the statistical approach of regression allows to smooth some measurement errors. It also provides estimations of model inaccuracy based on model statistics. This is particularly valuable in practice, since no reference values are available before the secant piles are installed. These inaccuracy measures based on statistics reflect site-specific features, like borings locations, borings number, secant piles locations and the real top of rock surface variability. The examined multiple linear regression models show good performance, in general notably better, than interpolations. Regression seems to provide good means for TOR approximation from borings.

The case study confirmed that ±5 ft inaccuracy interval is sufficient for the purpose of TOR determination at 35 ft shaft and for inclusion in the contract documents. It is worth to note, that the interval width is only partially resulting from the elevation measurement error, rock surface variation and unevenness or model inaccuracy. In large extend it is related to indistinct TOR definitions and identification criteria.

Our recommendation for the GBR writers would be to drill at least 5 borings in the each shaft vicinity, assign TOR with regression method and specify the real TOR as located within ±5 ft interval. The necessary interval can possibly be smaller, depending on the specific regression results, and the number of available borings.

REFERENCES

Chatterjee, S. & Hadi, A.S. 1986. Influential Observations, High Leverage Points, and Outliers in Linear Regression. *Statistical Science* 1(3): 379–416.
Drapper, N.R. & Smith, H. 1966. *Applied Regression Analysis*. John Wiley & Sons.
Johnson, H.L. 1982. *An Introduction to Error Analysis. The Study of Uncertainties in Physical Measurements*. Oxford University Press, University Science Books.
Mooney, J. & Stypulkowski. J. 2008. Two HDD crossings of the Harlem River in New York City. In C. Madryas, B. Przybyla & A. Szot. (eds), *Underground Infrastructure of Urban Areas. Wroclaw University of Technology.*: 213–223. CRC Press. Taylor & Francis Group
MTA 2010. *C-26008 GBR 86th St Station*. New York: Metropolitan Transportation Authority.
Stypulkowski, J., Mooney, J. & Lacy, H. 2009. Tunneling under Harlem River. In *RETC – Rapid Excavation & Tunneling Conference & Exhibit*, June 14–17, Las Vegas, Nevada. pp 48–57.
Stypulkowski, J., Villani, L.N. & Forsyth, G.F., Lacy, H, 2010. Drill and blast tunnel in Inwood Marble under Harlem River in New York City. In *EUROCK10 ISRM European Rock Mechanics Symposium*: 479–482.

Underground Infrastructure of Urban Areas 2 – Madryas, Nienartowicz & Szot (eds)
© 2012 Taylor & Francis Group, London, ISBN 978-0-415-68394-4

A case study of shield TBM tunnel faced to fault zone under Han River

B.S. Kim, Y.K. Kim & W.I. Jung
Korea Water Resources Corporation, Korea

S.W. Lee
Korea Institute of Construction Technology, Korea

Y.S. Yang
Sambu Construction Company, Korea

ABSTRACT: The Gimpo river-bed tunnel under Han River was planned to provide a stable water supply to Gimpo city and is under construction by using a small diameter slurry shield TBM to minimize environmental damages. The geological survey was conducted and showed the mixed layer of weathered rock and soft rock. However, the shield TBM was faced to unexpected geological conditions. This paper deals with the case study of a water transfer tunnel by slurry shield TBM. The project outline and unexpected geological conditions are described. The troubleshooting and suggestions to solve this difficulty are discussed. Especially the stabilization of tunnel face and the improvement of the TBM driving ratio by controlling mixture ratio of slurry contents are highlighted.

1 INTRODUCTION

The shield TBM has been widely used for tunneling in difficult ground conditions, such as high groundwater pressure and fractured zone (Jurbin 2008, Shirlaw & Boones 2008, Zhao et al. 2007, Lee & Kim 2010, Lee et al. 2011). The Gimpo river-bed tunnel under Han River was planned to provide a stable water supply to Gimpo city. During the design step, the geological survey was conducted. The results showed the mixed layer of weathered rock and soft rock. The use of a small diameter slurry shield TBM was decided to minimize environmental damages. During the construction, however, the shield TBM was faced to unexpected geological conditions of fractured zones consisting of sand, gravel and clay.

This paper deals with the case study of a water transfer tunnel by slurry shield TBM, especially for Segment #6, Stage 1 of waterway system at the lower reaches of Han River (L = 1.2 km, Haengjuwe-dong, Goyang~Gochon-eup, Gimpo city). The project outline and unexpected geological conditions are described. The troubleshooting and suggestions to solve the difficulty are discussed.

2 PLANNING AND CONSTRUCTION OF TUNNEL

2.1 *Project outline*

The project was planned to provide a stable water supply to Gimpo city by distributing 175,000 m^3/day among the extra capacity of 847,000 m^3/day in Metropolitan area. The outline of the projects is as follows (Figure 1, Kim et al. 2011):

- Location: Haengjuwe-dong, Goyang ~ Boundary between Gyehwa-dong, Gangseo-gu and Gochon-eup, Gimpo.

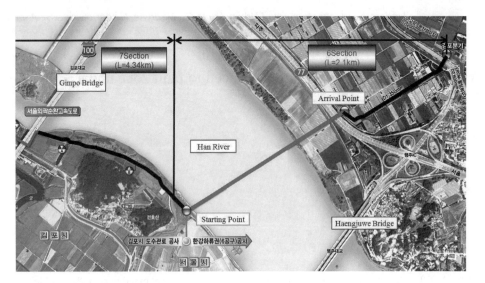

Figure 1.   Outline of the project.

Table 1.   General description of shield TBM.

Classification	Description	Classification	Description
–Advance method	Slurry pressure	–Length	13,080 mm
–Rock strength	~150 MPa	–Steel thickness	25~50mm
–OD	2,220 mm	–Cylinder length	1,550 mm
–No of cutter	15 (4 twin discs)	–Traction force	920 ton (115t x8)

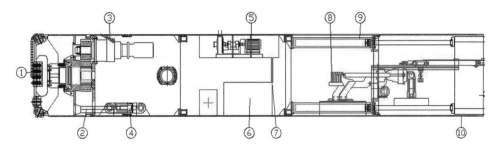

Figure 2.   Section of TBM equipment.

- Scope: Steel pipe D1350, 2.1 km (including 1.2 km of underwater tunnel), 2 maintenance manholes.
- Major work: Underwater tunnel (OD 2.2 m, ID 1.8 m)
- Construction period: Sep 2008~Dec 2011

2.2   Shield TBM equipment

The tunneling method is usually determined by considering ground condition, groundwater level, neighboring structure and underground utilities, traffic control on the ground, construction cost, construction schedule, constructability and stability, impact on surround environment, etc. A shield TBM, which offers the benefit for stable excavation without requiring separate waterstop measure, was adopted. Table 1 summarizes the general description of the used shield TBM and Figure 2 shows the section of TBM equipment.

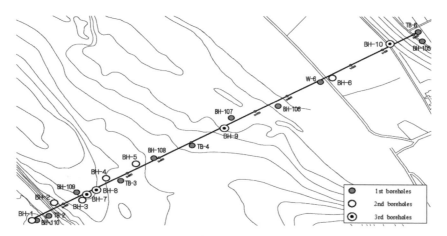

Figure 3. Location of boreholes.

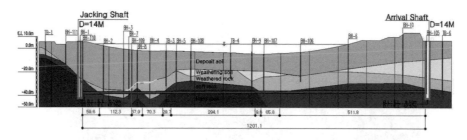

Figure 4. Geological condition through the tunnel (during design step).

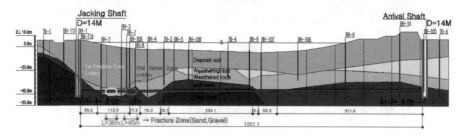

Figure 5. Two fracture zones through the tunnel (during construction).

2.3 *Construction*

The construction work sequence is as follows (Kim et al. 2011): (1) excavation for slurry wall, (2) reaction wall and opening entrance, (3) mobilization of support equipment, (4) advancement: advancement → segment fabrication → backfill grouting, (5) arrival at receiving pit: arrival → equipment disassembly → lifted out.

3 PROBLEMS AND COUNTERMEASURES

3.1 *Geological survey*

During the design step, the geological survey was conducted at 20 locations over a 1.2 km-long segment. Figure 3 shows the location of the boreholes and Figure 4 shows the geological condition through the tunnel. The tunnel section comprised the weathered rock and soft rock.

During the construction, two fractured zones comprising of sand, gravel, weathered rock, soft rock and hard rock were encountered as shown in Figure 5. The first fractured zone was encountered

Table 2. Comparison of soil and rock classification.

During the design		After the construction	
Advance	Classification	Advance	Classification
87.2 m	Soft rock	48.4 m	Soft rock
		75.4 m	Soft rock & Weathered rock
		90.4 m	Weathered rock
		94.4 m	Sand & Gravel
		99.4 m	Clay & Silt
		103.4 m	Weathered rock
		105.4 m	Clay & Silt
		125.4 m	Weathered rock
		138.4 m	Soft rock
		145.65 m	Weathered rock
		148.65 m	Weathered rock
		153.9 m	Weathered rock
'		161.4 m	Gravel
		171.4 m	Weathered soil
		257.4 m	Weathered rock
		258.4 m	Soft rock
341.8 m	Weathered rock	378.0 m	Hard rock

Table 3. Soil/rock samples from the first fractured zone.

90 m after the drilling. The total length of fracture zones was 81 m: 36 m for the first fractured zone and 45 m for the second. Table 2 summarizes the comparison of soil/rock conditions during the design step and after construction for the fracture zones.

3.2 *First fractured zone*

The soil and rock samples taken from the first fractured zone are shown in Table 3. The excavated material produced by slurry shield TBM can be pumped away from the face to the surface by a

Table 4. Several mixture trials.

	Design	1st mix	2nd mix	3rd mix	4th mix
Water (*l*)	900	800	950	950	800
Bentonite (kg)	60	40	35	40	100
CMC (kg)	1.3	1	1	1	1
Clay (kg)	–	200	–	120	200
MAK DP (kg)	–	2	3	3	2
Poly (kg)	–	–	200	40	20
MAK stopper (kg)	–	–	35	5	5
Slurry		25 m^3	50 m^3	75 m^3	75 m^3
Return rate		0%	0%	50%	80%

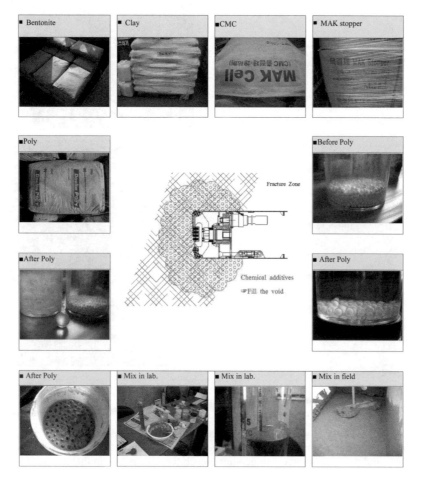

Figure 6. Final slurry mixture.

series of pumps and a suitable media returned to the TBM on the same circuit. In this project, the slurry mixture of water, bentonite and CMC were first used. However, the loss of slurry was occurred in sand and gravel zone. This led to decrease of face stability, increase of water inflow and block the route to the suction line.

In order to solve this problem, several slurry mixture, aided by chemical additives, were tried, as summarized in Table 4. After checking the slurry return rate, the final slurry mixture was decided, as shown in Figure 6.

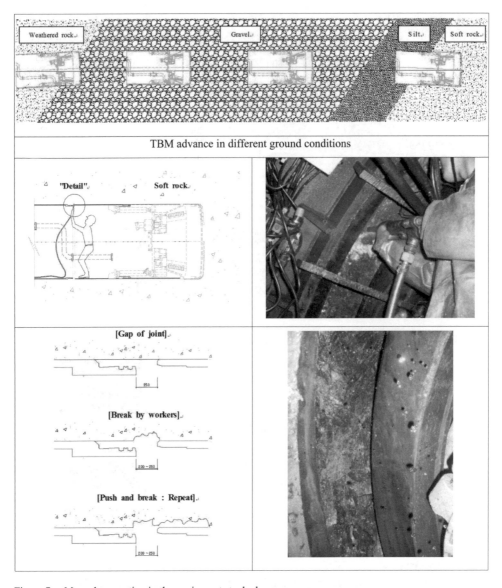

TBM advance in different ground conditions

Figure 7. Manual excavation in the equipment stocked zone.

Together with the loss of slurry, the cutter head was not rotated in the fractured zone. To make the cutter head rotation, a slurry feed/suction pipe was added. The cutter head was reactivated after 40-day delay and the excavation continued until reach the soft rock with the average advance rate of 0.87 m/day.

3.3 Between first and second fractured zones

After passing through the first fractured zone, the soft rock was encountered. The equipment maintenance and tool replacement were necessary.

Due to the excessive bit wear, the TBM equipment was stocked into the rock. This difficulty was solved by manually braking the rock by the workers (Figure 7). Also, the groundwater ingress made the bit replacement difficult. The Urethane and ARC were used to minimize the water ingress (Figure 8).

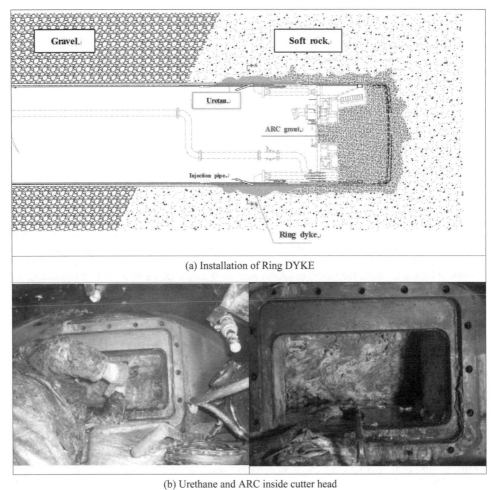

(a) Installation of Ring DYKE

(b) Urethane and ARC inside cutter head

Figure 8.   Installation of ring dyke for bit replacement.

Table 5.   Soil and rock samples from the second fractured zone with TBM advance.

| 155m | 158m | 160m | 162m |

After the bit replacement, the excavation continued with the average advance rate of 0.44 m/day.

3.4   *Second fractured zone*

The second fractured zone comprised mainly gravel. The soil and rock samples taken from the second fractured zone with the TBM advance are shown in Table 5.

Table 6. Summary of slurry mixture.

	1st fractured zone	2nd fractured zone	Difference
Water (*l*)	800	800	–
Bentonite (kg)	100	50	−50
CMC (kg)	1	1	–
Clay (kg)	200	300	100
MAK DP (kg)	2	2	–
Poly (kg)	20	50	30
MAK stopper (kg)	5	10	5
Slurry	$75\,m^3$	$75\,m^3$	
Return rate	80%	80%	

The new slurry mixture, as summarized in Table 6, was used. A serious concern was arose in the excavation of the second fractured zone. If the gravel zone exceed the equipment length of 13 m, the head of TBM may be down and the tunnel line be seriously changed. It was decided to take additional boring to check the length of gravel zone after consultation with the TFT (including the consulting team) and the experts. The boring result showed less than 13 m of the gravel zone.

4 CONCLUSIONS

The case study of a water transfer tunnel by slurry shield TBM has been described. The troubleshooting and suggestions to solve the unexpected geological condition in the project was discussed. Especially the stabilization of tunnel face and the improvement of the TBM driving ratio by controlling mixture ratio of slurry contents were highlighted.

REFERENCES

Jurbin, T. 2008. TBM drive for clean energy from Ashlu Creek. *Tunnels & Tunnelling International* March, 2008: 17–18.

Kim, B.S., Kim, Y.K., Jung, W.I. & Lee, S.W. 2011. A case study of water transfer tunnel under Han River by small diameter shield TBM. *1st International Congress on Tunnels and Underground Structures in South-east Europe*, April 7–9, Dubrovnik, Croatia, 2011.

Lee, S.W. & Kim, C.Y. 2010. State-of-the-arts of TBM tunnel constructions and performance in Korea. *Mini-Symposium on Dams and Mountain Tunnelling*, August 3–4, Chiangmai, Thailand, 2010, 75–84.

Lee, S.W., Chang S.H., Park K.H. & Kim C.Y. 2011. TBM performance and development state in Korea. *12th East Asia-Pacific Conference on Structural Engineering and Construction (EASEC-12)*, January 26-28, Hong Kong, China, 2011, 860–861.

Shirlaw, J.N. & Boones, S. Risk mitigation for slurry TBMs. *Tunnels & Tunnelling International* April, 2008: 26–30.

Zhao, J., Gong Q.M., Eisensten Z. 2007. Tunnelling through a frequently changing and mixed ground: a case history in Singapore. *Tunnelling and Underground Space Technology*, 22: 388–400.

Underground Infrastructure of Urban Areas 2 – Madryas, Nienartowicz & Szot (eds)
© 2012 Taylor & Francis Group, London, ISBN 978-0-415-68394-4

Investigations on fatigue strength of polymer concrete pipes in the light of current guidelines and standards

A. Kmita & M. Musiał
Institute of Building Engineering, Wroclaw University of Technology, Wroclaw, Poland

ABSTRACT: The issues connected with the investigative procedure of a fatigue strength of pipes according to Eurocodes requirements are discussed in this paper. Special attention was paid to representation of assumed static scheme in reference to work of whole element – pipe. The assessment of tensile strength of polymer concrete as fundamental criteria for determination of the load amplitude in fatigue test was emphasized as well. Strain distribution in mid-span cross-section and its change in function of number of load cycles were analyzed. The influence of dimensional imperfections of the specimens connected with its preparation process on state of strain and stress was taken into account. Selected results of the tests conducted on polymer concrete pipes DN 1000 are included in this paper.

1 INTRODUCTION

Basic loads acting on collectors, consist of precast pipes, as self-weight, weight of transported material, soil, hydrostatic lift of underground water have a character of static loads. Additionally, changing loads induced with railway or tram rolling stock, fleet of motor vehicles or even aeroplanes (in case of airstrips) should be considered. By the relatively thin soil cover of about 1.5 m (Hornung & Kittel 1988, ATV-A 161E 1990) the influence of moving loads is significant and must be taken into account in static analysis of pipes. Therefore, from one point we have total static load of pipes and connected with it values of internal forces and from the second one by the low thickness of soil cover and dynamic loads we have to conduct the analysis of fatigue effort of material of which the pipe is performed.

For the most of applied materials (steel, concrete, polymer concrete) proper standards (EN 1916 2002, EN 14636-1 2009) give the maximum effort level of these materials subjected to fatigue loads. In the further part of the paper the way of assessment of the strength parameters for fatigue loads and corresponding to them load level for polymer concrete pipes for micro tunneling was discussed.

2 INVESTIGATIONS ON FATIGUE STRENGTH OF POLYMER CONCRETE PIPES

According to the Eurocode recommendations (EN 14636-1 2009) the procedure of fatigue strength assessment for pipes is as follows:

– in the first phase the tensile strength of polymer concrete should be estimated in three points bending test of rectangular prism beams with the appropriate dimensional proportions according to (Rilem 1975),
– on the basis of obtained results the lower and the upper load values for investigated specimen should be estimated in such a manner as to not exceed the allowable value of the stress in polymer concrete during fatigue test.

The results of experimental studies on the material properties are listed in table 1. The Young modulus was taken from investigations of the pipes producer.

Table 1. The properties of investigated polymer concrete.

Property	Value
Mean compressive strength	103.50 MPa
Mean splitting tensile strength	8.38 MPa
Mean bending tensile strength	21.20 MPa
Mean Young modulus	31.29 GPa

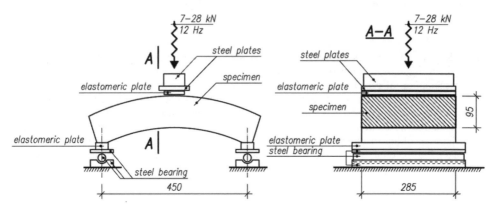

Figure 1. Experimental set – up.

The value of the upper and lower load to be applied in the fatigue test is calculated using the appropriate one of the proper following equations (EN 14636-1 2009):

$$P_{calc,up} = \sigma_{rb,min} \times \frac{2 \times b \times e^2}{3 \times l_b \times f_{corr}} \times f_{up} \tag{1}$$

$$P_{calc,low} = \sigma_{rb,min} \times \frac{2 \times b \times e^2}{3 \times l_b \times f_{corr}} \times f_{low} \tag{2}$$

where $P_{calc,up}$, $P_{calc,low}$ = respectively the calculated upper or lower applied load, f_{up}, f_{low} = respectively the factor for the upper (0,4) or the lower (0,1) load, $\sigma_{rb,min}$ = the minimum ring bending tensile stress, l_b = the spacing between the centres of the bearing beams, b = the width of the test piece, e = the wall thickness of the test piece, f_{corr} = the correction factor to allow for stress distribution in the curved beam (in considered case 1,057).

The minimum ring bending tensile stress value was assessed with the own investigations performed on the beam specimens with dimension $40 \times 40 \times 160$ mm. It was assumed as 5% quantile calculated for twelve single tests. The range of loads in the fatigue test was estimated as 7–28 kN. The investigations were performed in experimental set – up (fig. 1) built according to the standard recommendations (EN 14636-1 2009). The photograph from the experimental studies is shown below as well (fig. 2). It presents the shell in the experimental set – up and the instrumentation (dial displacement gauges and electric resistance wire strain gauges).

If the specimen passes through fatigue test (2×10^6 cycles) its fatigue strength is calculated with following formula (EN 14636-1 2009):

$$\sigma_{fat} = \sigma_{rb,min} \times (f_{up} - f_{low}) \tag{3}$$

where σ_{fat} = the calculated fatigue strength.

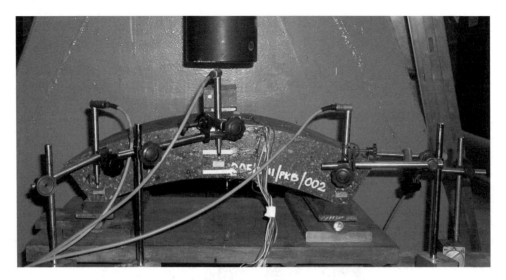

Figure 2.   The shell in the experimental set – up.

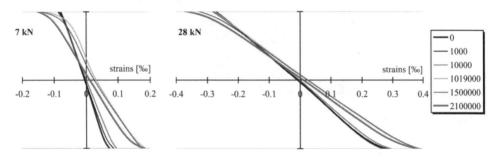

Figure 3.   Strain distribution for the specimen 2.

3   ANALYSIS OF THE EXPERIMENTAL STUDIES RESULTS

The main purpose of the investigations was to verify whether the selected specimens (shells) will pass the fatigue test with the number of cycles 2×10^6 for the load range assessed in the previous paragraph. In the considered case each specimen passed through the test. It means that their fatigue strength assessed according to the procedure is sufficient.

Additionally, during the experiment many other values, allowing to describe the changes of the strains and stress in the investigated specimens in function of load cycles number, were registered.

The strain distribution for the lower and upper load level in mid-span cross section for two selected shells is shown in figures 3 and 4.

In these figures the change of the strains for the selected load cycles is presented. The most significant strains increase was observed between the cycles 10^4 and 2×10^6. Moreover, the change of the localization of the neutral axis induced with the cyclic loads is noticeable.

During the investigations the state of the specimen preparation for the experiment was considered as well. For "nonlinear load" (dimensional imperfection) at the supports the deformations of the element at the mid-span cross-section differed from correct support case significantly. Three cases were considered:

– the specimen prepared properly with the horizontal support edges,
– the specimen with the imperfection $\Delta h = 1$ mm (with the diagonal support edges – falling gradient 0.35 %),

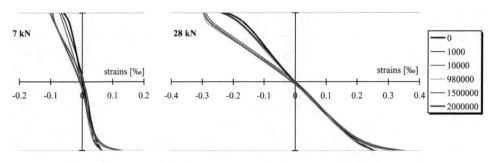

Figure 4.   Strain distribution for the specimen 3.

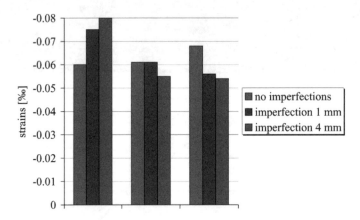

Figure 5.   Strain distribution at the compressed surface.

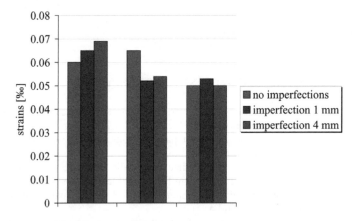

Figure 6.   Strain distribution at the tensioned surface.

– the specimen with the imperfection $\Delta h = 4$ mm (with the diagonal support edges – falling gradient 1.4 %).

The set geometric imperfections caused swaying of the specimens at the supports by the diagonal plane between the steel support bearings. The strains distributions are show in figures 5 and 6. The measurements were carried out for the load of 7 kN at the mid-span in three places (in the middle and by the edges) at the upper and lower surface.

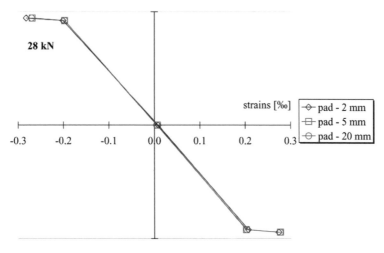

Figure 7.   Strains for the different pad thickness.

During the investigations the thickness of the elastomeric pads at the supports was changed. However, the significant influence of their thickness on the strains distribution in the middle cross-section was not observed. The exemplary strains diagram for three different pads thicknesses and the load level of 28 kN is shown in figure 7.

4   CONCLUSIONS

On the basis of the performed investigations it was founded that crucial element of the studies is the assessment of the bending tensile strength of the polymer concrete. The specimens should be taken from the products (pipes) with the proper proportions according to the Rilem guidelines (Rilem 1975). Moreover, the maximum dimension of an aggregate should be taken into account.

The very important is correct preparation of specimens for experiment. It was proved that dimensional precision and proper supporting of the specimens are of great importance. Thick elastomeric plates cannot replace correct preparation of specimens for experimental studies.

A curious issue is the strain distribution in the specimens in the function of the number of cycles and in the function of the effort level of the material as well. The authors plan further investigations in this field.

It is very important to verify the state of strains (stress) in the tensioned area of specimen in the initial phase of the investigations. It influences on the accuracy of the fatigue investigations. During the fatigue load, especially with high frequencies, in the whole investigative system the higher impacts than it was planned can be induced. It may be connected with the interference of loading press and specimen characteristics.

REFERENCES

ATV-A 161E 1990. *Structural Calculation of Driven Pipes*. Hennef: GFA.
EN 1916 2002. *Concrete pipes and fittings, unreinforced, steel fibre and reinforced*. Brussels: CEN.
EN 14636-1 2009. *Plastics piping systems for non-pressure drainage and sewerage – Polyester resin concrete (PRC) – Part 1: Pipes and fittings with flexible joints*. Brussels: CEN.
Hornung, K. & Kittel D. 1988. *Structural Analysis of Buried Pipes*. Wiesbaden und Berlin: Bauverlag GmbH.
RILEM TC14 – CPC5. 1975. *Flexural test on concrete*. Bagneux: RILEM Publications SARL.

Underground Infrastructure of Urban Areas 2 – Madryas, Nienartowicz & Szot (eds)
© 2012 Taylor & Francis Group, London, ISBN 978-0-415-68394-4

Strength properties of basalt pipes intended for trenchless installations of sewers

A. Kolonko & L. Wysocki
Institute of Civil Engineering, Wrocław University of Technology, Wrocław, Poland

ABSTRACT: The results of research on the physical parameters of pipes made of basalt have been described in the article. Possible areas of application of the pipes have been recommended due to their specific characteristics, especially the possibilities of usage in trenchless technology of constructing sewage systems. In this field, due to their presented advantages they are remarkably useful.

1 INTRODUCTION

Due to their advantages, trenchless methods of constructing underground pipelines find increasingly wider application in municipal infrastructure development. Classification of trenchless methods applicable for construction of sewers, as set forth in PN-EN 12889 "Trenchless construction and testing of drains and sewers" (Kolonko & Kolonko 2005) is illustrated in Figure 1.

In case of sanitary drainage network with gravitational flow, directional drillings are important trenchless method of constructing the pipelines, especially for short lengths (e.g. passages under streets). This technology allows to install circular cross-section sewers in wide range of nominal diameters DN200–DN1200. Detailed description of the technology is available in (Madryas et al. 2006). A very accurate guidance of pipe axis (as specified in design documentation) is the good point of directional drillings resulting from inherent controlling capabilities. It is of great importance as the design downgrades of gravitational sewers are extremely small. This method can be used to construct long sewerage lines in particular sections determined by location of inspection chambers. Generally, the spacing of these chambers amounts to 40–50 metres. It is especially essential when the works are run in compact settlement where closing the streets creates large communication disturbances.

Decision about selecting the trenchless method for constructing sewer line is often the outcome of analyzing various aspects of the investment project to be executed. Among them, there can be distinguished the following:

– economic viewpoints, especially in case of sewers planned at large depths in difficult subsurface and hydrological conditions, in close vicinity with other buried utilities which could be easily damaged when works are run with traditional methods. Under such conditions, the costs of excavations, dewatering, area occupancy and reproducing the roadway are often higher than those for trenchless technologies;
– social aspects can occur especially in case of town centres where such works made in open excavations cause essential hindrances in pedestrian and motor traffic, the more oppressive as in most Polish towns the communication systems are working badly even without any additional disturbances. Social conditions are also related to higher ecological consciousness of public and decision-makers who more and more frequently demand trenchless technologies;
– ecological aspects related to works run in excavations include, but are not limited to interfering in natural environment due to reduction of ground water level (plant withering), potential hazard of contamination, possible necessity of tree cutting, damage to their roots or enlarged emission of car exhausts from diverted traffic. These hazards to environment could be avoided by means of trenchless methods;

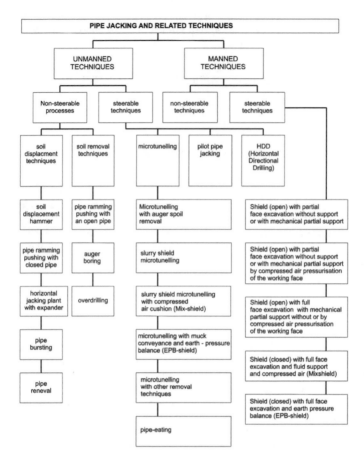

Figure 1. Classification of trenchless methods applicable in constructing sewers.

– technical points of view are related e.g. with safety hazard of civil structures when works are made in excavations due to worse foundation conditions (differential settlement). This refers to building structures located close to deep excavations, especially when it would be necessary to lower the ground water level. Additional problems are the conflicts with existing buried utilities causing hindrances for works in excavations and threat of damaging these utilities.

2 MATERIALS USED FOR EARTH PIERCING PIPES

Pipes used for directional drilling are generally known as earth piercing (casing) pipes. They are designed to carry the loads existing during operation (e.g. soil load, communication load, water head pressure of ground water) and during execution – forces caused by piercing (mostly the longitudinal forces). The common feature of earth piercing pipes is the construction of joints, which must be flush with side surface.

Now, earth piercing pipes are fabricated out of such materials as: steel, reinforced concrete, stoneware, polymer concrete, GRP (pipes out of polyester resins reinforced with glass fibre) and molten basalt.

Molten basalt pipes are the newest structural material not quite well known, however due to their strength properties and chemical resistance they should be more widely used in designing micro-tunnel passages for diameters DN100 to DN600. Such series of molten basalt pipe types is now available on the market. The investment projects including earth piercing pipes out of molten basalt is increasingly popular in the last years. Such implementations have been also completed in Poland (Ostrowski et al. 2010). Examples of projects where earth piercing basalt pipes were applied are:

- Łódź – pipes DN 200 – contractor: Inkop
- Warszawa – pipes DN 200 – contractor: Pol-Aqua
- Lewin Brzeski – rury DN 300 – contractor: Nawitel
- Sulików – pipes DN 300 – contractor: Sobet
- Łódź – pipes DN 200 – contractor: WUPRINŻ
- Wrocław – pipes DN 250/300 – contractor: Nawitel
- Opole – pipes DN 200 – contractor: Sobet
- Opole – pipes DN 350 – contractor: Nawitel
- Kraków pipes DN 250/300 – contractor: Hamer
- Lewin Brzeski – pipes DN 200/250/400 – contractor: Sobet
- Stara Miłosna – pipes DN 300 – contractor: Telbial

For the sake of some similarity with vitrified clay pipes, the basalt pipes are tested in accordance with Standard (PN-EN 295-3: 2002). Further in the study, both these types of pipes, having similar applicability, will be compared to each other.

3 SELECTING CRITERIA FOR STRUCTURAL MATERIAL

When selecting the pipes to construct sewer line with micro-tunnelling method, the following fundamental criteria related to:

hydraulics,
structural analysis/strength properties,
performance characteristics,
environmental protection, and
economy
shall be considered.

Hydraulic criteria take into account the volume of effluents on the basis of current and forecast consumption of water over the area serviced by the section of sewer under design (in case of combined sewerage system, rainwater inflow need to be considered). This criterion shall be the basis for determining the diameter of sewer planned.

Static/strength criteria allow to select appropriate thickness of pipe wall (for particular structural material) so as to carry all loads existing both during installation and operation stages. In case of earth piercing pipes, most often the wall thickness results from necessity to withstand large longitudinal forces.

Environmental criteria refer to effects of aggressive ground waters and aggressive substances generated as a result of biochemical processes taking place in flowing effluents, especially when there is no effective ventilation of sewer. Ground water could be, first and foremost, the alkali aggressiveness. Furthermore, ground water may also have acidic and acidic-carbonic corrosive power (ground water may include humic acids, inorganic acids and salts of strong acids – acid rains). Domestic sewage represents heavily contaminated water. This water always include some chlorides, sulphates, sulphites, sodium carbonate, detergents, fat and large amount of organic matter. Aggressiveness of domestic sewage for concrete and many other materials is not high. However, digestion processes may occur in the sewer causing generation of hydrogen sulphide and carbon dioxide. Hydrogen sulphide is a source of sulphur for Thiobacillus thiooxidans bacterium genus that may oxidize it to sulphurous acid. According to examinations carried out by the authors, the pH value of digested sediments and condensates deposited on sewer wall may be even as high as 1.5.

In case of concrete pipes, especially dangerous are sulphates. Proper material shall be also selected for gaskets, which may appear the weakest point of the whole system. Commonly used polymers, EPDM and SBR, are of low resistance to oil derivatives and fats, whereas the NBR polymer being resistant to fats has low resistivity to acidic medium.

Figure 2. Earth piercing pipes out of molten basalt on construction site.

In Poland economic criteria often come down to investment costs. Obviously, more rational and wider approach shall also include durability of the solution selected and the costs calculated for such period. Certainly, service life of even a hundred years can be considered when good and durable pipes are selected and proper installation is ensured. It is worth noting that when sewer is installed with micro-tunnelling method, the risk of execution errors is lower than that when sewer is laid down in excavation where essential is to ensure careful soil compacting and often the soil need to be, even completely, replaced.

4 BASALT PIPES

The first basalt pipes were fabricated in early 1950s. They were 180 mm in diameter and 330 mm long. Implementations with molten basalt pipes were not numerous. These pipes are manufactured by casting molten basalt in steel or sand moulds. Basalt material of proper quality is melt down in special furnaces at temperature of 1,280°C. When poured into moulds, basalt temperature is about 1,200°C. Detailed description of production process for pipes and fittings out of molten basalt is included in (Kolonko & Kolonko 2005). Fundamental properties of molten basalt pipes, as specified by the manufacturer (Materials of EUTIT s.r.o.), are summarized below. The values for vitrified clay are given in parenthesizes (Materials of KERAMO – STEINZEUG N.V.).

- Specific gravity: 29–30 kN/m^3 (22 kN/m^3),
- Compression strength: 300–450 MPa (150 MPa),
- Tensile strength at bending: (min.) 45 MPa (10–20 MPa),
- Coefficient of thermal expansion for temperature range from 0°C to 100°C: $8 \cdot 10^{-6}$ K^{-1},
- Chemical resistance: pH 1–pH 14,
- Absorbability: 0%
- Hardness in Mohs scale: min. 8
- Resistance to temperature jumps: up to 150°C

In practice, extensive experience and good raw material with appropriate parameters are required to fabricate good quality basalt pipes. Excessive content of undesirable additives may cause that given basalt deposit is inappropriate for such purposes. Elements fabricated from molten basalt subject to quality control and qualified as class 1 or class 2 products, or else as discards that could be reprocessed. Exemplary earth piercing pipes out of molten basalt on construction site are shown in Fig. 2 (Materials of EUTIT s.r.o.). The joint solution allowing their installation with trenchless methods is shown. The band at pipe joint is made of stainless steel while the elastomeric O-ring gasket is placed in a groove on so called bare end.

Table 1. Fundamental design parameters for basalt earth piercing pipes

DN [mm]	Crushing strength FN [kN/m]	Unit mass [kg/m]	Admissible piercing force N [kN]
100	32	22	830
150	42	30	1,200
200	40	39	1,500
250	30	51	1,900
300/25	48	73	3,000
300/33	60	97	4,100
350	42	99	4,300
400	48	128	5,800
500	60	205	9,900
600	95	254	11,600

Table 2. Fundamental design parameters for stoneware earth piercing pipes

DN [mm]	Crushing strength FN [kN/m]	Unit mass [kg/m]	Admissible piercing force N [kN]
150	58	36	170
200	80	60	350
250	110	105	810
300	120	125	1,000
400	160	240	2,200
500	140	295	2,700
600	120	350	3,100

Fundamental design parameters for basalt pipes are summarized in Table 1 (Materials of EUTIT s.r.o.) and those for vitrified clay pipes – in Table 2 (Materials of KERAMO – STEINZEUG N.V.).

5 BASALT PIPE TESTING

5.1 *Purpose of testing*

The tests were carried out to verify the strength parameters of basalt pipes declared by manufacturer. Moreover, it was essential to determine the modulus of elasticity E, which was not specified in data sheets and which is often necessary for detailed static and strength analyses.

5.2 *Crushing strength*

As aforementioned, due to some resemblance to stoneware pipes, the basalt pipes were tested in accordance with (PN-EN 295-3 2002) Standard. Testing for crushing strength was run for load scheme shown in Figure 3. The testing stand is illustrated in Figue 4.

Testing was made for pipe sections $L = 0.3$ m long. The values of breaking force, F_N, for $L = 0.3$ m and corresponding standard values for $L = 1.0$ m are given below.

Testing results

$F_{N1} = 22.44$ kN $\rightarrow$ $F_{N1} = 74.67$ kN/m
$F_{N2} = 23.02$ kN $\rightarrow$ $F_{N2} = 76.73$ kN/m
$F_{N3} = 23.00$ kN $\rightarrow$ $F_{N3} = 76.67$ kN/m

Analysis of testing results.

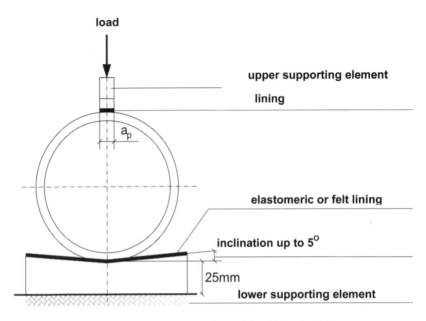

Figure 3. Scheme of crushing strength testing according to (PN-EN 295-3: 2002).

Figure 4. View of testing stand for determining the crushing strength acc to (PN-EN 295-3:2002).

The minimum value, $F_{Nmin} = 74.67\,kN/m$, is much higher than $F_N = 40\,kN/m$ declared by manufacturer for basalt pipes of diameter DN 200 (Materials of EUTIT s.r.o.). Hence, it should be corrected, the reserves are excessive, the more that the factor of safety according to German recommendations (ATV-DVWK A-127 2000) for the safety class A, most commonly used, is 2.2.

5.3 *Tensile strength at bending*

Testing for tensile strength at bending was also performed according to PN-EN 295-3 (PN-EN 295-3: 2002) for the load scheme shown in Figure 5. Testing stand is shown in Figure 6.

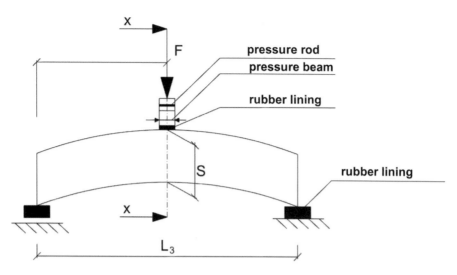

Figure 5. Scheme of loads for determining tensile strength at bending according to PN-EN 295-3.

Figure 6. View of testing stand for determining tensile strength at bending according to (PN-EN 295-3: 2002).

Tensile strength at bending determined on the basis of force applied at sample fracture can be calculated from the equation:

$$\sigma_{bz} = M_b \cdot \alpha_k / W$$

$$\sigma_{bz} = \frac{1000 F_m \cdot L_3}{4 b_3} \cdot \frac{6}{s_1^2} \cdot \alpha_k$$

where:

$$\alpha_k = \frac{3 d_1 + 5 s_1}{3 d_1 + 3 s_1}$$

87

Table 3. Tensile strength at bending for samples tested

Test No.	σ_{bz} [N/mm^2]
1	56.37
2	49.89
3	48.79
4	49.61
5	46.46
6	62.63

Table 4. Values of modulus of elasticity for samples under testing

Test No.	σ_{bz} [N/mm^2]	E [MPa]
1	56.37	119681
2	49.89	110620
3	48.79	124782
4	49.61	110244
5	46.46	165338
6	62.63	104209

where: σ_{bz} – tensile strength [N/mm^2]; F_m – force applied at the moment of sample fracture [kN]; L_3 – spacing of support axes [mm]; b_3 – sample width [mm]; d_1 – internal diameter of pipe tested [mm]; s_1 – sample thickness [mm]; α_k – correction coefficient [mm]; M_b – bending moment [Nmm]; W – bending index [mm^3]

Six samples were tested. Results from testing are summarized in Table 3.

5.4 *Modulus of elasticity E*

The values of the modulus of elasticity were determined from measurement of strain during testing for tensile strength at bending. The modulus of elasticity can be derived from the equation:

$$E = \frac{\sigma}{\varepsilon}$$

where: σ – tensile stress [N/mm^2]; ε – strain at given stress [μm/m]

The values of the modulus of elasticity are given in Table 4:

Upon reviewing the results, a conclusion can be drawn that the lowest value of the modulus of elasticity as measured during testing was E = 104,209 MPa. Hence, for design calculations the nominal value E = 100,000 MPa can be assumed. This is two times higher than that for stoneware pipes which E = 40 000 ÷ 50 000 MPa according to the manufacturer.

6   ADVANTAGES OF EARTH PIERCING BASALT PIPES

The most important advantages of basalt pipes are:

- Pipe deadweight.
  Data sheets of the two pipe manufacturers (compare Tables 1 and 2) show that unit mass of basalt pipes DN 200 is 39 kg/m and that of stoneware pipes 60 kg/m. Thus, earth piercing basalt pipes of the same performance values are evidently lighter than stoneware pipes, so their installation is clearly facilitated.

- Resistance to corrosion.

In case of basalt pipes, high resistance to corrosion results both from the material itself and from homogenous structure of pipe wall that is also free of any pores.

- Admissible value of longitudinal force.

Analysis of maximum admissible longitudinal force which may be applied during micro-tunnelling can be done on the basis of data sheets available for the two manufacturers. These data show that the admissible force for stoneware pipes (DN200) is $N_{dop} = 350\,kN$, while respective value for basalt pipes is $N_{dop} = 1500\,kN$; hence the latter is more than four times higher than that for stoneware pipes despite thinner walls in basalt pipes. Hence, there are potentials to reduce the wall thickness in basalt pipes. Here, the production process may impose some limitations. High value of admissible piercing force may be additionally valuable for especially difficult soil conditions and when longer sections of pipelines are to be made from a single starting chamber.

- Simplified installation.

As opposed to stoneware pipes which durability is dependent on the glaze layer, the basalt pipes are homogenous across their full section. As the basalt pipes are free of glaze on the wall surface, they can be installed using horizontal drilling without an intermediate stage, i.e. piercing temporary steel pipes including screw conveyors that could damage the glaze. Screw conveyor may be placed directly inside basalt pipes without any threat of damage. Such simplified solution can be applicable especially in sandy soils.

- Consistency with conditions of sustainable development.

Apart from obvious advantages of molten basalt pipes like high strength, resistance to corrosion, lack of absorbability and long durability, their additional value lies in a problem-free reuse of worn pipes. Crushed basalt is absolutely neutral for natural environment and may be used for various purposes, and the most obvious include: for pipe fabrication after melting, in road construction, e.g. as a road base, in building material fabrication, e.g. as an aggregate for concrete mixtures.

Hence, basalt completely meets the conditions of sustainable development.

7 SUMMARY

In the face of numerous advantages, the most important among them being very high resistance to corrosion resulting in durability of a given solution and withstanding very large piercing forces, the basalt pipes should find wider application in trenchless technology used for constructing sewer lines.

REFERENCES

ATV-DVWK A-127. 2000. *Structural calculation of buried pipes.*

Kolonko A., Kolonko A.: Rury i elementy z topionego bazaltu w zastosowaniu do budowy i renowacji przewodów kanalizacyjnych. Gaz, Woda i Technika Sanitarna. 2005 t. 78, nr 6.

Madryas C., Kolonko A., Szot A., Wysocki L.: Mikrotunelowanie: Dolnośląskie Wydawnictwo Edukacyjne, Wrocław 2006.

Materials of EUTIT s.r.o.

Materials of KERAMO – STEINZEUG N.V.

Ostrowski K., Zyguła M., Dalewski G.: Ekstremalne roboty inżynieryjne; Inżynieria Bezwykopowa Nr 1/2010.

PN-EN 12889: 2003 „Bezwykopowa budowa i badanie przewodów kanalizacyjnych".

PN-EN 295-3 2002 *Vitrified clay pipes and fittings and pipe joints for drains and sewers. Test methods.*

Underground Infrastructure of Urban Areas 2 – Madryas, Nienartowicz & Szot (eds)
© 2012 Taylor & Francis Group, London, ISBN 978-0-415-68394-4

Release of heavy metals from construction materials

A. Król
Opole University of Technology, Poland

ABSTRACT: The civil engineering structures and elements, wherein concrete makes the principal component, have been in common use and can be found all around us. Since numerous waste materials and industrial by-products have been used in the production of cement and concrete, attention should be paid to the environmental impacts of contemporary construction materials. Within that area, it is especially important to evaluate the release of heavy metals to the aquatic environment and/or to the soil. The paper presents the test methods which are applicable in the assessment of the leaching level of heavy metals from the construction materials. Also, the rate of release was determined for chromium, copper, zinc, lead and nickel from concrete elements which had been produced with the use of Portland cement CEM I and slag cement CEM III/B. The theoretical model for the leaching (release) profile of heavy metals from concrete composites versus time was provided, too.

1 INTRODUCTION

Wastes and/or by-products from other industrial processes have been more and more commonly used in the cement and concrete manufacturing processes. Those additives comprise predominantly fly ash, granulated blast-furnace slag, flue gas desulfurisation products (reagips) and/or silica dust which is generated in the production of ferroalloys. They can find their outlets in the cement industry as time-of-set control agents, or as the principal and/or secondary components in the cement formulation, provided that they satisfy the requirements of EN 197-1:2002. They may also be used as type II additives (EN 206-1:2003) in the composition of the concrete mix (Giergiczny et al. 2002).

The permanent price increase for conventional fuels and the ability to recover energy which is contained in combustible components of industrial and municipal wastes attracted the interest of the cement industry in the alternative fuels. They make substitutes for the fossil fuel (i.e. coal) in the sintering process of the basic clinker components for the production of the Portland cement. The optional fuels which are most frequently used comprise: tyres, wastes from the rubber industry, textiles, paper which is not suitable for recycling, wood wastes, spent oils, paints and solvents, dehydrated sewage sludge, etc. (Wzorek et al. 2007).

A question is: whether such a broad utilisation of industrial by-products and wastes may be responsible for increased leachability of heavy metals from cement composites, and whether it may have any negative environmental impact (transferred drinking water, soil, surface and ground water). That problem is critical when concrete is used for the drinking water reservoirs and drinking water transfer/distribution systems.

In accordance with the Construction Products Directive (89/106/CEE), the construction materials must satisfy the requirements of its Annex I, clause 3, within hygiene, health and the environment. The rate of emission of dangerous substances from those materials in their commercial use, working life, recycling and/or disposal is evaluated as one of parameters for compliance with those requirements.

The key question is if there is a potential unacceptable release of contaminants from the construction product to soil and groundwater, and how this can be determined. This question will be dealt

Table 1. Description of the different release scenarios (Dijkstra et al. 2005).

No.	Scenario description
1	Granular products placed on soil
2	Monolithic products placed on soil
3	Runoff (wet/dry) from monolithic product
4	Unbound granular materials e.g. construction debris (varying particle size; end-of-life)[1]
5	Pipes (e.g. drinking water pipes)[2]
6	Monolithic products in water (e.g., coastal works)
7	Runoff from metal plates

[1] End-of-life stage is not under the scope of the CPD, but may be important in judging acceptability of materials for use.
[2] Release to both the transported water as well as the surrounding soil.

with by addressing the following topics (van der Sloot et al. 2000, Dijkstra et al. 2005, CEN/TC 51/WG12/TG6/2005):

- Construction products are used in different configurations and exposure conditions, called "application scenarios". However, recent investigations have shown that only a limited set of chemical/physical factors is responsible for the release for a wide range of (primary and secondary) construction products, and in practice, only a few of those factors dominate.
- For each product or product group, which of these scenarios will be relevant for practice?
- How should widely different construction products be tested on their release potential? Are different tests needed for each different construction product or application scenario?
- How is the relation established between the test results and the regulatory criteria to be developed? Useful criteria can only be derived when they are based on the basis of an (expected) soil and/or groundwater quality on a certain distance of the construction product in its application ("point of compliance", which might be chosen anywhere). This relationship between the test result and soil/groundwater quality can only be established by setting up model scenarios for each of the application scenarios.

When selecting a method for evaluation of leaching, one should consider the form of the construction material under investigation since different release mechanisms of heavy metals may be involved. In case of monolithic elements, for example, heavy metals will be released as the result of surface wash-off, diffusion and/or dissolution processes (EA NEN 7375:2004).

Various factors need to be considered under test conditions since they will affect the performance of individual heavy metals. Structural changes due to external impacts (changes in temperature, changes in pH, contact with water) may give escalation in the release of heavy metals to the environment (Helms et al. 2009). The oxidation and carbonation processes, as well as other corrosive impacts of aggressive media, involve the need of understanding the effects of individual factors on the release mechanisms, with consideration given to the whole service cycle of the construction material (civil structure). Hence, the Dutch TANK tests, for example, makes allowance for the impacts of changing pH of the leaching liquid on the tested material (pH from 4 to 12) (PrCEN/TS 14429:2004). The external conditions also include the service conditions of the material (structure), the liquid-to-solid ratio (L/S), time of exposure to the leaching medium, type of exposure, pH value, temperature, and mechanical impacts (abrasion, erosion, frost impact) (PrCEN/TC 351/WG1).

The internal factors which may be specific to the test material comprise: porosity, thermal conductivity, shape, extended surface, size of element and reactivity of its material (susceptibility to carbonation, alkalinity), and the length of its service.

Hence, various possible application scenarios (and their combinations) for the construction materials have been reported in the professional literature (Dijkstra et al. 2005). These have been quoted in Table 1.

Table 2. The leaching scenarios for dangerous substances from construction materials, and suggested test methods (PrCEN/TC 351/WG1).

Scenario		Suggested test methods		Examples of materials
I.	Impermeable materials	Dynamic Surface Leaching Test (DSLT)	Method which tests leaching of dangerous substances versus pH value (pH – test dependence)	Metal plates, glass items, etc.
II.	Low permeability materials	DSLT method. A method for broken up material should additionally be used.		Concrete items, bricks, mortar, etc.
III.	Permeable materials	Percolation method (column test)		Soil, broken up material e.g. construction and demolition debris, etc.

A different approach to the application scenarios (Tab. 2) has been provided in the draft standard, in the documents developed by the European Technical Committee TC 351 (PrCEN/TC 351/WG1):

- scenario I – for impermeable products which are placed underground or underwater, and/or the surface of which is washed by moving water, e.g. metal plates, metal strips, roofing-tiles, glass elements, bituminous products;
- scenario II – that is specific for low permeability products, where water is transferred inside their matrices through capillary pores. The soluble substances are transported outside the matrix due to advection and diffusion, e.g. bricks, concrete elements, mortar, concrete tubes;
- scenario III – permeable (porous) products, through which water finds it easy to penetrate just by gravity, e.g. soil, high porosity materials, construction and demolition debris.

2 SELECTED TEST METHODS FOR EVALUATION OF HEAVY METALS RELEASED FROM CEMENT COMPOSITES

Various test methods can be used to evaluate how much heavy metals are released from the construction materials (concrete, building materials) to the environment (Nagataki et al. 2002). Those methods may be classified from the viewpoint of:

- time of leaching test: long-term methods – like the TANK tests (EA NEN 7375:2004) as mentioned above, and short-term methods – e.g. method as per EN 12457-4:2002, which can be used to characterise granular wastes and sediments with regard to their content of leachable heavy metals;
- leaching dynamics: static methods – which make it possible to predict the performance of the hardened concrete under static conditions, as for example TANK tests, and which are based principally on EA NEN 7375; and dynamic methods – these prefer leaching tests under dynamic conditions, as for example the method which is provided in EN 12457-4:2002;
- sample pre-treatment – the samples may have the following forms: intact test piece, broken up sample, sample which was cut out from a monolithic block and then broken up. The TANK tests investigate monolithic test pieces which are placed in containers filled up with a leaching liquid (demineralised water) and kept there for a specified period of time. According to EN

Table 3. Examples of characterization and compliance test methods (van der Sloot et al. 2004).

Test	Brief description	Level
prEN14429	pH-dependence test, on granular or size-reduced products	characterization
prEN14405	column test, on granular products	characterization
EN12457	batch test (natural pH), granular products	compliance
wi292010	compliance tests, monoliths	compliance

12457-4:2002, on the other hand, a sample of 100 g is broken up to obtain the grain size below 10 mm, it is covered with a suitable volume of water (liquid-to-solid ratio $L/S = 10$), and it is shaken over 24 h;

- The column tests, in which the test material is subjected to comminution (grain size <2 mm) and packed into a column (in accordance with DIN 14405:2004), make us of the percolation process to leach out heavy metals from the construction material samples;
- reaction of leaching medium – neutral or acidic. According to EA NEN 7375, a liquid (distilled water) is employed with neutral pH which contacts a monolithic test piece, while according to the documents of the Technical Committee (CEN) the draft standard PrCEN/TS 14429:2004 suggests that the test piece should be subjected to the impact of eight different pH levels (from 4 to 12).

It is also an important issue to establish priorities for the use of leaching tests since not all known test methods are applicable to all construction materials. Three basic levels are distinguished in the testing hierarchy (van der Sloot et al. 2004):

1. Characterization tests – used for basic characterization (or initial testing) of the release of the product. All the relevant parameters for an impact assessment are measured. These data allows for the assessment of products by categories, based on common controlling mechanisms, and therefore reduce the amount of products within a product group that require characterization. This type of data is necessary for scenario modelling, but can be also for insight in variability of measurements, expected range under field conditions, quality improvement (relevant for producers), once the dominant mechanisms are clear.
2. Compliance tests – which have the purpose to check with a simple test whether a material (still) complies with the previous characterization and subsequently with criteria. Once the leaching behavior has been investigated by characterization, a simple test measuring the same property (e.g., leaching at a certain pH) suffices. A close relationship between characterization and compliance should be ensured.
3. On-site verification/quality control tests – are quick tests to see whether a material complies with earlier determined or expected behavior, in its practical application. In general, a simple chemical measurement (e.g. pH, redox, conductivity), an administrative or visual check may be done in this case. For a real confident chemical check, at least a compliance test should be done. The main advantage that such a testing hierarchy has, is that once a characterization step has been done, much more simplified testing on compliance level can be chosen to verify the con-sistency of subsequent data with the characterization test results. This is very time and cost efficient. Examples and a detailed description of (characterization level) test methods and how they can be used for several purposes is referred to van der Sloot et al. 2004. These are summarized in Table 3.

Those problems are important and it is advisable to take up research programmes within that field. The importance of the issues is confirmed by the fact that the Technical Committee TC 351 has been established within the European Committee for Standardisation (CEN), and its aim is to develop applicable regulations in this area which will be adhered to by all EU member states. At the same time, more and more reports are published not only on the need to evaluate the leaching process of dangerous substances from the construction materials but also on prediction of that process over a longer time horizon. In order to provide that capacity, it is necessary to develop a model for the leaching process versus time, and also versus other parameters (e.g. pH of extracting liquid) which affect the release of heavy metals and other pollutants (van Herck et al. 2000, Dijkstra et al. 2005).

Table 4. Chemical compositions of cements used in researches.

Chemical composition	Content, %	
	CEM I	CEM III/B
Loss on ignition	3.46	0.42
CaO	64.60	49.75
SiO_2	19.20	32.92
Al_2O_3	4.69	6.96
Fe_2O_3	3.04	1.80
MgO	1.22	5.04
SO_3	2.65	1.44
K_2O	0.81	0.83
Na_2O	0.09	0.28
Cl^-	0.047	0.067

Table 5. Content of heavy metals in cements used in researches.

Heavy metals	Content, mg/kg	
	CEM I	CEM III/B
Zn	316.0	105.2
Cr	54.0	30.9
Ni	18.0	10.9
Pb	24.0	39.1
Cu	60.0	27.3

3 MATERIALS AND METHODS USED IN TESTS

The conducted research involved the design and development of concretes with the use of: Portland cement CEM I and slag cement CEM III/B (75% of granulated ground blast furnace slag). The chemical compositions of the cements used in the tests are summarized in Table 4. Table 5 contains the contents of heavy metals (Zn, Cr, Pb, Ni, Cu) in the Portland cement CEM I and in the slag cement CEM III/B.

The composition of the concrete mixture was as follows: cement – $300.0\,kg/m^3$; sand – $685.2\,kg/m^3$; $2 \div 8$ mm gravel – $600.4\,kg/m^3$; $8 \div 16$ mm gravel – $628.6\,kg/m^3$; water – $180.0\,kg/m^3$; water/cement ratio (w/c) in the two mixtures was the same (0.6).

The resulting mixture serves as the material for formation of 10x10x10 cm cubes. After 24 hours the samples were removed from the mould and subsequently subjected to leaching tests.

The water extracts from concretes were taken by method in the standard EA NEN 7375:2004. This test involves the preparation of aqueous extract from a monolithic sample. The concrete cubes are placed on pillars, hence enabling them to remain in contact with the liquid across the entire surface of the sample (Fig. 1).

The volume of the liquid which is used to immerse the sample must not be smaller than twice the volume of the examined sample and is to be determined in accordance with the formula (1).

$$2 \times V_p \leq V \leq 5 \times V_p \tag{1}$$

where:

V_p – volume of sample,
V – volume of liquid.

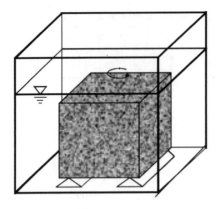

Figure 1. Principles of placing concrete sample in a container in accordance with the EA NEN 7375:2004 standard.

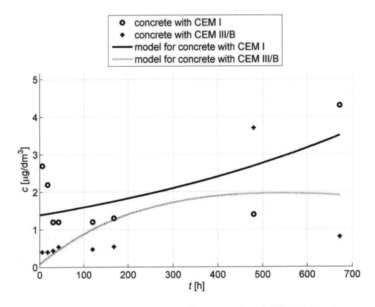

Figure 2. Time profiles for leachability of chromium from CEM I- and CEM III-based concretes.

In the conducted examinations the concrete sample was placed in a container holding 4 liters of distilled water. The total duration of the test amounted to 64 days and was divided into 8 research periods (0.25; 1; 2,25; 4; 9; 16; 36; 64 days). After each of the researched periods an aqueous extract was taken from the container and the liquid was replaced. The content of heavy metals was subsequently taken in the resulting extracts.

This test is a procedure to evaluate the release from monolithic material by predominantly diffusion control (e.g., exposure of structures to external influences).

4 TEST RESULTS

The leaching tests of monolithic materials which are in line with the methods presented in EA NEN 7375:2004 give 8 water extracts which are then analysed for the concentrations of heavy metals. Figures 2–6 show the concentration profiles for heavy metals in water extracts obtained for CEM I- and CEM III-based concretes at 8 time intervals, as represented by respective points.

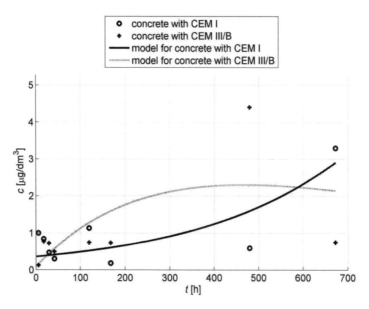

Figure 3.   Time profiles for leachability of copper from CEM I- and CEM III-based concretes.

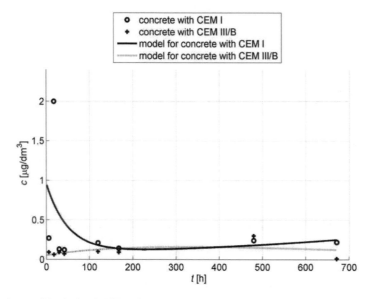

Figure 4.   Time profiles for leachability of nickel from CEM I- and CEM III-based concretes.

Papers can be found in journals which tend to establish both model leaching profiles for individual heavy metals from cement-based composites and to predict the leaching levels over long periods of time (Dijkstra et al. 2005).

In order to find trends for release of individual heavy metals versus time, as shown in Figures 2–6, the leaching levels were analysed for eight different time intervals, and the regression functions were also estimated for which the structural parameters were derived from the author's model (2) for five selected heavy metals, and for two concrete grades (based on CEM I and

97

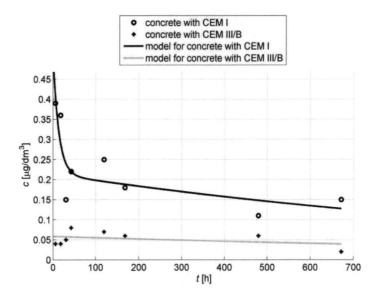

Figure 5.   Time profiles for leachability of lead from CEM I- and CEM III-based concretes.

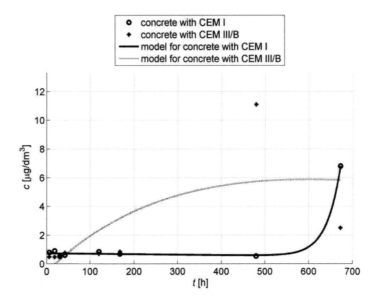

Figure 6.   Time profiles for leachability of zinc from CEM I- and CEM III-based concretes.

on CEM III). That model represents the time profile for leachability: $c(t)$.

$$c(t) = \sum_{i=1}^{k} a_i e^{b_i t}$$ (2)

where:

$c(t)$ – leachability $[\mu g/dm^3]$,
a_i, b_i – structural parameters of the regression function $[-]$,
k – number of exponential components $k = 2$ $[-]$,
t – time $[s]$.

The time-dependent profile of leachability $c(t)$ [$\mu g/dm^3$] may be approximated with the function $\hat{Y}(t)$ which comprises the sum k of exponential functions. The estimation process of the parameters in the function $\hat{Y}(t)$ involved the non-linear model for the regression function as described by equation (3) (Therrien 1992)

$$\hat{Y}(t) = f(X, P) + \varepsilon \qquad (3)$$

where:

$\hat{Y}$ – theoretical value of the regression function $\hat{Y}(t) = f(X, P)$,
X – set of variables for the regression function,
P – set of structural parameters for the regression function Y versus X,
ε – random component.

Estimated regression function $\hat{y}(t)$ has the form as presented by equation (4).

$$f(X, P) = \hat{y}(t) = \sum_{i-1}^{k} a_i e^{b_i t} \qquad (4)$$

where:

$\hat{y}(t)$ – approximated regression function,
a_i, b_i – estimated structural parameters of the regression function,
k – number of exponential components $k = 2$,
t – independent variable, time.

The structural parameters of the regression function a_i, b_i were evaluated with the use of the least squares method. They were then subjected to optimisation process which was based on the Nelder-Mead Simplex method (Lagarias et al. 1998). As criterion the least norm of residuals δ_r (5) was used (Ostrowski 1973). The number of exponential components in the regression function (2) $k = 2$ was chosen based on the visual analysis of the obtained estimation results.

$$\delta_r = \|\hat{y} - y\| = \sqrt{\sum_{i=1}^{n}(\hat{y}_i - y_i)} \qquad (5)$$

where:

δ_r – value the residuals norm $\|\hat{y} - y\|$,
$\hat{y}$ – approximated regression function,
y – empirical data,
$n = 8$ – number of data points in single time run.

The experimental findings from the leaching behaviour tests and from the developed model for the leaching level versus time show that leachability of Cr, Cu, Zn is increasing both for the concrete which is based on the Portland cement (CEM I) and for the concrete which is based on the slag cement CEM III (Figs 2, 3 and 6). In the long run, concentrations of those elements in water extracts may be higher than initially. The opposite trend is demonstrated by Ni and Pb in the case of CEM I concrete. The concentrations of those metals in water extracts decline over time, going down to very low levels of leaching. The release of Ni and Pb from the CEM III concrete is constant (during 64 days of testing) and it falls within the ranges 0.01 $\mu g/dm^3$ ÷ 0.26 $\mu g/dm^3$ and 0.02 $\mu g/dm^3$ ÷ 0.08 $\mu g/dm^3$, respectively.

The leaching levels for heavy metals from the tested concretes may also be specified as total values after 64 days of test (Table 6).

The dominant part of heavy metals which are contained in concrete elements has been firmly fixed in the structure of the Portland clinker, or of the phase components of granulated blast-furnace slag (principally in the glassy phase), and it cannot be leached out (van der Sloot 2000). Hence, the concentrations of heavy metals in water extracts from concrete matrices is very low. The

Table 6. Total leaching of heavy metals from concrete.

Heavy metal	Total leaching of heavy metal [mg/dm³] Concrete with cement		Permissible value acc. to (Regulation of Minister ...)
	CEM I	CEM III/B	
Chromium (Cr)	0.0155	0.00729	0.05
Copper (Cu)	0.00781	0.008714	2.0
Nickel (Ni)	0.00333	0.00081	0.02
Lead (Pb)	0.00181	0.00037	0.05
Zinc (Zn)	0.01169	0.01746	nn

nn – not specified in the quality standard for water intended for human consumption (Regulation of Minister ...)

concentrations of heavy metals, which were obtained after contacting concrete blocks with distilled water were compared to the most restrictive water quality standards – to those for water which is intended for human consumption (Regulation of Minister ...) (Table 6). It should be stresses that none of the allowable levels/concentrations of heavy metals was exceeded. Especially low leaching was identified for copper, as referred to the allowable quality of drinking water. The eluate copper concentrations make 0.5% of the permissible limit at maximum.

5 SUMMARY

The available test methods were analysed which make it possible to learn the release level of heavy metals to the aquatic environment. New ways were also mentioned for the efforts intended to unify the test methods with due consideration of various "application scenarios" for the construction materials.

The author's leaching tests were based on the method presented in EA NEN 7375:2004. The release of chromium, copper, zinc, lead and nickel was observed from concrete tests blocks which were composed and moulded with the use of Portland cement (CEM I) and blast-furnace cement (CEM III/B). The differences were revealed in the time profiles for the heavy metals leaching process. Within the analysed heavy metals, nickel and lead were found to offer the lowest leaching level from the CEM III/B concrete (the total leachability after 64 days was 0.00081 mg/dm³ and 0.00037 mg/dm³, respectively).

The conducted tests demonstrated differences in the release trends for different test periods (as described by the author's model). Irrespective of that, it may be concluded that the concentrations of heavy metals which are released from tested concretes to water are very low: lower than the permissible concentrations of heavy metals in drinking water. It is also worth highlighting that, within the tested concrete matrices, the CEM III/B concrete (with 75% of granulated blast furnace slag, i.e. by-product from the iron manufacturing process) makes a better matrix for heavy metals than the traditional Portland cement CEM I concrete – the concentrations of heavy metals leaching to the aquatic environment were much lower than for concrete on CEM I.

REFERENCES

CEN/TC 51/WG12/TG6/2005 Schemes for testing either 'new'/unapproved constituents of concrete and mortar or production concrete and mortar for release of regulated dangerous substances into soil, groundwater or surface water.
CEN TC 292 WG2 Monolith compliance leaching test WI292010
Dijkstra, J.J., van der Sloot, H.A., Spanka, G., Thielen, G. 2005. How to judge realease of dangerous substances from construction products to soil and groundwater. ECN-C-05-045.
DIN 14405:2004 Characterization of Waste. Leaching behavior tests. Up-flow percolation test.
EA NEN 7375:2004 Leaching characteristics of moulded or monolithic building and waste materials. Determination of leaching of inorganic components with the diffusion test. The Tank Test.

Ehrnsperger, R., Misch, W. 2006. *Implementation of Health and Environmental Criteria in Technical Specifications for Construction Products*. Deutsches Institut für Bautechnik (DIBt). Research Report 200 62 311/2006.

EN 12457-4:2002 *Characterisation of waste – Leaching Compliance test for leaching of granular waste materials and sludges – Part 4: One stage batch at a liquid to solid ratio of 10 l/kg for materials with particle size below 10 mm (without or with size reduction)*

EN 197-1:2002/A1:2005 *Cement –Part 1: Composition, specifications and conformity criteria for common cements*

EN 206-1: 2000 *Concrete – Part 1: Specification, performance, production and conformity.*

Gendebien, A., Leavens, A., Blackmore, K., Godley, A., Lewin, K., Whiting, K.J., Davis, R., Giegrich, J., Fehrnbach, H., Gromke, U., del Bufalo, N., Hogg, D. 2003. *Refuse Derived Fuel, Current Practice and Perspectives*. (B4-3040/2000/306517/MAR/E3). Report No. CO 5087-4, WRc Swindon, The United Kingdom.

Giergiczny, Z., Małolepszy, J., Szwabowski, J., Śliwiński, J. 2002. *Cements with mineral admixtures in the new generation concrete technology*. Instytut Śląski Publishers, Opole (89/106/EEC) *Construction Products Directive.*

Helms, G., Thorneloe, S. 2009. *Improved leach testing for evaluating fate of metals from coal ash.* Environmental Protection Agency, Lyon.

Lagarias, J.C., Reeds, J.A., Wright, M.H., Wright, P. E. 1998. *Convergence properties of the nelder-mead simplex method in low dimensions.* SIAM J. Optim., 9(1), pp. 112–147.

Nagataki, S., Yu Q., Hisada, M. 2002. *Effect of leaching conditions and curing time on the leaching of heavy metals in fly ash cement mortars.* Advances in Cement Research, vol. 14, no 2, pp. 71–38.

Ostrowski, A.M. 1973. *Solution of equations in Euclidean and Banach spaces.* AcademiPress.

PrCEN/TS 14429:2004 *Characterisation of waste – Leaching behaviour tests – Influence of pH on leaching with initial acid/base addition.*

PrCEN/TC 351/WG1 N 117 TS – 2 2009 – 01 – 19

PrCEN/TC 351/WG1 N 118 TS – 3 2009 – 01 – 19

PrCEN/TC 351/WG1 N 142 TS – 1 2009 – 03 – 15

Regulation of the Minister of Health, of 19 November 2002, *on quality requirements for water which is intended for human consumption.* Journal of Laws No 203, item 1718.

Therrien, C.W. 1992. *Discrete random signals and statistical signal processing.* Prentice-Hall.

van der Sloot, H.A. 2000. *Comparsion of the characteristic leaching behavior of cements using stan dard (EN 196-1) cement mortar and an assessment of their long-term environmental behavior in construction products during service life and recycling.* Cement and Concrete Researches vol. 30, pp. 1079–1096

van der Sloot, H.A., Dijkstra J.J. 2004. *Development of horizontally standardized leaching tests for construction materials: a material based or released based approach?* ECN-C–04-060, Energy Research Centre of the Netherlands (ECN): Petten, The Netherlands.

van Herck, P., Van der Bruggen, B., Vogels, G., Vandecasteele, C. 2000. *Application of computer modelling to predict the leaching behaviour of heavy metals from MSWI fly ash and comparison with a sequential extraction method.* Waste Management, 20, pp. 203–210.

Wzorek, M., Troniewski, L. 2007. *Application of sewage sludge as a component of alternative fuel.* Environmental Engineering. Taylor & Francis, New York, Singapore, pp. 311–316.

Underground Infrastructure of Urban Areas 2 – Madryas, Nienartowicz & Szot (eds)
© 2012 Taylor & Francis Group, London, ISBN 978-0-415-68394-4

Modernization of the Cracow non-circular sewers using glassfiber reinforced plastic panels

A. Kuliczkowski & D. Lichosik
Department of Civil and Environmental Engineering, Kielce University of Technology, Kielce, Poland

R. Langer
The Municipal Waterworks and Sewer Enterprise in Cracow, Cracow, Poland

A. Wojcik
HOBAS Poland, Dąbrowa Górnicza, Poland

ABSTRACT: The paper presents one of the largest renovation projects in Poland conducted in 2007–2009. It was a part of "Water and Sewage Management System in Cracow – Phase I" investment. The project consisted of rehabilitation of 5.3 km sanitary and combined sewers with dimensions from DN 800/1200 to DN 3000/2520. The rehabilitation used panels made of glassfiber reinforced plastics, being a combination of unsaturated polyester resin, chopped and hoop glass fiber and mineral reinforcing agents. The study describes innovative technology HOBAS NC Line which enables production of shaped elements tailored to the dimensions of the refurbish channels. Particular attention was paid to the production of special shaped panels. They are used in the technical refurbishment of profiles with dry weather flow channels located under Daszynskiego and Wanda streets. In addition, the assembly stage in the city center, in a dense infrastructure and with considerable intensity of traffic was described.

1 INTRODUCTION

Cracow municipal sewage system includes sanitary and combined sewers reaching a total length of 1477 km. It needs rehabilitation due to the occurrence of the leaks and damage, mainly related to the long period of maintenance. In 2005 The Municipal Waterworks and Sewer Enterprise in Cracow received co-financing from the Cohesion Fund for the project titled "Water and Sewage Management System in Cracow – Phase I". It was one of the largest rehabilitation projects in Poland conducted between 2007 and 2009 and also the second consecutively EU-funded project in the field of water and sewage infrastructure in Cracow. The total cost of the project reached about 38.1 Million Euros with approximately 21.5 Million Euros coming from the Cohesion Fund. The project consisted of four parts included in six contracts. One of the main elements of the project was rehabilitation of about 55 km of sanitary and combined sewers, man-entry and non- man-entry sewers.

An agreement for Contract II "Renovation of Cracow sewerage system, man-entry sewers" has been signed at the end of August 2007. The scope of work included rehabilitation a Total length of 5.3 km of man-entry sewers. Channels selected for rehabilitation with standard and custom sections (DN 800/1200-3000/2520) where placed under 13 Roads of city center. Rehabilitation has been carried out using polyester resin reinforced panels with glass fiber – HOBAS NC Line.

2 THE BASIC FACTORS TAKEN INTO ACCOUNT WHEN CHOOSING THE SEWERAGE SEWER MAINS REHABILITATION TECHNOLOGY

The decision on the selection of described trenchless technology has been influenced by economic, social and environmental factors.

Economic factors:

– laying pipes at great depths, in difficult soil and water conditions and simultaneous conflicts with other existing networks,
– high cost of occupying, restoration and reconstruction of lane or sidewalk (Góra 2010).

Social factors:

– necessity for sewer rehabilitation in areas of the city with heavy vehicular and pedestrians traffic (city center),
– occurrence of narrow roads in the works area and the associated inconvenience and danger to users of these sites (Góra 2010).

Ecological factors:

– preservation the existing natural environment,
– waste reduction and minimize environmental disturbance during the execution of the works,
– constant groundwater table (Góra 2010).

Concrete pipelines of renewed man-entry sewer section were built between 1900 and 1945, and their condition was poor in some instances. On the basis of existing documentation and a partial inventory which was carried out using sewer TV inspection, the state of sewers was assessed. Corrosion and infiltration were a problem as well as cracks and root intrusion. Used technology had to restore the load capacity of analyzed sewer mains. In addition, it should consider the limitations of place, technologies and materials. Since rehabilitations were carried out in the city center, it was important to minimize earthworks carried out in the lanes.

The HOBAS NC Line panels were selected for the technical implementation of pipelines rehabilitation using the relining method. They met customer requirements in terms of both quality and essential characteristics important during installation. These include, among others minimum amount of excavation works and sewage over pumping, and the installing panels possibility, regardless of the season and weather conditions. The contract was implemented based on the FIDIC Yellow Book and the contractor needed to prepare draft of the project taking into account Technical Specification requirements. Choosing the appropriate panels for rehabilitation of pipeline depended on information about its condition, shape and associated loads gathered during technical inspection. Overall calculations were accepted by the customer.

3 DESCRIPTION OF THE SEWER REHABILITATION TECHNOLOGY USING NC LINE PANELS

The HOBAS NC Line system is used for the trenchless pipeline rehabilitation of non-circular cross-sections. They can be made of a variety of materials: brick, stone, concrete. Depending on the shape of the existing pipeline, panels may have typical or special shape. Production panels takes place in specially constructed forms – with standard or non-standard cross-section shapes. Special metering and control systems are used. Forms rotate resulting creation of elements with uniform thickness and parameters around its circumference (Wojcik 2010).

HOBAS NC Line panels are made of: unsaturated polyester resin, the continuous and chopped glass fiber (E-CR type) and quartz sand (grain size max 1.0 mm). Wall structure created during winding process consists of the following four layers:

– internal layer with a thickness of 1.0 mm including a resin-rich internal layer with quartz sand as a filler, characterized by a high smoothness, chemical and abrasion resistance,
– barrier layer of resin and chopped glass fibers with a thickness of 1.5 mm,
– structural layer composed of resin, chopped and continuous fiberglass and quartz sand as a filler which gives panel suitable stiffness,
– external layer of highly sand-filled resin covered with coarse sand with a thickness of 0.5 mm (Wojcik 2010).

Depending on the route of the sewer main, different unit lengths are used. On the straight sections – panels of a length of 2.35 m, while on the curves – shorter elements, for example 0.5–1.5 m. All performed elements are equipped with an integrated socket. Spigot of the pipe after curing process is ground in order to settle an elastomeric seal. On customer demand it is also possible to implement bonded joints.

Choosing the appropriate NC Line shape element, its length and thickness of the wall, depends on data delivered by customer such as: shape of the existing or designed pipeline, its technical condition and also on the weight of the soil, surface water and surface traffic loads. Moreover, static calculations by finite element method, or based on the "ATV-M 127" guidelines are carried out for rehabilitated pipeline. The purpose of these calculations is to check the resistance of segment for:

- pressure generated during filling annular space between the existing structure and the liner with cement grout,
- external pressure exerted by groundwater after the restoration of leaking pipeline,
- the weight of the soil and surface traffic loads when existing structure losses its load capacity (Wojcik 2010).

Principal advantages of the NC Line system:

- Panels installation is possible while maintaining a low flow rate, which does not threaten the crew.
- The panels can be installed without bypassing these low flows which reduces costs related to over pumping. In this case, the work is most often performed from upstream to downstream.
- Low construction cost. Excavation work is limited to the creation of access shafts. Simple and rapid installation of the panels and the absence of costs related to traffic disruption help to reduce costs of the whole project.
- Work not influenced by outside weather conditions.
- No environmental impact and no impact on existing structures. Since it is not necessary to lower the level of the groundwater, the installation of the pipes has no negative influence on the stability of structures which are nearby the sewer to be rehabilitated.
- Improvement load capacity. The NC Line panels, the existing structure and the grout in the annular space create a composite system that is able to resist vertical loads, external water pressure and corrosive effluents, while maintaining high flow capacity.
- A system applicable to sewer pipes with variable dimensions and shapes. The production of the NC Line panels is based on the use of steel molds, both standard and custom, adapted to each project in accordance with the data collected during the preliminary inspection of the existing structure.

4 SELECTED ISSUES REGARDING THE PIPELINE REHABILITATION PROJECT

Preliminary panels dimensions were established based on Technical Specifications used for the tender. In the next stage, when the contract between Employer and Contractor was signed, an additional inspection was conducted. It finally confirmed the current shape, dimensions and condition of renewed pipelines. These data were used for the panel profiles selection, which on the one hand have to provide adequate flow capacity of the pipeline, on the other hand comfortable installation. Based on the static strength calculations, which take into account the material strength parameters and the type of loads acting on the structure during their installation and subsequent operation, in each case the appropriate thickness of the modules were determined. The parabolic, oval, arch and special shapes of panels with dimensions up to DN 600/900 DN 2677/2075 and wall thicknesses from 11 to 45 mm were selected.

While most renewed channels have shape allowing application of oval and parabolic standard elements, in two cases, the manufacturer of the NC Line panels was forced to seek a special solution. Pipelines located under Wanda and Daszynskiego Roads were built as DWFC (Dry Water Flow Channel) and the use of typical panels in these cases would result in a significant reduction in cross-section and channels flow capacity. Consequently, modules consisting of several separate elements combined using lamination were selected. Not only the special production place needed to be prepared, but also special structure which protected produced panels both during transportation and installation. The modules were equipped with special sockets joining system. Unusual shape

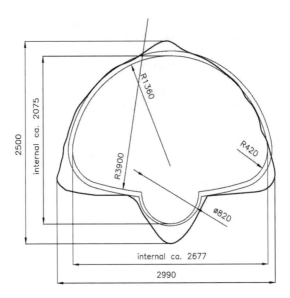

Figure 1. Parabolic cross-section resin panels with side shelf and semicircular invert.

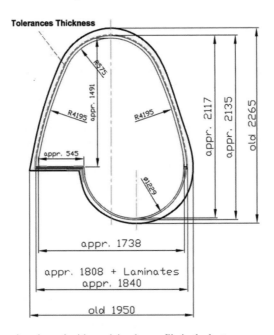

Figure 2. Arch cross-section channel with semicircular profile in the bottom.

of the panels excluded standard gasket joining and an epoxy adhesive joining system was applied. In the most difficult spots TEXTEC synthetic material (Wojcik 2010), which provided better adhesion to filler has been used.

Examples of non-standard-shaped elements are shown in Figures 1–2.

5 SELECTED REALIZATION ISSUES

Implementation of described project began in May 2008 and it was completed in September 2009. Initially, the rehabilitation of the pipeline was carried out on test section. Works on

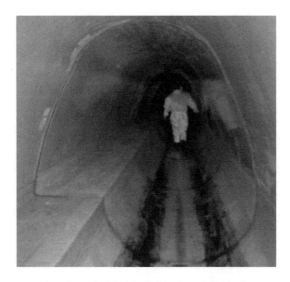

Figure 3.   Parabolic cross-section channel with side shelf and semicircular invert.

subsequent legs began after successful passing of all trials and tests. Moreover, project parameters set up by customer needed to be confirmed by laboratory research. Achieving the desired
results depended on many factors, including the proper conducting of preparatory work, removal
of all the obstacles (including collision with existing waterworks network), careful cleaning of
the existing channel walls and also on the established properties of materials used for panels
manufacturing.

New modules were inserted into the host pipe through a specially prepared access shafts with
dimensions adapted to the size of the modules. Then, using an appropriate equipment, such as
trolleys, the panels were placed into the depth of the channel, where they were fixed and combined.
In order to prevent buoyancy, modules were properly wedged during filling annular space with
cement grout. The filler, which is one of the most important elements of the applied technology,
ensured appropriate coordination between the existing channels and panels and also provides a
uniform loads distribution on new construction. Injection was carried out in several stages. Holes for
existing house connections were cut consecutively as the work progressed. The house connections
with the panel were combined using top hat or pipe stub, where the point of contact was sealed by
the laminate.

One of the most interesting task was the rehabilitation of sewer mains under Wandy and Daszynskiego Roads. Those sewer mains are the largest from all rehabilitated in Cracow but also they have
different shapes. Respectively, their dimensions are: 1950/2265 mm and 2990/2500 mm and they
were built as reinforced concrete structures. Requirement for this part of work was not to reduce
the cross-section of the channel after the rehabilitation above 10%. This goal has been achieved by
using special modules designed to match the shape of the channel profile.

Rehabilitated stretch in Wandy Road was 254-m-long. Because of its straight section, equal
(2.35-m-long) panels has been used. The access shaft was located in the middle of the rehabilitated
section and the installation took place in two directions. Around 12 to 15 modules were installed,
during the day. Pipe view after rehabilitation is shown in Figure 3.

The second non-standard 220-m-long sewer main was located under Daszynskiego Road. There
were used 82 sections (2.35-m-long) for the straight parts and 32 sections (0.85-m-long) for the
curves. The individual parts of the shaped sections were securely joined together with laminate. In
that case sewage needed to be pump.

View of existing duct after rehabilitation is shown in Figure 4. Assembly works under
Daszynskiego Road ended entire project in which for the first time in Poland channels with shelves
were rehabilitated.

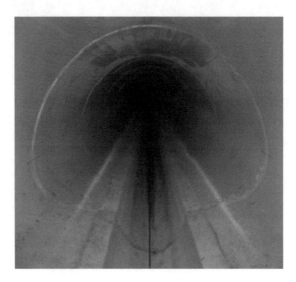

Figure 4.   Arch cross-section channel with semicircular profile in the bottom, after rehabilitation.

6   CONCLUSIONS

Adopted trenchless technology of sewer mains rehabilitation using HOBAS NC Line – resin rein-forced panels with glass fiber has provided a lot of benefits. Traffic disruption has been reduced by limitation of excavation work to create access shafts, which also lowered the costs of occupy-ing, disassembly and restoration of the pavement. Inconvenience such as detour designation and building footbridges across the trenches were avoided. The water table has not been lowered and installation of the pipes has had no negative influence on the stability of structures, which are located nearby the sewer. The risk of damage to other networks or cables has been eliminated (Kuliczkowski et al. 2010). An negative impact on environment has been reduced by groundwa-ter protection and limitation of waste, noise, CO_2 emission to the atmosphere (Kuliczkowska & Kuliczkowski 2011). Installation work has not depended on weather conditions.

The quality of used materials had significant importance for described project. High chemical protection, abrasion resistance and improved hydraulic parameters were obtained, thanks to using HOBAS NC Line technology. Renewed channel ensures mechanical resistance necessary to balance static and dynamic loads. Easy installation facilitated the work and shorten implementation period, so it was possible to achieve significant financial savings.

REFERENCES

Góra, M. 2010. Cracow MPWiK S.A. achievements in man-entry sewers restoration in historical city centers. Economic and engineering aspects: 6. Cracow.
Kuliczkowska E. & Kuliczkowski J. 2011. Trenchless technologies help reduce CO2 emissions. *Modern Construction Engineering, No. 1:* 68–70.
Kuliczkowski A., et al. 2010. *Trenchless technologies in environmental engineering.* Warsaw: Publisher Seidel – Przywecki.
Wojcik, A. 2005. New technologies in the environmental sector used in projects financed by the Cohesion and ISPA Fund: 7. Warsaw: HOBAS.
Wojcik, A. 2010. HOBAS pipes in all shapes and sizes. Renovation of sewer main in Cracow city center with panels HOBAS NC Line: 2. Cracow: HOBAS.

Underground Infrastructure of Urban Areas 2 – Madryas, Nienartowicz & Szot (eds)
© 2012 Taylor & Francis Group, London, ISBN 978-0-415-68394-4

Trends in application of materials to construction and rehabilitation of water supply networks in Poland after 1990

M. Kwietniewski & K. Miszta-Kruk

Warsaw University of Technology, Department of Water Supply and Wastewater Management, Poland

ABSTRACT: The paper outlines the trends which could be seen in development and modernisation of water supply networks in Poland after the economic transformation initiated in the 90-ties of the last century. On the basis of research efforts made in the past 20 years the changes in the material structure of the water supply networks and service lines in that period of time have been determined. Pipelines made of traditional materials (grey cast iron and steel) continue to dominate in water supply networks. For some time thermoplastic materials have been widely used for production of conduits. Moreover, the paper shows the range of technologies applied to rehabilitation of water supply networks and identifies the materials that the conduits before and after rehabilitation are made of. The pipelines subject to rehabilitation are made predominantly of traditional materials and, in most cases, they are replaced with PE or ductile cast iron conduits

1   INTRODUCTION

In Poland 267 332 km of water distribution networks (Fig. 1) were in operation in 2009. Since 1990 one could witness a steady development of the water supply networks, with the average annual increase by around 8,700 km in the period of 1990 to 2009 in the total network length, although in the past 5 years the network construction process slowed down to around 4,300 km a year on the average. This intense development of the networks leads to more and more people using collective water supply systems. Although the indicator illustrating the share of the general public benefiting from collective water supply networks is quite high 87.2% (in towns – 95.2% and in villages – 74.7%), there is still a substantial group of consumers who don't have access to an organised system of fresh water supplies for household purposes.

Water supply networks are made of various materials. The diversity of the materials is much bigger now than it was in the past due to a wide array of material solutions available on the market, the solutions which are permitted to be used in water supply systems. In the previous period of time, i.e. before 1990, especially in the 60-ties and 70-ties, (Technical Guidelines1964) mainly pipes made of asbestos cement or PVC for small-diameter conduits, or pipes made of reinforced concrete, grey cast iron or steel for large-diameter conduits were recommended for construction of water supply networks. It was so because implementation of water supply projects probably involved preferential treatment of certain branches of industry in Poland, conducive to development of, i.a., cement and asbestos industries.

Research into the range of application of various material solutions to construction of water supply networks in Poland and into the network renovation technologies has been conducted for 20 years by Department of Water Supply and Wastewater Management at Faculty of Environmental Engineering, Warsaw University of Technology in specific cycles, each more or less five year long. The last stage of the research, covering the period of 2005 to 2008, was conducted in cooperation with Chamber of Commerce in Polish Water (Kwietniewski et al. 2010).The research focused on water supply networks (distributing pipes and water mains) and water service lines. The research was conducted in two main subject areas, namely:

- Identify the scope of application of materials to construction of water supply networks and service lines, and criteria employed to select those materials;

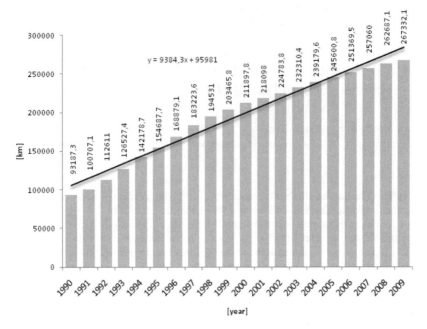

Figure 1.   Development of water distribution networks in Poland in the period of 1990 to 2009 (Prepared on the basis of: Infrastruktura komunalna w 2011r., Central Office of Statistics).

- Determine the scope of application of various technologies to rehabilitation of water supply pipelines, and criteria employed to make rehabilitation decisions, as well as identify the materials that the conduits subject to rehabilitation and the replaced conduits are made of.

This paper shows major research results obtained in the entire twenty-year-long period of observations, covering the changes in the structure of materials used for construction of water supply networks and service lines, as well as the material structure of rehabilitated conduits.

2   MATERIAL STRUCTURE OF EXAMINED WATER SUPPLY NETWORKS AND SERVICE LINES

Domestically, the water supply networks and service lines laterals are made predominantly of the materials listed below. The list includes abbreviations assigned to individual materials and used consistently in further result analyses and presentations:

- ST – steel;
- CI – grey cast iron;
- DI – ductile cast iron;
- PVC – non-plasticized polyvinyl chloride;
- PE – polyethylene;
- AC – asbestos cement
- PP – polypropylene
- Pb – lead.

The past 20 years witnessed a downward trend in application of traditional materials, i.e. grey cast iron and steel, to construction of water supply networks lat (Fig. 2). In that period of time the share of examined networks made of grey cast iron decreased by around 27.5%, and the share of steel networks decreased by about 6%.

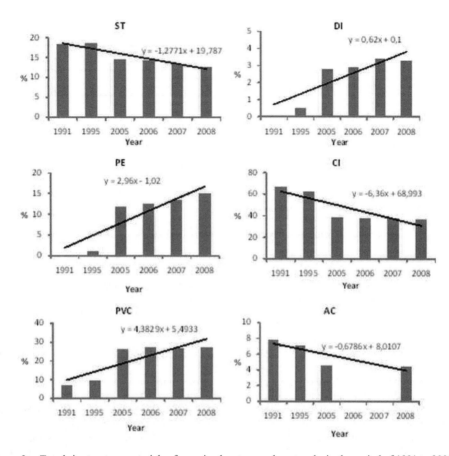

Figure 2.   Trends in structure materials of examined water supply networks in the period of 1991 to 2008.

A reverse trend is displayed by networks made of thermoplastic materials. The share of PVC conduits increased in the examined water supply networks by over 20% and the share of PE conduits increased by around 15%.

It is important to mention conduits made of asbestos cement. Their share in the examined water supply networks decreased in the analysed period of time by around 3.5%. It is so mainly because of liquidation of pipelines made of that material as a result of the conduit failures. The elimination of asbestos cement pipes is in line with a nationwide Programme of Elimination of Asbestos and Products Containing Asbestos Used in Poland adopted by the Cabinet of the Republic of Poland on 14 May 2002. The Programme provides for elimination of products containing asbestos in the period of 30 years, i.e. until by 2032. The revised version of the Programme entitled Asbestos Disposal Scheme for Poland for the Period of 2009 to 2032 developed by Ministry of Economy in 2009 (Program of the National Asbestos Cleaning for the years 2009–2032) allows for the possibility of leaving the pipes containing asbestos in the ground provided that they are documented in local plans and shown in the documentation of the given property. It is important to mention in this context that in our country there is a total of 15.466 million tonnes of products containing asbestos, including 14.866 million tonnes of asbestos cement cardboards and 0.6 million tonnes of pipes, the latter including water supply pipes, sewers, chimney shafts or rubbish chutes used in multi-storey buildings. However in the period of 2003 to 2008 around 1 million tonnes of the products were disposed of, accounting for around 6% of the total. Hence, there continues to be around 14.5 million tonnes of asbestos products in Poland (The program for remove asbestos and products containing asbestos used on Polish territory 2002, Program of the National Asbestos Cleaning for the years 2009–2032).

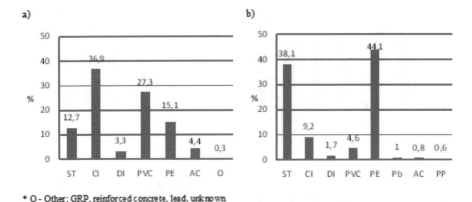

a)   b)

* O - Other: GRP, reinforced concrete, lead, unknown

Figure 3.   Trends in structure materials of examined water service lines in the period of 1994 to 2009.

Less research material was collected for service lines than for water supply networks. Neverthe-less one can see certain regularities in the analysed period of time (Fig. 3). A clear downward trend shown by steel service lines can be seen. The share of pipelines made of that material decreased in the analysed period of time by around 25%. But no dramatic changes in the use of grey cast iron and PVC for construction of service lines have been recorded.

The share of PE in the examined water service lines in the period of 14 years increased very clearly by around 24%.

Moreover, the use of ductile cast iron for construction of service lines has been recorded, although the extent to which that material is used remains rather small.

As a result of material development in the analysed period of time the material structures of the examined networks and service lines have been obtained as shown in Figures 4 and 5.

The material structure of the water supply networks (Fig. 4) continues to be dominated by pipelines made of traditional materials, namely grey cast iron and steel. The pipelines account for a total of 59.6% of the network length. But the share of conduits made of thermoplastic materials, i.e. PVC and PE (collectively they account for around 42.4% of the network length), which are gradually replacing traditional pipelines, has become quite substantial.

The share pf conduits made of ductile cast iron in the structure of the examined water supply networks is rather significant.

Two types of materials dominate definitely in the case of water service lines, namely PE and steel, with the polyethylene conduits outnumbering the steel ones (the former account for 44.1% of the length of the examined service lines) (Fig. 4).

The share of service lines made of grey cast iron and PVC is also quite significant.

3   MATERIAL STRUCTURE OF REHABILITATED WATER SUPPLY NETWORKS

Technologies applied to modernisation of water supply conduits in the analysed period of time have been divided into three groups, namely: traditional Excavation-based replacement, renovation and Trenchless replacement. It results from the analysis of the obtained data that in the entire period of time the traditional excavation-based replacement was employed definitely in the greatest number of cases. Application of that technology shows an upward trend, and it has been used for moderni-sation of around 81% of the examined water supply networks in recent years. Renovation was the technology which had played a significant role until 2005. In general, trenchless technologies show an upward trend and in the recent research period more pipelines were modernised using trenchless technologies than renovation (Fig. 5).

Due to a dominating share of the excavation-based technology the material structure of the conduits before and after modernisation based on that technology was analyzed (Table 1 and 2).

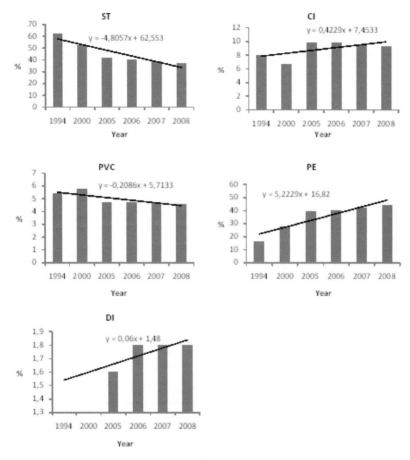

Figure 4.   Material structure of the examined water supply networks (a) and service lines (b) in Poland in 2008 (Kwietniewski et al. 2011).

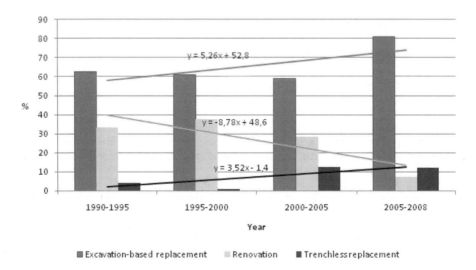

Figure 5.   Trends in application of technologies to rehabilitation of the examined water supply networks in the period of 1990 to 2008.

113

Table 1. Material structure of the examined water supply networks subject to rehabilitation using traditional excavation-based technique in the period of 1991 to 2008 (% of the length of rehabilitated pipelines).

Material of the pipelines subject to renovation	1991 to 1995	1995 to 2000	2005 to 2008
ST	7.0	51.6	54.9
CI	62.9	40.5	23.9
AC	2.5	7.1	16.2
PVC	N/A*	0.1	2.7
DI	N/A	N/A	2.0
PE	N/A	N/A	0.2
Lead	N/A	0.7	0.1
Other	27.5	–	–
TOTAL	100	100	100

*N/A-lack of data. **Other: reinforced concrete, some listed in the table, unknown.

Table 2. Material structure of the examined water supply networks after rehabilitation involving traditional excavation-based replacement technology in the period of 1995 to 2008 (% of the length of rehabilitated pipelines).

Material of conduit after replacement	1995 to 2000	2005 to 2008
ST	10.5	51.6
PVC	21.5	6.7
DI	4.0	14.4
PE	64.0	78.2
Other (unknown)	–	0.1
TOTAL	100	100

The results, as put together in Table 1, confirm the conclusions reached in the analysis of the material structure of the examined water supply networks (Fig. 2). Conduits made of traditional materials (steel and grey cast iron) were subject to excavation-based replacement. But with time more and more steel conduits and less and less grey cast iron ones were replaced.

The data shown in Table 2 indicate that old and worn out conduits are replaced mainly with PE pipelines. In the period of 2005 to 2008 they accounted for around 78% of the length of the examined water supply networks subject to excavation. Recently conduits made of ductile cast iron have been used quite frequently; the same applies, although to a smaller extent, to PVC conduits.

4 CONCLUSION

National water supply networks continue to be dominated by conduits made of steel or grey cast iron. At present pipelines made of those materials account for almost half the network. The share of the pipelines made of PE or PVC has become quite substantial, over 42% of the examined networks is made of one of those two materials.

The material structure of the water service lines is dominated by conduits made of PE or steel, with the polyethylene pipes clearly outnumbering the steel ones. Collectively they account for over 82% of the length of the examined service lines.

It can be observed that thermoplastic materials are extensively used for construction of water supply conduits. Especially PE pipes are extensively used for assembly of smallest diameter conduits (service lines). It is also important to mention that networks made of ductile cast iron have been increasing steadily.

Old pipelines made of traditional materials, i.e. grey cast iron or steel are replaced mainly with PE conduits.

It is important to point out in this recapitulation that at present there are many products (pipes, fittings), made of various materials and designed to be used for construction of water supply networks, available on the Polish market. Therefore, there is a problem of choice: which material satisfies the water supply network user's requirements best in given circumstances. The users include, first and foremost, water supply network operators who operate and maintain the networks. To solve the material selection problem in a sound and informed manner you need a thorough knowledge about technical and functional, and usability characteristics of the products, as well as an adequate methodology of how to proceed. The special selection criteria should be taken into account in such a process. Reaction material with water, and durability and pipeline failures there are the most important criteria (Dohnalik & Jędrzejewski 2004, Hotloś & Mielcarzewicz 1995, Kuliczkowski & Kuliczkowska 2010, Kusak et al. 2002, Miszta-Kruk et al. 2009, Pawełek & Wojdyna 2000). The elaboration (Kwietniewski et al. 2011) presents such a methodology based on appropriate and well-developed selection criteria and procedure. The methodology should facilitate the selection of material solutions to be used by water supply network operators and designers for construction of the networks.

REFERENCES

Dohnalik P. & Jędrzejewski A. 2004: *Effective operation of water supply systems*. Kraków.
Hotloś H.& Mielcarzewicz E. 1995: Failuring of water and wastewater systems in terms of mining operations. Gaz, Woda i Technika Sanitarna 12/1995, s. 429–433.
Kuliczkowski A. & Kuliczkowska E. 2010: Renewal startegy of water networks, Gaz, Woda i Technika Sanitarna nr. 2/2010r. Wyd. SIGMA. ss. 26–29.
Kusak M., Kwietniewski M. & Sudoł, M. 2002: Influence of different factors on the failuring of water supply pipes based on operational reliability tests. Gaz, Woda i Technika Sanitarna nr 10/2002 ss. 366–371.
Kwietniewski M., Tłoczek M., Ferszt E. & Sobierajski M. 2010: The research of materials structure and the scope of the renewal technology, water supply network in Poland in 2005–2008. Zeszyty Izby Gospodarczej „Wodociągi Polskie". S. 17–33. Rok. IV, nr 4/2010, ISSN 1734-0896.
Kwietniewski M., Tłoczek M. & Wysocki L. 2011: Rules for selection of material and structural solutions for the construction of water pipes. Wyd. Izba Gospodarcza "Wodociągi Polskie". Bydgoszcz 2011.
Miszta-Kruk K., Kwietniewski M., Osiecka A. & Parada J. 2009: Material structure of municipal networks In Poland in the period of 2000 to 2005. [w:] Madryas, C, Przybyła B., Szot A. (eds.): Underground Infrastructure of Urban Areas. CRC Press. Taylor & Francis Group. Boca Raton, London, New York, Leiden.
Pawełek J. & Wojdyna M. 2000: Analysis of the distribution pipes failure in a large water supply system. Gaz, Woda i Technika Sanitarna 2/2000, s. 49–54.
Program of the National Asbestos Cleaning for the years 2009 – 2032 , Rada Ministrów, 14 lipiec 2009r.
Technical Guidelines for the design of municipal water supply networks. Zarządzenie MGK z 17 stycznia 1964 r. (Dz. Bud. nr 8, poz. 26).
The program for remove asbestos and products containing asbestos used on Polish territory, Rada Ministrów, 14 maj 2002 r.

Underground Infrastructure of Urban Areas 2 – Madryas, Nienartowicz & Szot (eds)
© 2012 Taylor & Francis Group, London, ISBN 978-0-415-68394-4

The workability, porosity characteristics, frost resistance, durability coefficient and compressive strength of SCC incorporating different type of admixtures

B. Łaźniewska-Piekarczyk

Department of Building Processes, Faculty of Civil Engineering, Silesian University of Technology, Poland

ABSTRACT: The influence of superplasticizer (SP) type, air-entraining admixture (AEA), anti-foaming admixture (AFA) and viscosity modifying admixture (VMA) on air-content, workability and change of workability after time of self-compacting concrete (SCC) is analyzed in the paper. The compressive strength, frost resistance and DF durability coefficient, parameters of air-voids and total porosity of hardened SCC are also investigated. The research results indicate that the admixtures significantly change the analyzed properties of fresh and hardened SCC.

1 INTRODUCTION

The effect of the self-compaction depends on the values of rheological parameters, such as yield stress and plastic viscosity of cement paste. Technological tests, assessing the self-compactibility of the concrete mix (SCC), such as flow test (Tables 1 and 2), are carried out because the availability of the direct measurement of rheological properties is limited. The value of SCC flow diameter depends on the mix yield stress τ_{0m}, whereas SCC time flow depends on its plastic viscosity η_{pl}. The diameter and time flow of SCC should correspond with the classes presented in Tables 1 and 2. European guidelines for self-compacting concrete (European Project Group 2005) describe detailed outlines in respect of SCC classes and other technical tests of the self-compacting concrete mix depending on its purpose.

On the basis of different tests concerning self-compacting concrete mixtures, the authors found out too high air content in concrete volume, which was the result of high range of water reducer (SP) presence, in spite of fulfilling the criteria (European Project Group 2005). The molecules of SP should also modify the surface of solid particles in order to keep its hydrophilic character. The air bubbles can adhere only to hydrophobic surfaces. The presence of listed functional groups (oxygen in form of etheric group (-0-), hydroxyl group (-OH) and carboxyl group) produce water surface tension decrease, producing flocculation of associated molecules and increase in moisten of not only grains of cement but also the whole mineral framework (Kucharska 2000).

In the superplasticizers group there are ones that show only dispergration functioning not decreasing surface tension. They are: hydrocarboxylen acid salts, sulphonic melamine-formaldehygenic resins, formaldehygenic picodensats salts of beta-naphtalensulphonic acid.

Results of researches (Mosquet 2003) presented in Table 3, prove that new superplasticizers generations show air-entraining functioning, which was proved by the research results carried out by the authors. The research results showed that an excessive air-entrainment is caused – mostly – by the decrease of surface tension of liquid phase in paste by the PCP superplasticizer.

On the basis of author's tests results the influence of etheric and poly-carboxyl SP the SCC on porosity structure could be concluded. Tests results prove that SP made on the basis of poly-carboxyl ether generate considerable SCC air-entrainment. The air-entrainment is higher in case of higher w/c value and amounts even to 8.30%. According to authors' publication (Hanehara & Yamada 2008) PC usually has an air-entraining effect, but some types of PC drastically reduce the freezing/thawing resistance.

Table 1. Slump-Flow classes (European Project Group 2005).

Class	mm
SF1	from 550 to 650
SF2	from 660 to 750
SF3	from 760 to 850

Table 2. Viscosity classes (European Project Group 2005).

Class	sec. T_{500}	V-funnel
VS1/VF1	≤2	≤8
VS2/VF2	>2	from 9 to 25

Table 3. The influence of SP type on the concrete air-entrainment (Mosquet 2003).

				New Generation SP	
SP	LS	SNF	SMF	PCP	AAP
Air volume	++	+	0	++	++

Where: LS – Lignosulfian, SNF – Sulfonated Naphtalene Formaldehyde Condensate, SMF – Sulfonated Melamine Formaldehyde Condensate, PCP – PolyCarboxylate Polyoxyethylene, AAP - Amino Phosphonate Polyoxyethylene.

Considering mentioned above tests results, it is evident that certain SP of new generation produce an excessive air-entrainment remaining in the volume of the fresh mix and concrete, although the mix meets commonly accepted criteria of technical tests (Tables 1 and 2). Desired suitable fluidity of the fresh mix, essential for its effective self-compacting, is not included in any commonly used technical tests. Accepted criteria for such tests are insufficient in this scope and do not guarantee effective self-compacting. It can be obtained by increasing the fluidity of the fresh mix with the SP, however it may generate its segregation. In order to prevent the presence of the excessive air-entrainment, the SP should not only be compatible with cement, but also do not create air-entraining effect in the paste.

Thus, in case of non-air-entrained SCC, achieving low air-content might became a slightly problematic task (Khayat 2000, Kobayashi et al. 1981, Litvan 1983, Szwabowski & Łaźniewska-Piekarczyk 2008b). Some SP types cause too high air-content in SCC (Szwabowski & Łaźniewska-Piekarczyk 2008b), in spite of the fact that the SP contains (according to producers' information) the anti-foaming admixtures (AFA). AFA decreases the air-content in SCC effectively (Łaźniewska-Piekarczyk 2010). Moreover, AFA increases the level of mix consistency. Thus, it is necessary to reduce the dosage of SP. In case of segregation of the mix, viscosity modifying admixture (VMA) should be used.

As in the case with non-air-entraining SCC, in case of air-entraining SCC, achieving suitable air void characteristics is also a difficult task (Kamal H et al. 2002, Khayat 2000, Szwabowski & Łaźniewska-Piekarczyk 2008a). Considerable fluidity of self-compacting concrete mix, air bubbles, presented in air entrained concrete mix, can be unstable because of floating and coalescence of air bubbles or fading of bubbles with a diameter less than 0.10 mm (Kamal et al. 2002, Khayat 2000). Intentionally introduced air bubbles are unstable due to the high level of consistency of self-compacting mix. In case of too high level of mix consistency, which encourages the segregation of the mix, VMA should be used.

Table 4. Composition of SCC, kg/m^3.

CEM II B-S 32.5R	Lime stone	w/c	w/b	Sand 0–2 mm	Gravel 0–8 mm	Volume of paste, %
442.40	190.00	0.45	0.31	693.20	866.49	41.10

Table 5. Type of admixtures used in SCC.

Admixture type	The main aim of use of the admixture:
SP1 (with air-entraining side effect)	Air-entrained SCC (as a results of sidle effect of SP1)
SP1+AFA	Elimination of too high air-content (as a results of sidle effect of SP1) in SCC
SP1 + AFA + VMA	Prevention of a SCC segregation as a results of SP1 and AFA action
SP2 (without air-entraining side effect)	Non air-entrained SCC
SP2+VMA	Prevention of a SCC segregation as a results of SP2 action
SP2+AEA	Intentionally air-entrained SCC
SP2+AEA+VMA	Prevention of a SCC segregation as a results of SP2 and AEA action

Main base of admixtures: SP1; SP2 – polycarboxyl ether; VMA – synthetic copolymer; AFA – polyalcohol; AEA – synthetic tensid.

The main purpose of the paper is a comparison of the influence of SP, VMA, AEA and AFA on the properties of the self-compacting concrete mix, porosity characteristics of SCC that affect the compressive strength and frost-resistance.

2 RESEARCH METHODS

Investigated SCCs (Table 4) were prepared from CEM II B-S 32.5R, limestone, gravel 2–8 mm, sand 0–2 mm. In the research, the following admixtures were used (Table 5): SP1 (with air-entraining side effect), SP2 (without air-entraining side effect), AFA and VMA, AEA. Because the consistency of concrete mix influences the air-content in SCC (Szwabowski & Łaźniewska-Piekarczyk 2008b), the dosages of the admixtures were conformed to the same slump flow class (SF2) of SCC.

The properties of self-compacting concrete mix were tested by means of the following method: slump flow tests according to (European Project Group 2005) air-content in mix according to EN 12350-7:2001. The slump flow tests of SCC were carried out after 20 and 60 min.

The compressive strength of SCC was measured according to EN 12350-3:2001.

The DF durability of SCC coefficient (ASTM C 666, 1991) was also investigated. With the aim to estimate DF coefficient, the modulus elasticity of concrete beams ($10 \times 10 \times 40$ cm) was measured according to CEN/TR 15177 – beam test. The values of modulus elasticity of SCC beams were estimated after 25, 50, 75, 100, 150, 200, 250, 300 freezing-thawing cycles. According to (ASTM C 666, 1991) concrete is frost-resistant when DF > 80%, and concrete is not frost-resistant when DF < 60%.

The frost-resistance of SCC was investigated in accordance to PN-88/B-0625. According to the PN-88/B-0625, concrete is frost-resistant when the decrease of compressive strength after n cycles in less than 20% and decrease of mass is less than 5%.

The values of air-voids parameters were estimated according to EN 480-11. The measurements of total porosity of SCC were also carried out. Total porosity of SCC was estimated by the means of porosimeter. The pores with diameter from 3 nm to 250 μm were measured with pressure from 1 kPa to 400 MPa. The measurement was carried out by a continuous method (scan mode). The changes of mercury volume were estimated during pressure increase and decrease.

Table 6. Test results of SCC mix.

Symbol	D [mm]	T_{500} [s]	Ac [%]	D* [mm]	T_{500}^{*} [s]
B1	730	3	8.0	730	3
B1A	705	2	2.7	705	2
B1VA	710	5	3.1	710	5
B2	715	2	2.1	685	4
B2V	710	4	2.5	780	6
B2A	640	2	5.0	540	3
B2AV	690	4	5.0	730	3

*After one hour after start mixing of SCC.

3 RESEARCH RESULTS AND DISCUSSION

The research results of concrete mix properties are presented in Table 6. Because of the dosages of admixtures were conformed to the constant slump flow class (SF), the diameters flow of concrete mix are similar (Table 6). The main aim of this step of research is to compress the influence of admixtures with air-content in SCC (with similar SF class). The analysis of data presented in Table 6 results in the following conclusions.

SP type significantly influences the air-content in self-compacting mix with similar flow diameter. The air-content amounts to 8% in spite of the fact that the flow diameter amounts to 730 mm.

The application of AFA causes a considerable decrease of the air-content in SCC. The use of higher dosage of SP causes mix segregation. The introduction of AFA does not result in segregation of concrete mix. In this case, the dosage was decreased to achieve the SF2 class flow of SCC. Normally, flow diameter of SCC incorporating AFA is higher than the flow diameter not incorporating AFA and the same amount of SP. The T_{500} time flow of concrete mix that contains AFA is similar to time flow of concrete mix without AFA. Moreover, mix with AFA adheres less to Abram's cone until conducting the test of consistency (European Project Group 2005).

The introduction of VMA into SCC with AFA (and into SCC without AFA but with SP without air-entraining effect) results in increase of the time flow. Fortunately, the previously decreased air-content as a result of AFA acting is not significantly increased. The self-compacting mix incorporating AFA is more resistant to segregation.

Deliberate air entraining contributes to the decrease of diameter flow of self-compacting concrete mix (Table 6) more than air-content as a result of SP side effect. It should be emphasized that in the same publications are decreased contradictory conclusions concerning the increase of air-entrainment on modification of flow diameter of concrete mix. The other author's research results prove that the character of the influence of air-entrainment depends on amount of introduced AEA into self-compacting concrete mix. Initially, small dosage of AFA may results in increase of flow diameter of SCC as a result of AEA acting. However, successive addition of AEA into self-compacting concrete mix causes the decrease of flow diameter due to the interaction between air bubbles and concrete mix particles (Szwabowski & Łaźniewska-Piekarczyk 2008b).

Research results (Lachemia et al. 2004) in Figure 2 indicate that VMA results in the decrease of air-content in previously air-entraining concrete mix. Khayat's publication (Khayat 2000) also confirms that the character of VMA influence the decrease of air-content in air-entrained concrete mix. However, research results in Table 3 and 5 indicate that VMA not always decrease air-entrainment of concrete mix.

The investigated admixtures (Table 3) influence the workability of SCC after time in fundamental way. The biggest workability loss of SCC is in case of intentional air-entrained and as a side effect of SP action. VMA allows SCC to maintain initial consistency of SCC. Other author's research results indicated that cement pastes contain SP and VMA set for a longer time than cement pastes without VMA.

In data in Table 7 the total porosity characteristics by the means of porosimeter and compressive strength of SCC are presented. The pores with diameter from 3 nm to 250 μm were measured. There

Table 7. The research results of porosity characteristics and compressive strength of SCC.

Symbol	Volume of pores [cm³/g]	Volume of pores [%]	Specific surface of pores [m²/g]	f_{cm} [MPa]
B1	0.0728	15.46	8.2	61.7
B1A	0.0543	11.85	6.4	61.7
B1VA	0.0596	12.94	8.2	57.5
B2	0.0492	11.16	5.7	74.6
B2V	0.0613	13.37	7.4	73.1
B2A	0.0597	13.15	6.4	55.9
B2AV	0.0597	13.16	5.6	65.1

Table 8. The air-voids characteristics and decrease of compressive strength after 300 freezing-thawing cycles f SCC.

Symbol	A [%]	L [mm]	α [mm⁻¹]	A_{300} [%]	Decrease of f_{cm} after 300 freeze-thawing cycles [%]
B1	4.47	0.29	20.83	1.55	−3.0 (increase)
B1A	2.10	0.58	15.04	0.25	−0.4 (increase)
B1VA	2.56	0.43	18.34	0.42	2.6
B2	1.86	0.84	10.88	0.22	49.3
B2V	3.14	0.99	7.16	0.16	32.6
B2A	3.80	0.33	20.21	1.39	−3.3 (increase)
B2AV	3.72	0.32	20.71	1.54	3.1

is no correlation between the compressive strength and volume of pores in SCC. The influence of used admixtures probably disrupts the relationships.

VMA does not significantly influence on the compressive strength of SCC (Table 8), compare B2 and B2V). Admittedly, research results (Fu & Chung 1996, Gołaszewski 2009, Kamal & Khayat 1998, Lachemia et al. 2004, Leemann & Winnfield 2007, Rols et al. 1999, Saric-Coric et al. 2003,) suggest that VMA has a positive or negative compressive influence on strength of concrete, however, publication (Şahmaran et al. 2006) indicate that the influence of VMA on compressive strength depends on VMA and SP type.

It is a common knowledge that air-entraining of concrete results in a decrease of compressive strength. In this case, air-entrainment, as a result of AEA acting, causes decrease in compressive strength of SCC. Differences between the compressive strength of air-entrained and non-air-entrained SCC amount to 20%, while the difference between total air content in analyzed SCC amounts to 1,94%. The introduction of VMA into previously air-entrained self-compacting concrete mix results in the increase of SCC compressive strength. The increase of the compressive strength of SCC amounts to 10 MPa.

The comparison of data in Table 6 and 8 suggests that it is possible to predict the air-content in SCC on the basis of air-content in self-compacting concrete mix. However, there are significant differences between air-content in concrete mix, such as air-entraining side effect of SP and air-content in hardened SCC. It indicates that the air-content as a side effect of SP is very instable. On the other side, research results cited in publications (Szwabowski & Łaźniewska-Piekarczyk 2008b) indicate that air-content in hardened SCC, as a side effect of SP acting, may amounts to even 8%.

In Table 8 the air voids parameters research results are presented. The test results suggest the significantly influence of admixtures on values of porosity parameters. SP type significantly influences the values of air-voids parameters. The air-void factor in case of SP2 is almost three times bigger than SCC with SP1. The specific surface of air-voids in case of SCC with "air-entraining" SP is almost twice bigger than in case of SCC with SP without air-entraining side effect. The

Table 9. Research results of DF coefficient measurements after freezing-thawing cycles, [%].

Symbol	0 cycles	25 cycles	75 cycles	100 cycles	150 cycles	250 cycles	300 cycles
B1	99	104	104	107	104	107	107
B1A	102	101	98	99	101	100	101
B1AV	100	104	103	102	103	103	104
B2	99	102	102	103	100	81	65
B2V	100	101	100	99	101	99	101
B2A	100	102	103	101	103	106	105
B2AV	100	103	104	103	105	103	103

volume of air-voids with diameter smaller than 300 μm is bigger seven times in case of SCC with "air-entraining" SP.

The research results in Table 8 indicate that adding of AFA into SCC results in significantly decrease of air-content in SCC. The air-voids parameters research results suggest the adverse effect of AFA on frost-resistance of SCC.

The application of VMA into SCC with SP and AFA does not result in changes in values of air-voids parameters besides volume of the smallest air-voids (Table 8). Only the content of air voids with diameters smaller than 300 μm is increased. This rise is beneficial to frost resistance of SCC.

VMA influences the air-content in SCC. Total air-content in SCC is higher in case of SCC with VMA. Other parameters of air-voids of SCC with and without VMA are slightly different.

There is no correlation between air void content (Table 8) and compressive strength of SCC. Powers' equation and other formulas for calculating air-voids parameters according to EN 480-11 were empirically verified for air-entrained concrete. Little is known about the application of Powers' equations in case of non-air entrained concrete with small content of air voids and small air voids specific surface. This is why the comparison of the air void content and compressive strength of all types of concrete is not adequate in a large extent. Moreover, the methodology of testing of air voids parameters according to EN 480-11 does not cover the air-voids with diameter smaller than 5 μm. Thus, considerable air voids content is not measured. The interactions between aggregates and cement paste, micro cracking and micro porosity influence largely the compressive strength of SCC.

The air-content as a side effect of SP acting secures the frost-resistance of SCC. However, the research result analyzed in publication (Kobayashi et al. 1981, Szwabowski & Łaźniewska-Piekarczyk 2008a) indicate that "air-entrained" SCC, as a results of side effect of SP acting, was not frost-resistant. There must be an explanation for such a different influence of SP on frost-resistance of SCC. Probably, various types of SP cause different resistance of SCC to freeze-thaw cycles.

Research results of DF coefficient durability testing) are presented in Table 9. SCC with SP2 is frost-resistant until 250 cycles (DF ≥ 80%). After 300 cycles the SCC with SP2 is not frost resistant (DF < 60%). The research result indicates that the type of SP is very important for frost-resistance of SCC.

The research results in Table 9 suggest that the VMA influences frost resistance of SCC. The research results, indicated in publication 0, suggest that the same admixtures significantly influence the relationship between the frost-resistance of concrete and air voids space factor. Non air-entrained SCC with VMA retains the DF = 100% after 300 freeze-thawing cycles. The research results also indicate the beneficial influence of VMA to frost-resistance of concrete and value of air-voids space factor. However, the data in Table 8 indicate the negative influence of VMA on values of air-voids parameters regardless of concrete frost-resistance. On the other side, the decrease of compressive strength after 300 freeze-thawing cycles of SCC with VMA is smaller than SCC without VMA. Research results in Tables 8 and 9 also indicate that SCC is frost-resistant even though the values of parameters of air voids are different from recommendations of standards (presented in Table 10).

Table 10. The recommendations for air-voids parameters values by different standards.

Standard	Requirements	Class of frost exposure			
		XF1	XF2	XF3	XF4
Standard EN 206-1	Minimum air content in the mixture [%]	–	4.0	4.0	4.0
Polish Standard PN-B-06265	–	–	–	–	–
Austrian Standard	Minimum air content in the mixture [%]	–	2.5	2.5	4.0
ÖNORM B 4710-1	Minimum content of micro-voids A_{300} [%]	–	1.0	1.0	1.8
	Maximum void spacing factor L [mm]	–	–	–	0.18
Danish Standard DS. 2426	Minimum air content in the mixture [%]	–	4.5	4.5	4.5
	Minimum air A content in the mixture [%]	–	3.5	3.5	3.5
	Maximum void spacing factor $\bar{L}$ [mm]	–	0.20	0.20	0.20
	Concrete resistance on surface scaling	–	good	good	good
German Standard	Minimum air A content in the mixture [%]	–	3.5% at $d_{max} = 63$ mm		
			4.0% at $d_{max} = 32$ mm		
			4.5% at $d_{max} = 16$ mm		
			5.5% at $d_{max} = 8$ mm		
German Federal Ministry of Transport ZTV Beton-StB 01	Concrete with air-entrained and plasticizing or streamlining admixture	Concrete for road surfaces			
	Minimum air content in the mixture [%] (daily average)	5.0		4.0	
	Minimum content of micro-voids A_{300} [%]	1.5		1.8	
	Maximum void spacing factor $\bar{L}$ [mm]	0.20		0.20	

Symbol:

L – factor of air voids spacing in hardened concrete acc. to PN-EN 480-11,

A – air voids content in hardened concrete acc. to PN-EN 480-11,

A_{300} – the content of micro-voids with a diagonal diameter of less than 0.3 mm in hardened concrete acc. to PN-EN 480-11.

4 CONCLUSIONS

In the range of investigated of SCC, used admixtures and received research results it was indicate that:

- The type of SP very significantly influences the values of air-voids parameters and frost-resistance of SCC. The "air-entrainment", as a result of side effect of SP acting, secures the frost-resistance of SCC, which is also indicated by DF coefficient durability research results. SCC made of "non air-entraining" SP is not frost-resistant, also according to DF research results. Nevertheless, the air-entrainment, because of AEA acting, results in the best values of air-voids parameters regardless of frost-resistance of SCC.
- The application of VMA results in increase of content, particularly the volume of air voids with diameter smaller than 300 µm. Moreover, adding of VMA does not cause the decrease of air volume as a result of AEA acting. The changes of porosity characteristics are important for frost-resistance of SCC. The VMA cause the improvement of SCC frost-resistance. DF research results also indicate that conclusion. Nevertheless, the influence of VMA on frost-resistance of SCC still needs further research.
- AFA effectively decreases the air-content in SCC. Unfortunately, the use of AFA results in adverse effect on values of air-voids parameters of SCC, regardless of frost-resistance of SCC. Nevertheless, SCC is frost-resistant, which is also indicated by DF research results. The unexpectedly positive influence of AFA on frost-resistant needs further research.
- The application of the VMA in case of SCC with AFA beneficial decreases the size of air voids. The other parameters of air voids were changed because of decrease of air voids size: L was decreased and α i A_{300} was increased. SCC with AFA i VMA is frost-resistant, which was also indicated by DF research results.

- SP, VMA, AEA and AEA significantly influence the relationship between air-content of all pores and compressive strength of SCC. The issue needs further research.
- Examined admixtures significantly affect the properties of fresh self-compacting concrete. AFA improves the workability of SCC and protects concrete mix against segregation. Moreover, the analyzed admixtures influence the change of workability of SCC after time in a fundamental way. The air-entrained SCC is characterized by the highest decrease of workability after time. The use of VMA results in decrease of workability changes after time.

REFERENCES

ASTM C 666. Standard Test Method for Resistance of Concrete to Rapid Freezing and Thawing, Annual Book of ASTM Standards. 1991.

European Project Group. 2005. The European guidelines for self-compacting concrete: specification, production and use.

Fu X. & Chung D.D.L. 1996. Effect of methylcellulose on the mechanical properties of cement. *Cement and Concrete Research*. Vol. 26, No. 4/1996: 535–538.

Gołaszewski J. 2009. Influence of viscosity enhancing agent on rheology and compressive strength of superplasticized mortars. *Journal of Civil Engineering and Management International Research and Achievements, Vilnius: Technika*. Vol. 15, No 2/2009: 181–188.

Hanehara S. & Yamada K. 2008. Rheology and early age properties of cement systems. *Cement and Concrete Research* 21/2008: 175–195.

Kamal H. & Khayat K.H. & Assaad J. 2002. Air-Void Stability in Self-Consolidating Concrete. *ACI Materials Journal*. V. 99. No. 4. July–August/2002: 408–416.

Kamal H. & Khayat K.H. 1998. Viscosity-Enhancing Admixture for Cement-Based Materials – An Overview. *Cement and Concrete Composites* 20/1998: 171–188.

Khayat K. H. 2000. Optimization and performance of the air-entrained self-consolidating concrete. *ACI Materials Journal*. Vol. 97/2000. No. 5: 526–535.

Kobayashi M. & Nakakuro E. & Kodama K. & Negami S. 1981. Frost resistance of superplasticized concrete. *ACI SP-68*: 269–282.

Kucharska L. 2000: Traditional and modern water reducing concrete admixtures, Cement, Wapno, Beton, 2/2000: 46–61.

Lachemia M. & Hossaina K.M.A. & Lambrosa V. & Nkinamubanzib P.-C. & Bouzoubaâb N. 2004. Self-consolidating concrete incorporating new viscosity modifying admixtures. *Cement and Concrete Research* 34/2004: 917–926

Łaźniewska-Piekarczyk B. 2010. The influence of anti-foaming admixtures on properties self-compacting mix and concrete. *Cement-Wapno-Beton 3/2010*: 164–168.

Leemann A. & Winnfield F. 2007. The effect of viscosity modifying agents on mortar and concrete. *Cement & Concrete Composites* 29/2007: 341–349.

Litvan G. 1983. Air entrainment in the presence of superplasticizers. *ACI Journal*. Vol. 80. No. 4/1983: 326–331.

Mosquet M. 2003. The New generation admixtures. *Budownictwo Technologie Architektura*, SP/2003: . . .

Rols S. & Ambroise J. & Péra J. 1999. Effects of different viscosity agents on the properties of self-leveling concrete. *Cement and Concrete Research* 29/1999: 261–266.

Şahmaran M. & Christianto H.A. & Yaman İ.Ö. 2006. Effect of chemical and mineral admixtures on the fresh properties of self-compacting mortars. *Cement and Concrete Composites*, Vol. 28, Issue 5, May 2006: 432–440.

Saric-Coric M. & Khayat K. H. & Tagnit-Hamou A. 2003. Performance characteristics of cement grouts made with various combinations of high-range water reducer and cellulose-based viscosity modifier. *Cement and Concrete Research* 33/2003: 1999–2008.

Szwabowski J. & Łaźniewska-Piekarczyk B. 2008. The requirements for porosity parameters of frost-resistance self-compacting concrete (SCC). *Cement-Wapno-Beton*. No. 3/2008: 155–165.

Szwabowski J. & Łaźniewska-Piekarczyk B. 2008. The increase of air-content in mix under influence of carboxylate superplasticizers acting. *Cement-Wapno-Beton*. No.4/2008: 205–215.

Underground Infrastructure of Urban Areas 2 – Madryas, Nienartowicz & Szot (eds)
© 2012 Taylor & Francis Group, London, ISBN 978-0-415-68394-4

Estimation of hydraulic efficiency of linear drainage systems

J. Machajski
Instytute of Geotechnics and Hydrotechnics, Wrocław University of Technology, Wrocław, Poland

D. Olearczyk
Instytute of Environmental Engineering, Wrocław University of Environmental Life Sciences, Wrocław, Poland

ABSTRACT: Linear drainage systems are more and commonly applied, particularly there, where the beauty of drained area becomes essential. Unfortunately, producers often forget about hydraulic efficiency of a system understood as ability of collecting the precipitation waters and transport ability of fine pollutants flowing into a drainage flume. The producers of some available drainage systems usually define conditions for their choice, but they are not always compatible with real conditions of working, especially if it concern a hydraulic work. The authors, based on their experience in dimensioning of surface drainage systems, including in this linear drainage systems, give the principles for estimation of hydraulic efficiency of these systems, considering a capacity ability of every component of system separately. Paper can be treated as a trial to systematize the work conditions of such a type systems, and also recommendations for computational procedure application for determining the possibilities of collecting a given discharge from precipitation waters and transport of impurities flowing to a system. On the basis of carried out investigations connected with elaboration of analytical methods of capacity ability verification of linear drainage systems, a procedure of dimensioning an individual elements of that system is given, in this a grate, drainage flume, trash box and conduit carrying away water from system. Authors considered every component of system separately and also in their general configuration. It is important for proper protection against precipitation waters, for example pedestrian subways, entries into road tunnels or parking places.

1 INTRODUCTION

Linear drainage systems are included in surface drainage systems, collecting precipitations waters from the most often hardened parking places, yards of logistic centres, entries into road tunnels, pedestrian subways or locally from area entries into garages (Madryas et al. 2009, Edel 2006). These systems considerably improve a beauty of drained areas, however, to obtain this effect the components of system should be made in such a way that does not change an existing or adopted elevation configuration of drained surface (Madryas et al. 2009, Edel 2006, Geiger et al. 1995). On the other hand, a functional basis of all types of drainage systems is to give to drained areas both the longitudinal slopes and crosswise slopes. Hence, it is difficult to reconcile these two questions without functional loss of system, which is understood as capacity ability. An essential advantage of linear drainage system is in addition an uniform collecting of water from drained area, whereas a drawback is a necessity of frequent carrying away waters from a system and draining them to a recipient, for example to a storm water drainage or more and more often used a retention-seepage systems (Geiger et al. 1995).

2 SYSTEMS OF LINEAR DRAINAGE

System of linear drainage consists of a few basic elements, including: drainage flume, grate closing a flume from above and trash box with conduit which connects a trash box to water recipient in

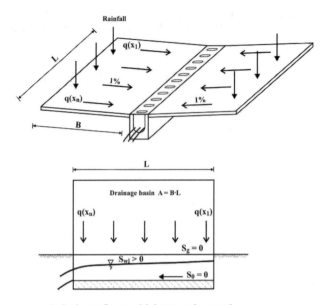

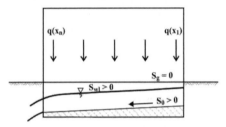

a) drainage flume with bottom slope and
   ground slope equal zero

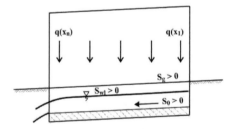

b) drainage flume with bottom slope greater
   than zero, ground slope equals zero

c) drainage flume with bottom slope and
   ground slope greater than zero

Figure 1. Possible schemes of drainage flumes built-in a drained area.

certain area (Madryas et al. 2009, Machajski et al. 2011, Edel 2006, Geiger et al. 1995). Efficiency of surface drained systems depends on their capacity ability, determined on the basis of properly assumed computational schemes, provided that an exact adaptation to the existing conditions is made. The possibilities of carrying out the fine pollutants, that can be present in rainfall waters becomes also an important issue. Because such systems usually work under free flow conditions, a longitudinal slope of a bottom of drainage flume becomes also important (Machajski et al. 2011).

Presented in Figure 1 schemes of possible drainage flume built-in a ground, were elaborated to obtain a maximum capacity ability, what assure not only the parameters of grate meshes, parameters

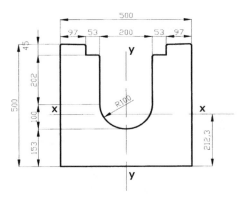

Figure 2. Cross-section and dimensions of drainage flume.

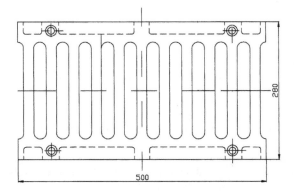

Figure 3. Example of grate cross-section with the length of 0.50 m.

of cross-sectional area of flow of drainage flume, kind of material of flume, but also a longitudinal slope of bottom flume. On the other hand, it is important to laid a flume in such a way that a necessity of change a longitudinal slope of ground will not occur; it could impact on area beauty and assumed usable functions.

Task of a drainage flume of linear drainage system is to take over a whole outflow from rainfall from a drainage basin (Fig. 1), hence its dimensions (Fig. 2) should guarantee a maintenance of appropriate capacity ability, similarly also a longitudinal slope of bottom line, which is important as regards removal of fine pollutions from flume cross-section.

A basic task of grate closing a drainage flume from above is en entry of rainfall waters into a flume, with simultaneous effective arrestment of bigger impurities washing from drainage area in such way in order to not influence on grate capacity ability. Therefore it becomes important a section of one grate gap both in respect of its length, width as well and the numbers of gaps on one running meter of drainage flume length. A general view of grate closing a drainage flume is shown in Figure 3.

The main task of trash box (Fig. 4), assembled usually in half of drainage flume section, is water reception from flume and to carry away to the nearest recipient by conduit with properly chosen diameter and appropriate longitudinal slope of bottom line. The same in trash box computations it is important not only a hole diameter selection but also its height location in relation to water level at the outlet of drainage flume. Hence, dimensions of trash box should be chosen in such way that outlet is placed on such depth, which give a possibility of water damming above it to height that allow to create an inlet velocity giving a guarantee of effective water draining off from a system.

A recipient of rainfall water in the case of linear drainage systems application is usually the nearest located conduit of storm water drainage, and in the case of its lack at present more frequently used

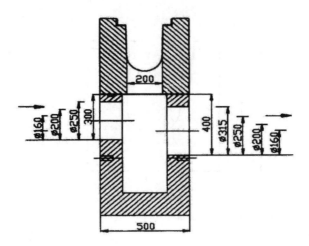

Figure 4. Example of cross-section of trash box.

retention – seepage systems that allow in convenient soil – water conditions to collect part of rainfall by subsoil of drained area (Madryas et al. 2009, Machajski et al. 2011, Geiger et al. 1995).

3 ESTIMATION OF HYDRAULIC EFFICIENCY OF LINEAR DRAINAGE SYSTEM

Systems of linear drainage have number of tasks, that should be fulfilled, because only then it can be told about certain system that is hydraulically efficient (Madryas et al. 2009, Machajski et al. 2011, Edel 2006, Geiger et al. 1995). These tasks are:

– taking over a whole waters from precipitation, flowing crosswise in relation to drainage flume with outlet to recipient,
– to obtain the highest possible under certain conditions hydraulic efficiency, as a result of flow cross-section's size, its smoothness and longitudinal slope,
– to obtain an ability to create and maintain a self-purification velocity, that guarantee a taking out of fine impurities washed out by precipitation waters from drained area and carried in a drainage system's direction,
– to assure a maximum possible hydraulic efficiency of a grate closing a drainage flume, that guarantee a reception of entire flowing precipitation's waters and effective penetration of fine impurities washed out from drained area without their stoppage at grate surface.

Carried out an assessment of hydraulic efficiency of individual elements of linear drainage system had the aim to indicate the most profitable, with attention on requirements, built-in conditions of this system. Hence, it was determined: grate inlet ability, capacity ability of drainage flume in different cases of bottom longitudinal slope and ground slope, inlet ability of trash box and capacity ability of conduit taking away precipitation waters from system. For this purpose the available in hydraulics equations were applied, chosen according to working conditions of certain device.

3.1 Estimation of grate inlet ability

Estimation of grate inlet ability, closing drainage flume of linear drainage system was carried out using a below giving formula (Rogala et al. 1991) and for fixed, usually according to work conditions of flume, precipitation waters level that fill a space over an grate's inlet edge:

$$Q_s = \alpha \ A \ \varphi \sqrt{2 \, g \, h}$$ (1)

where Q_s – inlet ability of single grate's mesh over a flume (m^3 s^{-1}); α – coefficient of side contraction; A – area of single grate's mesh (m^2); φ – velocity coefficient; h – thickness of water layer over grate's mesh (m).

Thickness of water layer over a grate inlet edge has the most important significance for grate's inlet ability, because from this results a velocity of flowing in water, and in connection with grate meshes parameters – an expected capacity ability. Unfortunately, in the regions of drained areas by linear drainage systems the conditions of build in drainage flume do not allow to create too high depth h (Geiger et al. 1995), hence it varies between $0.01 \div 0.03$ m.

3.2 Estimation of capacity ability of drainage flume

Estimation of capacity ability of drainage flume was carried out for three possible built in situations on section that needs to be drained. The first case represents a flume with both longitudinal slope of bottom and longitudinal slope of ground equal zero ($S_o = 0$, $S_g = 0$), shown in Figure 1a. The second one is for flume with longitudinal slope of bottom greater than zero and longitudinal slope of ground equals zero, shown in Figure 1b. The third case represents a flume with longitudinal slope of bottom greater than zero, adapted to ground inclination also greater than zero, shown in Figure 1c.

Estimation of capacity ability of drainage flume for chosen computational case is carried out applying a differential equation describing a lay-out of water level lines along a drainage flume, then a slope of water level line will impact on water movement in a drainage flume, apart from bottom longitudinal slope of a flume. A differential equation could be derived in a way given below (Machajski et al. 2011, Novak et al. 2007, Osman 2006, Ghosh 2006).

Conditions of flowing precipitation waters to drainage flume cross-section lead to form in its reach a specific motion, defined as spatially variable (Novak et al. 2007, Osman 2006, Ghosh 2006, Rogala et al. 1991). This motion occurs when in the reach of conduit a lateral inflow q_L is present, and a particular feature of that motion is a gradual rise of discharge in conduit, consistent with the flow direction. It should be emphasized that such a motion occurs in conduit independently of whether the slope of flume bottom is longitudinal or horizontal – as it often occurs for drainage flumes of linear drainage systems. The best solution for described problem is an application of the principle of conservation of momentum (Osman 2006, Rogala et al. 1991), based on the procedure scheme presented in Figure 5, for two cases of longitudinal slope of flume bottom: greater than zero and horizontal one.

An section of drainage flume is taken with a length dx, between two cross-sections (1) – upstream and (2) – downstream. It assumed that a flow velocity at upstream cross-section is equal v and discharge equals Q, at downstream cross-section it can be written that velocity and discharge are equal respectively $v + dv$ and $Q + dQ$, where dQ means a growth of discharge along a computational distance dx. Assuming that a drainage flume is conducted with a slope equals S_o, then the water weight component between two cross-sections, being consistent with a flow direction, marked in Figure 5b as $W \sin \Theta$, can be written in the following way:

$$\gamma S_o \left(A + \frac{dA}{2} \right) dx \cong S_o \, A \, dx \qquad (2)$$

where A – upstream cross-sectional area of flow; γ – water specific gravity.

Simultaneously, between upstream and downstream cross-sections, a loss of energy occurs which results from the roughness of material of drainage flume (Fig. 5). In a general notation that loss can be expressed as:

$$dy_f = S_{fm} \, dx \qquad (3)$$

where: S_{fm} – average energy line decline, determined on a length dx, for instance from modified Manning's formula (Osman 2006, Rogala et al. 1991). Then it can be expressed as (Fig. 5):

$$S_{fm} = \frac{1}{2}(S_{f1} + S_{f2}) = \frac{1}{2} \left(\frac{n_1^2 \, v_1^2}{R_{h1}^{4/3}} + \frac{n_2^2 \, v_2^2}{R_{h2}^{4/3}} \right) \qquad (4)$$

129

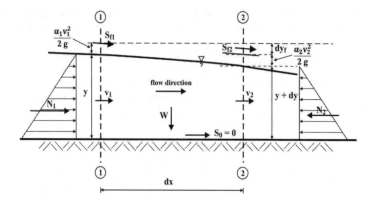

b) drainage flume with bottom slope equals zero

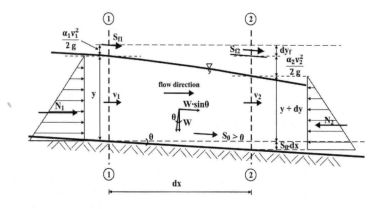

a) drainage flume with bottom slope greater than zero

Figure 5.   Possible procedure schemes.

where n – roughness coefficient for Manning's formula; R_h – hydraulic radius of cross-sectional area of flow (Osman 2006, Rogala et al. 1991).

Friction force F_f along the walls of drainage flume on a section length dx can be described as:

$$F_f = \gamma\,(A + \frac{1}{2}dA)\,S_f\;dx \cong \gamma\,A\,S_f\;dx \qquad (5)$$

The hydrostatic forces N_1 and N_2 act at the upstream cross-section and downstream cross-section, respectively. The force N_1 can be calculated as follows:

$$N_1 = \gamma\,\bar{z}\,A \qquad (6)$$

where $\bar{z}$ is the position of the centre of gravity of the area A in relation to water level line in drainage flume.

The hydrostatic force N_2 acting at the downstream cross-section can be determined from formula:

$$N_2 = \gamma\,(\bar{z} + dy)\,A + \frac{\gamma}{2}\,dA\,dy = \gamma\,(\bar{z} + dy)\,A \qquad (7)$$

Hence, a resultant force acting on a water volume between the upstream and downstream cross-sections is equal to:

$$N_1 - N_2 = -\gamma\, A\, dy \tag{8}$$

Between the upstream and downstream cross-sections there is a change momentum which can be written in the form (Osman 2006, Ghosh 2006):

$$[\rho\,(Q + dQ)\,(v + dv) - \rho\,Q\,v] = \rho\,[Q\,dv + (v + dv)\,dQ] \tag{9}$$

In accordance with the principle of momentum conservation (Osman 2006, Ghosh 2006, Rogala et al. 1991), the momentum should be equal to the sum of external forces acting on a water volume contained between two cross-sections, i.e., a pressure and resultant hydrostatic force and friction forces, hence:

$$\rho\,[Q\,dv + (v + dv)\,dQ] = -\gamma\,A\,dy + \gamma\,S_o\,A\,dx - \gamma\,A\,S_f\,dx \tag{10}$$

Substituting in a place of differentials the finite differences an equation (10) can be written as:

$$\frac{\gamma}{g}[Q\,\Delta v + (v + \Delta v)\Delta Q] = -\gamma \int_0^{\Delta y} A\,dy + \gamma S_o \int_0^{\Delta x} A\,dx - \gamma S_f \int_0^{\Delta x} A\,dx =$$

$$= -\gamma \overline{A}\,\Delta y + \gamma S_o\,\overline{A}\,\Delta x - \gamma S_f\,\overline{A}\,\Delta x \tag{11}$$

Because the discharge in a drainage flume changes (increase) with its length, a mean area of flume cross-section can be written as:

$$\overline{A} = \frac{Q_1 + Q_2}{v_1 + v_2} \tag{12}$$

Assuming simultaneously that $Q = Q_1$ and $v + \Delta v = v_2$, after some transformations an equation (11) can be written in the form (Osman 2006, Ghosh 2006):

$$\Delta y = -\frac{Q_1\,(v_1 + v_2)}{g\,(Q_1 + Q_2)}\left\{\Delta v + \frac{v_2}{Q_1}\,\Delta Q\right\} + S_o\,\Delta x - S_f\,\Delta x \tag{13}$$

The first term of equation (13) shows a change in a filling of a drainage flume cross-section, the second one – a drop of flume bottom line on a section Δx and the third one – an energy losses as a result of friction force action. This equation was solved numerically in the example at the end of this paper. Because a drainage flume can also be laid with a longitudinal bottom slope equals, there will be no impact of water weight component in the section dx, being consistent with a flow direction. Equation (13) for a case of drainage flume conducted horizontally will slightly change its form to (Osman 2006, Ghosh 2006):

$$\Delta y = -\frac{Q_1\,(v_1 + v_2)}{g\,(Q_1 + Q_2)}\left\{\Delta v + \frac{v_2}{Q_1}\,\Delta Q\right\} - S_f\,\Delta x \tag{14}$$

3.3 *Evaluation of efficiency of precipitation waters reception system*

The efficiency of precipitation waters reception system is evaluated based on a determination the possibility of collecting precipitation waters, flowing to the final component of drainage system, i.e. trash box. Trash box is placed somewhat below a bottom line of drainage flume with a hole of a required diameter. Possibility of precipitation waters reception is of fundamental importance for the efficiency of entire surface drainage system of traffic structures, understood as a system that

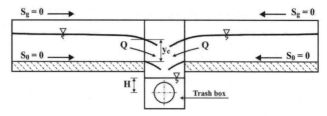

b) drainage flume with bottom slope equals zero

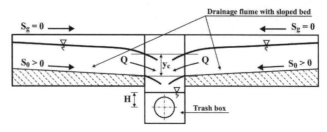

a) drainage flume with bottom slope greater than zero

Figure 6. Schemes for trash box calculations.

offers a possibility of taking over a whole precipitation waters flowing from belonging area and allows their fast removal from a system.

Capacity ability of a hole cross-section on a side-wall of trash box (Fig. 4) can be determined applying the formula which requires a precise determination of water height H at inlet, measured with respect to axis of outflow conduit (Rogala et al. 1991) (Fig. 6).

Hence, a capacity ability of trash box is given by:

$$Q_z = \alpha \ A \ \varphi \sqrt{2 \ g \ H} \tag{15}$$

where Q_z – capacity ability of a hole on a side-wall of trash box (m^3 s^{-1}); α – coefficient of side contraction; A – area of a hole on side-wall (m^2); φ – velocity coefficient; H – thickness of water layer above an axis of outlet, necessary to create an entry velocity (m).

3.4 *Estimation of possibility of water reception by outflow conduit to sewage system*

Calculations, checking a possibility of precipitation water reception by outflow conduit, are carried out on the basis of nomogram elaborated for conduits with a circular cross-section, according to the recommended in professional literature the Darcy-Weisbach and Colebrook-White formulas, for an assumed strict roughness of conduit's walls. Conditions of free surface work of this conduit are assumed, and the same a required longitudinal slope of conduit bottom (Rogala et al. 1991).

3.5 *Computational example*

The comparative calculations of capacity ability are carried out for a linear drainage system, collecting waters from rain and snow melt from both sides of drainage area of size: length $L = 200$ m, width $B = 60$ m (Fig. 1). Drainage area is a hard surface covered with concrete plates with a slope towards draining devices equals 1.0 %. To drain a given area a drainage flume as in Figure 2 is used, for two cases: bottom slope equals $S_d = 0$ and a bottom slope greater than zero, i.e. $S_o = 0.0015$, with a trash box of size as in Figure 4 installed in the middle of its length (L/2). A drainage flume is closed by a grate with dimensions as in Figure 3 and the a following size of mesh: width $b = 0.025$ m and length $l = 0.20$ m. Discharge determined on the basis of hydrological calculations (Madryas et al. 2009, Edel 2006) from one side of linear drainage system equals $Q = 20\,\mathrm{dm^3\,s^{-1}}$. Hence, a computational discharge on one running metre of the length of flume from both sides will be equal to $2\,Q/(L/2) = 40/100 = 0.40\,\mathrm{dm^3\,s^{-1}}$.

Step 1
Determination of grate inlet ability; a discharge of single grate mesh of the area $A = b \cdot l = 0.025 \cdot 0.20 = 0.005\,\mathrm{m^2}$ is taken into account. It is assumed that the height of water level over a grate is equal to $h = 0.01$ m, the velocity coefficient $\phi = 0.75$ and the side contraction coefficient $\alpha = 0.85$. Equation (1) is applied, hence:

$$Q_s = \alpha \ A \ \varphi \sqrt{2\,g\,h} = 0.85 \cdot 0.005 \cdot 0.75 \sqrt{2 \cdot 9.81 \cdot 0.01} = 0.00141\,m^3\ s^{-1}$$

Because for one running metre of drainage flume length a grate have twenty meshes, a discharge of 1 m grate length equals:

$$0.00141 \cdot 20 = 0.0282\ m^3\ s^{-1}$$

There is a possibility of some mesh choking, so the factor of safety equals 2 is taken for calculations (Madryas et al. 2009, Edel 2006) and a final grate discharge for one running metre is given by:

$$0.0282 : 2 = 0.0141\ m^3\ s^{-1}$$

Calculated meshes capacity ability exceed considerably the discharge from drainage area for one running metre of flume length:

$$0.0141\ m^3\ s^{-1} > 0.00040\ m^3\ s^{-1}$$

Step 2
Calculation of capacity ability of drainage flume along the length $L/2 = 100$ m, using an equation (13) for flume with bottom slope greater than zero and equation (14) for flume conducted horizontally. Computer model for calculation both cases was elaborated by the Authors. It was assumed simultaneously that along one running metre of drainage flume a discharge is equal $\Delta Q = 0.40\,\mathrm{dm^3\,s^{-1}}$. In calculations, the direction of procedure is important; for supercritical flow in a flume this procedure should progress upstream, for subcritical flow in a flume the calculation procedure should be reverse (Chadwick et al. 2004, Osman 2006, Ghosh 2006, Rogala et al. 1991). For computational example, for both cases the calculation procedure should progress upstream.

In every case, an initial depth is important for properly carried out calculations. For a given example under supercritical flow conditions, it is a depth that results from free fall of water in a joint cross-section of flume with trash box (Fig. 6). In that cross-section, for computational assumptions, it is a critical depth y_c, which can be calculated from critical flow condition (Chadwick et al. 2004, Osman 2006, Rogala et al. 1991):

$$\frac{A^3}{B} = \frac{\alpha\,Q^2}{g} \qquad (16)$$

where A – cross-sectional area of flow (m²); B – width measured at a height of water level (m); α – the Saint-Venant coefficient; Q – computational discharge (m³ s⁻¹); g – acceleration of gravity (m s⁻²).

For drainage flume parameters shown in Figure 2, when discharge equals $Q_o = 0.040\,\mathrm{m^3\,s^{-1}}$ reaches the outlet cross-section, a critical depth y_c determined from equation (16) equals 0.189 m.

Results of calculations for discharge passage through drainage flume are given in a Table 1 for both cases of bottom slope of a drainage flume and are shown in Figure 7.

Table 1. Hydraulic characteristics of drainage flume with bottom longitudinal slopes equals zero and greater than zero.

Flume length Δx [m]	Bottom slope $S_0 = 0$						Bottom slope $S_0 = 0,0015$						Depth differences for two computational cases [m]
	Depth y [m]	Cross-sectional area of flow A [m²]	Sum of discharge $q(x)$ [m³s⁻¹]	Flow velocity v [ms⁻¹]	Energy line slope S_f		Depth y [m]	Cross-sectional area of flow A [m²]	Sum of discharge $q(x)$ [m³s⁻¹]	Flow velocity v [ms⁻¹]	Energy line slope S_f		
0	0.189	0.0335	0.0400	1.194	0.0073813		0.189	0.0335	0.0400	1.194	0.0073813		0.000
5	0.254	0.0465	0.0380	0.818	0.0030614		0.245	0.0447	0.0380	0.851	0.0033562		0.009
10	0.279	0.0515	0.0360	0.699	0.0021641		0.264	0.0485	0.0360	0.742	0.0024860		0.015
15	0.296	0.0550	0.0340	0.619	0.0016602		0.276	0.0508	0.0340	0.669	0.0019862		0.021
20	0.310	0.0576	0.0320	0.555	0.0013196		0.284	0.0524	0.0320	0.611	0.0016402		0.026
25	0.320	0.0598	0.0300	0.502	0.0010676		0.289	0.0535	0.0300	0.561	0.0013775		0.031
30	0.329	0.0615	0.0280	0.455	0.0008711		0.292	0.0541	0.0280	0.517	0.0011667		0.037
35	0.336	0.0629	0.0260	0.413	0.0007125		0.294	0.0545	0.0260	0.477	0.0009909		0.042
40	0.342	0.0642	0.0240	0.374	0.0005815		0.294	0.0546	0.0240	0.440	0.0008405		0.048
45	0.347	0.0652	0.0220	0.338	0.0004716		0.294	0.0545	0.0220	0.404	0.0007093		0.053
50	0.352	0.0660	0.0200	0.303	0.0003785		0.292	0.0542	0.0200	0.369	0.0005933		0.059
55	0.355	0.0667	0.0180	0.270	0.0002993		0.290	0.0537	0.0180	0.335	0.0004899		0.065
60	0.358	0.0673	0.0160	0.238	0.0002319		0.287	0.0531	0.0160	0.301	0.0003971		0.071
65	0.360	0.0678	0.0140	0.207	0.0001747		0.284	0.0524	0.0140	0.267	0.0003139		0.077
70	0.362	0.0682	0.0120	0.176	0.0001268		0.279	0.0516	0.0120	0.233	0.0002395		0.083
75	0.364	0.0685	0.0100	0.146	0.0000872		0.274	0.0506	0.0100	0.198	0.0001738		0.089
80	0.365	0.0687	0.0080	0.116	0.0000554		0.269	0.0495	0.0080	0.162	0.0001169		0.096
85	0.366	0.0689	0.0060	0.087	0.0000310		0.263	0.0483	0.0060	0.124	0.0000696		0.103
90	0.366	0.0690	0.0040	0.058	0.0000137		0.257	0.0471	0.0040	0.085	0.0000329		0.110
95	0.367	0.0690	0.0020	0.029	0.0000034		0.250	0.0457	0.0020	0.044	0.0000088		0.117
100	0.367	0.0690	0.0000	0.000	0.0000000		0.243	0.0442	0.0000	0.000	0.0000000		0.124

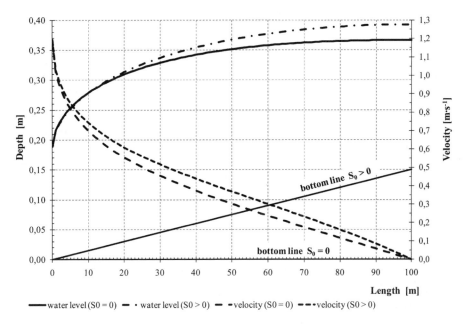

Figure 7. Hydraulic characteristics of drainage flume with bottom longitudinal slopes $S_0 = 0$ and $S_0 > 0$.

Step 3
Determination a possibility of collecting precipitation waters from drainage flume by trash box. Calculation procedure is the same for both analyzed cases: bottom slope greater than and equals zero. The aim of calculation is to determine a hole diameter adjusted to a water height over its axis. An equation (15) is applied in which H is replaced with a given height of water layer at trash box cross-section, measured in relation to outlet axis, hence assuming initially an outlet diameter $D = 0.30$ m, a discharge is equal to:

$$Q_z = \alpha \; A \; \varphi \sqrt{2 \, g \, H} = 0.90 \cdot 0.0707 \cdot 0.85 \sqrt{2 \cdot 9.81 \cdot 0.20} = 0.1071 \, m^3 \, s^{-1}$$

Because an inflow to an outlet equals $2 \cdot Q_o = 40 \cdot 2 = 80$ dm^3 s^{-1} (0.080 m^3 s^{-1}), hence for an advisable water damming up over an axis of conduit with diameter DN 0.300 there is a possibility of free carrying away a computational waters from a system.

Step 4
Determination of the cross-section and the longitudinal slope of conduit taking away the precipitation waters from trash box and draining them to the nearest well of storm water drainage. Ability of precipitation waters reception was determined assuming a conduit diameter DN 0.300 and imposed longitudinal slope of conduit bottom equal $S_o = 2.6‰$; for such assumptions the discharge Q_z equals 0.08884 m^3 s^{-1} (88.84 dm^3 s^{-1}).

Step 5
Verification of system ability to remove a fine pollutants from flume cross-section. Verification is carried out by comparison the flow velocities and their variability along a flume. Results, given in table 1, show that only from 25 m and 30 m of flume length, respectively for bottom slopes equals zero and greater than zero, there will be possible to remove easy the fine impurities from a drainage flume ($v \geq 0.50$ m s^{-1}).

4 SUMMARY

The producers of linear drainage systems rarely give their capacity ability which results in defective working, particularly concerning a drainage flume and also a grate. In the paper Authors define every component of system from the point of view of hydraulic conditions of their operation. It has an important impact on ability reception of waters from precipitation and snow melt flowing from surrounding area, included in a drainage project. This is reflected in the beauty of drained areas and in the safety of vehicles and pedestrians using these areas. The significant improvement of hydraulic efficiency of drainage flume, laid with longitudinal slope greater than zero, in enclosed computational example pays attention. For the same initial conditions a smaller depths, about 0.12 m, were obtained. It allows to elongate a drainage section and also to minimize a number of necessary outlets of precipitation waters using trash boxes. In every of computational case it is necessary to confirm by calculations a possibility of taking over a precipitation waters by system, in which the rules for estimation of its hydraulic efficiency, given in the paper could be helpful.

REFERENCES

Chadwick, A., Morfett, J. & Borthwick, M. 2004. *Hydraulics in Civil and Environmental Engineering*. New York: Taylor & Francis Group.

Edel, R. 2006. *Odwodnienie dróg*. Warszawa: Wydawnictwo Komunikacji i Łączności.

Geiger W. & Dreiseitl, H. 1995. *Neue Wege für das Regenwasser. Handbuch zum Rückhalt Und zur Versickerung von Regenwasser in Baugebieten*. München – Wien: R. Oldenbourg Verlag.

Ghosh, S.N. 2006. *Flood Control and Drainage Engineering*. London: Taylor & Francis

Machajski, J. & Olearczyk, D. 2011. Dimensioning of linear drainage systems. *Studia Geotechnica et Mechanica*. Vol. XXXIII, No. 1, 2011, pp. 1–15.

Madryas, C. Machajski, J. Olearczyk, D. Kolonko, A. & Wysocki, L. 2009. *Zalecenia projektowania budowy i utrzymania, odwodnienia tuneli samochodowych, przejść podziemnych i przepustów*. Warszawa: Generalna Dyrekcja Dróg Krajowych i Autostrad.

Novak, P. Moffat, A.I.B. Nalluri, C. & Narayanan, R. 2007. *Hydraulic Structures*. New York: Taylor & Francis Publishers.

Osman Akan, A. 2006. *Open Channel Hydraulics*. London: Elsevier Ltd.

Rogala, R. Machajski, J. & Rędowicz, W. 1991. *Hydraulika stosowana. Przykłady obliczeń*. Wrocław: Wydawnictwo Politechniki Wrocławskiej.

Underground Infrastructure of Urban Areas 2 – Madryas, Nienartowicz & Szot (eds)
© 2012 Taylor & Francis Group, London, ISBN 978-0-415-68394-4

Utility Tunnels – Old fashion or necessity? – Analysis of environmental engineering factors creating potential for growth

C. Madryas
Wrocław University of Technology, Wrocław, Poland

L. Skomorowski & R. Strużyński
HOBAS System Polska Sp. z o. o., Dąbrowa Górnicza, Poland

V. Vladimirov
HOBAS Engineering GmbH, Klagenfurt, Austria

ABSTRACT: Underground infrastructure is a very wide-ranging term. With the help of various kinds of networks we can ensure the provision of diverse sorts of utilities such as, potable water, energy, sewage, precipitation water as well as communication services. This paper has been divided into five parts that sequentially describe the reasons for which multi-conduit tunnels came into being and their types. The aspects of how such tunnels, and the process of building them, may impact the environment, based on CO_2 emissions, based on an example of a project in Warsaw are also discussed in the paper. The diverse methods of constructing the tunnels, the advantages and disadvantages of the methods, as well as the social costs borne during the constructions of multi-conduit tunnels are described. There are several example projects performed in the United States and Europe presented. It was indicated that the number of such tunnels constructed in cities, by means of modern methods, increases because they allow for scraping up space for additional conduits. At the end it includes conclusions and a vision for coordinating the activities of the services supervising individual utility networks in urban areas.

1 INTRODUCTION

Underground infrastructure is a very wide-ranging term. All types of buildings, both housing and facility ones, have to be constructed with diverse kinds of services like water and energy supply, communications services provision, waste and precipitation water disposal available. These services are mostly provided with the use of various kinds of networks of conduits. In general, there are two main methods of laying such conduits – directly in the ground and intermediately in the ground. In the first case, the conduit is laid in an open trench or by means of trenchless technologies allowing for the transfer of only one medium. Such situations mostly takes place when passing a conduit underneath various kinds of obstacles, such as rivers, roads and railway tracks. The passage of the Jamal pipeline under the Warta River in the town of Oborniki Śląskie in Poland, performed with the use of CC-GRP casing pipe in 1998 is a good example.

In the second case it is actually a set of conduits in one casing (thereof protective casings, i.e. both accessible and non-accessible multi-conduit tunnels as well as all kinds of large-diameter objects, i.e. conduit galleries and slab foundations. Because of the considerable concentration of conduits, very large financial costs (estimated at billions of PLN) and social costs arise in the case of failure of conduits, as well as the necessity of digging trenches and the difficulties resulting from it (blockage of streets, especially in the centres of large cities) (Dziennik Rzeczpospolita 14 stycznia 2011) make the strive for building multi-conduit tunnels, or service galleries, logical. Laid conduits have the following advantages: easy access to the conduits without the necessity of digging trenches in order to replace, repair or maintain the parts of the conduits, or their fittings, ordered arrangement of the underground infrastructure elements, and large volume of free space underground that can be used for other purposes.

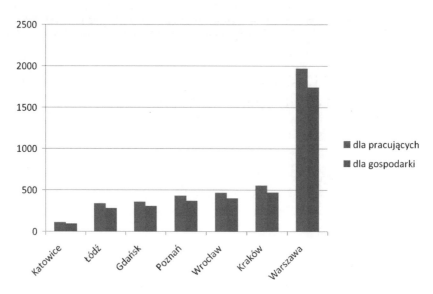

Figure 1.   Annual cost of congestion in millions zlotys according to Deloitte (Dziennik Rzeczpospolita 14 stycznia 2011).

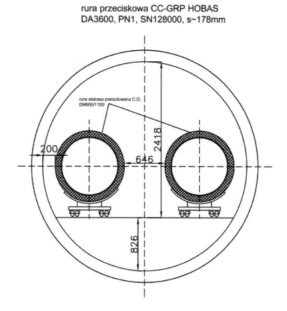

Figure 2.   An example of proposed tunnel scheme of district heating.

In recent years, especially the past decade, a considerable increase in the awareness of the decision-makers in the field in Poland has increasingly risen, especially regarding the necessity of taking into account the social costs, as well as, better organisation of the processes in large cities. The awareness is strengthened by the new engineering accomplishments in the area of performance methods and materials used. Trenchless technologies have been well developed in Poland and are of special importance as they can make the multi-utility tunnels a common solution used in the intensely developing Polish agglomerations, and support the development of utility systems. The application of the solution may not only minimise the costs of those investments but also the social costs resulting from the indispensable changes to street traffic, which cause traffic disorganisation, as well as a considerable increase in the time, and fuel, wasted in additional traffic jams. It can also reduce or eliminate other risks, e.g. an increase in the number of traffic accidents caused by temporary traffic bypasses, and more importantly, it can minimise the maintenance costs of the networks. It may cause a considerable decrease in serious failures of the electricity grid, caused by the overloading of overhead lines by wet snow or heavy winds which may be attained by replacing them with underground cables. The landscape would also be improved by eliminating heat-transfer installations and overhead lines of the electricity grid. It also eliminates the contact of conduits with the surrounding ground and so diminishes the destructive effect of mining damage, corrosion cased by the ground-and-water environment and mechanical loading by ground movements. When the tunnels are constructed of the appropriate materials it is also possible to eliminate the damaging impact of stray currents.

2 TYPES OF TUNNELS

The first multi-utility tunnels adapted to this function were the sewage collectors in London, Paris and Madrid. They were used to house water supply pipelines, electrical cables, communications and pneumatic tube transport. In the multi-utility tunnels constructed later and currently, it is possible to place all kinds of utility networks, i.e. water supply pipelines, electrical cables, gas supply pipelines, heat supply pipelines, pressure and gravitational sewage pipelines. Although, in the past, there were many negative opinions on placing gas supply pipelines in multi-conduit tunnels they proved to be ungrounded in light of the experience that has been gained from their exploitation in such tunnels in Great Britain for over one hundred years. Yet, these opinions caused that tunnel designs were developed there with gas supply pipelines located in the tunnel ceiling, separated from the tunnel's interior (e.g. the multi-utility tunnel in the town of Suhl, Germany (Kuliczkowski & Madryas 1991)). At present, there are regulations in many countries, and in Poland, that allow for the laying of gas supply pipelines inside the tunnels, provided that they are appropriately ventilated (Kuliczkowski & Madryas 1991).

The cross-section of a multi-utility tunnel may be of any geometry, but in most cases it may not violate the minimum value of the coefficient of free space (rectangular cross-section), or the installation rules (circular cross-section – microtunnelling). The accessible multi-tunnel tunnels can be constructed by means of three methods: open trench, semi-trenchless and trenchless. In the first case, the construction may be carried out on site (of brick, stone, concrete or reinforced concrete) or of pre-fabricated elements – most often with a rectangular cross-section. Yet, it is necessary to pay special attention to the way of the segments are joined. The essence of the semi-trenchless method lies in the fastest possible restoration of the destroyed on-ground facilities above the tunnel. The side walls of the tunnel, in this technology, are formed directly in the ground by means of screw piles, hollow walls or driven walls. The ceiling is constructed of pre-fabricated elements or is formed directly on site. The bottom slab and other inside furnishing elements are produced after the on-ground facilities and their functions have been restored and the ground has been excavated from below the ceiling. The trenchless method includes the tunnelling by mining, hydraulic jacking and microtunnelling. The mining methods consist of driving a tunnel like a traditional drift. Next, the side walls and ceiling of the drift are preliminary secured by means of section-steel supports and pre-fabricated reinforced-concrete lagging elements. Afterwards, a permanent lining is built in the tunnel, most often vaulted in its upper part. Nowadays, mining technologies are used very rarely. The method of hydraulic jacking is a good choice and is currently often applied in constructing short tunnels. The method consists of pushing lining elements, with either a rectangular or circular cross-section, from the jacking shaft to the receiving shaft. When the tunnels are longer the method of microtunnelling is used most often. In the latter case for the shape of the drilling head, and shield, the tunnels are most often given a circular cross-section. There are practically no limits with regard to the ground properties and depth at which the tunnel is driven. The variety of drilling heads available makes it possible to effectively execute the microtunnelling work at very high groundwater levels and in diverse types of ground. The method of microtunnelling is a very precise method. The most important issue is to choose an appropriate shield and the materials to be used in constructing the tunnel. The materials have to meet the requirements of the technology applied and later maintenance requirements.

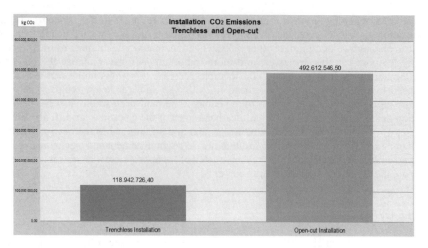

Figure 3.   Comparison of CO_2 emissions for the open-cut and trenchless scenario.

The method and materials to be applied should be chosen while taking into account not only the ground conditions and investment costs but also the social costs (the road traffic impediments) and the environmental noxiousness of the undertaking.

3   COMPARISON OF CO_2 EMISSION FROM TRENCHLESS AND OPEN-CUT INSTALLATION METHODS IN EXAMPLE TO INSTALLATION OF CC-GRP PIPE OD3000 IN WARSAW

Traditional open cut methods are usually obtrusive to the environment (i.e. digging out and trans-porting large amounts of earth), require increased consumption of resources both directly (i.e. fuel, electricity) and indirectly (for instance, thorough traffic delays). Thus, such methods are rather highly polluting and also tend to take longer amounts of time until completion, as compared with no-dig methods.

Trenchless technology provides a sustainable alternative more than often. Using trenchless as an environmentally conscious choice ensures less air pollution (less CO_2 emissions, less dust, less noise), reduce traffic disruptions and protect the natural habitat (i.e. preserving the trees that would otherwise be torn down for open-cut).

Installation calculations in this chapter refer to Czajka I (right side of the river) which represents most of the pipe line (5.7 km). Installation refers to the preparatory work for the installation of pipes and manholes, the actual installation as well as the closing works once installation is finished. CO_2 emissions were determined based on calculations according to the Polish Catalogue of Capital Expenditures (KNR).

The trenchless installation model is further divided in four calculation elements. These include the preparation of the chambers, the dewatering, the micro-tunneling and the handling of vegetation (trees). The open-cut installation model is further divided in five calculation elements. These include the preparation of the chambers, the dewatering, the actual installation of pipes and chambers, the rehabilitation after installation as well as the handling of vegetation (trees). It can be noticed that the open-cut calculation includes an additional element as compared with the trenchless scenario. This resides in the differences between the installation methods: while trenchless activities are focused on the underground, open-cut installations are focused on the surface, hence greater disturbances and need of rehabilitation of damaged landscape and infrastructure. Installation CO_2 emissions represent the most significant emission source for the Czajka I project

As shown in Figure 3 CO_2 emissions for the open-cut scenario are significantly higher than those for the trenchless scenario. Open-cut would have generated more than four times the emissions from the trenchless installation.

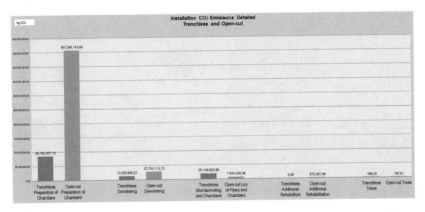

Figure 4.   More detailed assessment of the individual calculation elements of open-cut and trenchless scenario.

The figure above presents a more detailed assessment of the individual calculation elements of both scenarios.

Emissions in case of open-cut are significantly higher for the preparation of chambers, dewatering, rehabilitation after installation and trees. CO_2 emissions in case of trenchless are higher for the actual installation. This is because higher energy is needed to push the pipes underground as compared to laying the pipes above ground. However, much more energy is needed overall for open-cut. In addition, it should be noted that avoiding excavation (open-cut) also implies less CO_2 emissions and less effort to rehabilitate the adjacent structures. According to Tardiff, digging trenches near a paved surface will reduce its lifespan by at least 30 percent (Tadriff 2009). According to Jung and Shina, pavement life can be reduced by even 40% in case of open-cut (Yeun et al. 2004).

4   CASE STUDIES

4.1   *Tunnel under the Vistula River, Poland*

The tunnel under the Vistula river is an element of the "siphon" constructed in order to carry the sewage, which consists of the following elements:

- starting chamber located in the area of left-bank Warsaw,
- collecting tunnel (utility tunnel) with a diameter of 4500 mm built under the Vistula river,
- two sewers with a diameter of 1600 mm inside the multi-duct tunnel,
- exit chamber located in the right-bank part of Warsaw including chambers of gate valves, and expanding and connecting chambers, located inside.

The tunnel was designed to lead inside it also another kind of wires (e.g. fiber). The siphon was designed by a consortium consisting of four companies: DHV Poland Ltd. (leader of the consortium), Prokom Ltd., Grontmij Poland Ltd. and ILF Consulting Engineers Poland Ltd. Design materials made by the consortium constituted the basic source material for the preparation of this paper. At the stage of prospective solutions, technical opinions on the purposefulness of tunnel construction under the Vistula river were made by the Wroclaw University of Technology (directed by Prof. C. Madryas) and by the Kielce University of Technology (directed by Prof. A. Kuliczkowski). The arrangement of the described system components is shown in figure 5.

The tunnel with an inner diameter of 4.5 m and length of 1305 m, intended for installing CC-GRP pipes, will be constructed under the Vistula river in the area comprising two geomorphic units, i.e. the area of denuded erosive-and-accumulating terrace (so-called "błonski terrace") and the Vistula valley (Geoteko 2008). Along the route of the tunnel currently planned the ground and water conditions are not very changeable. From the starting chamber, along approximately 210 m of the route, silts and tertiary clays occur, with numerous watered sandy interbeddings. Further on (a section of 210–310 m) the tunnel runs through sandy-argillaceous-dusty soils, watered soils

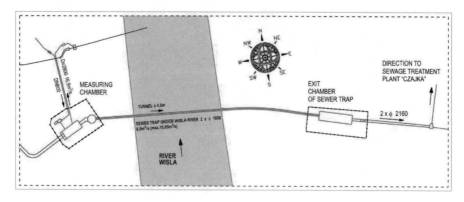

Figure 5.   Diagram of the siphon with the supply and exit sewers.

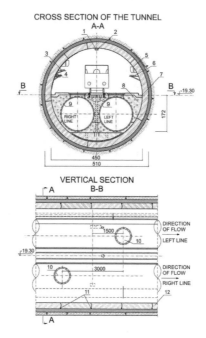

Figure 6.   Tunnel diagram with distributed ducts of the "siphon".
1 – main and emergency lighting, 2 – security camera, 3 – technical cables, 4 – water pipe, 5 – set of renovation sockets, 6 – electric cables, 7 – floor, 8 – lightweight concrete, 9 – sewer pipe made of CC-GRP, 10 – manhole d = 700 mm, 11 – sealing made of EPDM, 12 – tubing, fiber wire is not shown at this diagram.

occurring within an erosive trough. Further still, on a section of 370 up to 500 m, watered river sands occur, which turn into silts and tertiary clays. At 940 m, counting from the starting chamber, the sandy-argillaceous-dusty deposits described above appear again. From 1070 m (counting from the starting chamber) to the end of the route the tunnel runs again within watered dusty sands.

The building has been designed as a reinforced concrete structure made of tubing using a TBM (made by Herrenknecht AG). This is the first implementation in Poland in which this technology will be used. Figure 6 shows the tunnel diagram

4.2   *Deep Tunnel Force Main, US*

It has been an ongoing priority of the State of New Jersey to improve its infrastructure for their 8.6 million resident. Construction of Tunnel and Edison Force Main which was one from third phases

Figure 7.   View on the whole tunnel under the Raritan River.

Figure 8.   Installation of inside CC-GRP pipeline.

was part of overall goal to provide a redundant means for sewage conveyance from Middlesex County Utilities Authority's (MCUA) Edison Pump Station to their Central Wastewater Treatment Plant. The project consists of two parallel 1.2 km long CC-GRP pipelines constructed in a tunnel beneath the 900 m wide Raritan River and some shorter connections constructed in open trench excavations on the pump station and treatment plant grounds. This tunnel in diameter 4.72 m has several uses with the most important being the connection of the Edison Pump Station, one of five contributing pump station, to the Central Treatment Plant. Within the primary tunnel are two parallel force mains. The new dual force main will replace MCUA's existing DN1500 Arsenal Force Main, which was installed in 1969. The MCUA decided to leave the tunnel partially open to permit inspection of the pipelines and to permit future utilities to use the corridor to cross the Raritan River. In the final construction two additional pipelines in diameter DN400 were installed to convey landfill gas across the river.

4.3   *City of Sports in Rome, Italy*

Major works are under way in the district TorVergata where Rome's so called Second University is based. With the establishment of the City of Sport a substantial aim of the project is to improve the

Figure 9.   Installation of inside sewer pipeline from CC-GRP.

area – not least with an impressive design by the architect Santiago Calatrava. Stormwater from the sports facilities is collected in a water main DN2700 that has been installed by microtunneling. This pipeline, however, serves as host pipe for other service lines as well. It for instance holds a sewer pipe that runs on steel saddles which are fixed to traverses perpendicularly mounted to the tunnel every 3 m. A steel grid as walkway for easy inspection and maintenance of the tunnel is placed above these traverses. It was decided to make assurance double sure and installed double 760 m long sewer pipeline inside in diameter DN600 and DN800. The purpose is to allow keeping an eye on possible leakage and collect the water from the inner line in the clearance between the pipes in case of such.

4.4   *Water supply line in Modrany, Prague, Czech Republic*

A project to renovate the water supply line of Modrany, a fringe area of Prague that lies right at the Vltava River, was launched in 2010. The old steel pipe was installed in a tunnel and was fixed to steel supports and bituminous insulation every two to four meters. It had deteriorated over the course of time and was seriously damaged. To ensure a safe water supply for the population the authorities in charge were presented with two material options the line's replacement: cast iron or glass fiber reinforcement plastic pipes. The many advantages unique to CC-GRP Pipes convinced the investor. Installation was conducted half a meter above the tunnel ground replacing not only the pipelines but also its supports that hold the pipes at their couplings and at the center of their length.

4.5   *Jelenia Góra – tunnel under the mountain massif Chmielnik, Poland*

Jelenia Góra is a city situated in the south-west part of Poland and with 88 thousand inhabitants it is placed in $108.4\,km^2$ area. The location of the city in the valley, in the middle of mountains, and

Figure 10. View of the inside of tunnel in Prague.

Figure 11. View of the inside of tunnel in Prague.

presence of numerous monuments makes it attractive for tourists. The project which was covered in 60% by the cohesion funds from European Union was divided into a few jobs regarding building the pipeline connected water intake in Sosnówka tank and water treatment plant, the water mains supplying water from the WTP to the water networks of Jelenia Góra, building the sewage system and modernization of the existing wastewater treatment plant. A very interesting part of the project was constructing the tunnel under the mountain massif of Chmielnik. It was made as the arch profile 3 m wide and 3.2 m high, the length of it was 392.5 m. Inside the tunnel 2 CC-GRP pipelines DN 500 PN 6 transporting treated water from the WWTP to the surge tanks were installed on the steel supports.

Several other successful examples from Stockholm, Moscow, Switzerland, Scandinavia or Germany of use of microtunnel (GRP, concrete) confirm growing demand of such applications.

5   CONCLUSION

The application of the multi-utility tunnels puts order on the subsurface utility networks, as well as considerably lowers the costs calculated for the whole lifetime of the installations. Multi-utility

Figure 12.   View of the inside of tunnel in Jelenia Góra.

tunnels ensure better work conditions for the conduits (appropriate temperature and humidity of air), eliminate the static loads exerted by the surrounding ground and the dynamic ones caused by transportation means, limit the effects of mining damage and corrosive impact of the surrounding ground and groundwater. They, of course, require some additional engineering furnishings connected with the necessity of ensuring ventilation, health and safety-at-work, systems for monitoring tightness of the conduits, etc. The development of the trenchless methods, especially microtunnelling, allows for performing new utility investments in a non-arduous way, passing natural barriers (e.g. rivers) and using the existing tunnels for many purposes, also by adapting them to purposes that were not envisaged earlier. All that makes multi-utility tunnels attractive and opens a variety of applications which the persons managing the utility systems in large cities may find beneficial and worth implementing for a larger scale.

REFERENCES

Dziennik Rzeczpospolita 14 stycznia 2011.
Geoteko. 2008
HOBAS® 2011a, Comparison of CO_2 emissions from trenchless and open-cut installation methods. Installation of OD 3000 mm diameter pipes for Project Czajka, Warsaw, Poland.
HOBAS® 2011b, Pipeline January, Wietersdorf, Austria.
HOBAS®, materiały własne.
Kamler W. 1976, Ciepłownictwo, Państwowe Wydawnictwo Naukowe, Warszawa.
Kuliczkowski A., Madryas C. 1991, Tunele wieloprzewodowe, Wydawnictwa Politechniki Świętokrzyskiej, Kielce.
Kuliczkowski A. 2010, Współczesne tunele wieloprzewodowe budowane metodami bezwykopowymi, Nowoczesne Budownictwo Inżynieryjne, Styczeń – Luty, Kraków.
Madryas C., Strużyński R. 2009, Application of Composite Pipeline from Glass Reinforced Plastic for Tunnel Drainage, ITA- AITES 2009 World Tunnel Congress, Budapest, Hungary.
Madryas C., Wysokowski A., Gaertig M., Skomorowski L. 2011, Innovative Tunnelling and Microtunnelling Technologies of Record Parameters Used in the Construction of the sewage Transfer System Connected to the Czajka Sewage Treatment Plant in Warsaw, International. No-Dig 2011–29th International Conference and Exhibition, Berlin, Germany.

Motyczka A., Noga A. 2009, *Media w tunelach wieloprzewodowych*, www.edroga.pl.

Pawłowski J. 2011, Mikrotunneling, *Builder.* June 2011, p. 90–91.

Tardiff, I., Center of Expertise and Research on Infrastructure in Urban Areas, Montreal, Canada. 2009, *How to take advantage of green thinking to get more funding for infrastructure*, NASTT and ISTT No-Dig Show 2009,

Yeun J. Jung, Sunil K. Sinha. 2004, Trenchless Technology: An Efficient and Environmentally Sound Approach for Underground Municipal Pipeline Infrastructure Systems, No Dig 2004, New Orleans, p. 1.

Underground Infrastructure of Urban Areas 2 – Madryas, Nienartowicz & Szot (eds)
© 2012 Taylor & Francis Group, London, ISBN 978-0-415-68394-4

Modern systems and methodology for the technical examination of concrete underground infrastructure

C. Madryas & L. Wysocki
Institute of Civil Engineering, Wrocław University of Technology, Wrocław, Poland

A. Moczko
Institute of Building Engineering, Wrocław University of Technology, Wrocław, Poland

ABSTRACT: The paper presents an overview of modern systems and methodology addressed for technical examinations of underground infrastructures with particular attention paid to sewerage pipes and channels. Typical defects, specific conditions and testing procedures for "in-situ" examinations of such structures have been discussed. The main principles of the selected testing methods are also briefly presented. Among other things, the following testing techniques are introduced: core testing, "pull-out" and "pull-off" measurements, Rainbow-test, determination of water absorbability, internal inspection by means of Close Circuit Television (CCTV) and Ground Penetrating Radar (GPR).

1 INTRODUCTION

The underground infrastructure in contemporary cities consists of numerous networks (water, sewage, central heating, gas, telecommunications, power lines) and transport infrastructure (subways, tunnels, tube lines, garages). Tools and methodology used for making expert opinions on underground transport infrastructure do not substantially diverge from those adopted for buildings. However, elements considered in expert opinions concerning underground networks are essentially different, most of all because of their difficult accessibility and hence also working conditions for experts. For example, in the most complicated network system represented by a sewage system the following elements can be distinguished: sewer network, pumping stations, sewage treatment plants and sewage outlets. A network consists of sewers and reinforcements including: inspection chambers on small-diameter sewers (from 100 to 800 mm), manholes on large-diameter sewers (over 800 mm), side entries, ventilators, sewer flushing tanks, sewer section closures (flaps, gate valves, doors), as well as special objects, such as: network tanks, overflow chambers, sewer traps, aqueducts and other structures at intersections with engineering objects. These elements are equally important, but in terms of quantity sewers constitute the largest parts of networks. Therefore, the discussion presented here concentrates mainly on this part of the sewage system. At this point it should be explicitly emphasized that these structures are not classified among hydrotechnical constructions, etc. (sometimes such an erroneous classification is made), which from an ecological point of view is not acceptable.

2 CAUSES OF DAMAGE IN UNDERGROUND NETWORKS

Structural materials of underground sewers are exposed to numerous destructive chemical influences, both on their external side (ground waters and ground) and internal side (sewage). Domestic sewage is heavily polluted waste water. It always contains certain amounts of chlorides, nitrates, sulphates, sulphides, sodium carbonates, detergents, fats and large amounts of organic substances. Typical domestic sewage is characterized by: pH value from 6.5 to 7.0, content of sulphates from 100 to 250 mg/l, chlorides from 20 to 150 mg/l and nitrates from 30 to 200 mg/l. As can be seen, the composition of typical sewage indicates that its aggressiveness towards concrete is insignificant; hence the environment of domestic sewage can be classified under the XA1 exposure class according to the EN 206-1 European standard.

In sewers, however, putrefaction may occur, resulting in formation of hydrogen sulphide and carbon dioxide. Carbon dioxide causes the carbonatization of cement grout (strongly alkaline calcium hydroxide contained in the grout turns into inert calcium carbonate), which, for example, reduces the pH value of concrete; however, it should be noted that carbonatization of wet concrete is very slow (the concrete in the structure of sewers is always wet). In the case of possible decomposition of organic substances contained in the sewage, in the sewer environment hydrogen sulphide is formed, which represents a source of sulphur for *Thiobacillus thiooxidans* bacteria. Hydrogen sulphide may undergo oxidation to sulphur, which deposits on the concrete surface. Then the thiobacillus bacteria oxidize sulphur to sulphuric acid, which causes corrosion of materials on the cement matrix. This type of corrosion is called 'biological corrosion'. At the first stage sulphuric acid reacts with calcium hydroxide:

$$Ca(OH)_2 + H_2SO_4 \rightarrow CaSO_4 \times 2H_2O$$

As a result of this reaction calcium sulphate (gypsum) is formed, which crystallizes with two water molecules, increasing its volume simultaneously by 130%. At the second stage gypsum may combine with tricalcium aluminate in the following reaction:

$$3CaSO_4 \times 2\,H_2O + C_3A \times nH_2O \rightarrow C_3A \times 3CaSO_4 \times 32H_2O$$

thus creating cement bacillus, which crystallizes with a simultaneous increase in volume by 227%. The crystallizing gypsum and cement bacillus cause internal stresses in the concrete, which at the first stage result in cracks and fractures, and then total destruction of the concrete structure. Crystallization pressure during transition from $CaSO_4$ into $CaSO_4 \times 2H_2O$ is approximately 110 MPa (the tensile strength of concrete usually does not exceed 5.0 MPa).

The concrete corrosion rate depends on its quality (porosity, tightness), type of cement used, condition of cracks and content of sulphates. Cracks, caverns and porous structure facilitate concentration of aggressive substances and their penetration into the structure, which substantially accelerates the rate of corrosion.

The authors' own tests indicate that a particularly aggressive environment (exposure class XA3) is formed in expansion chambers and in sewers below such chambers. A very aggressive environment may also occur in sewers with a small slope, in which ventilation is ineffective. Too small slopes cause collection of deposits, which in the case of inadequate ventilation readily undergo fermentation. Such situations often occur in sewer mains carrying sewage to treatment stations, because such sewers usually are not equipped with house sewers ensuring ventilation. Fermented deposits or drips taken from such sewers show a pH value of approximately 2.0 (and even 1.5), and sulphates content may exceed 5%.

Tests carried out for many years in many Polish cities by the authors of this study show that sulphates are in practice the main corrosion hazard in sewers made of concrete or brick structures on cement mortar. Sporadically an increased content of chlorides can be found, which poses a threat to manhole steps, inspection chamber covers, sewer reinforcement rods and other ferroconcrete structures.

In sewers made of plastics, especially thermoplastics (PEHD, PVC, polypropylene), the hazard results however not from chemical substances, but from mechanical and thermal factors. Stones and other objects occurring in the filling exert spot pressure on the sewer wall, causing slow material flow. This results in local deformation of a sewer, reducing its peripheral rigidity and resistance to the loss of stability. A similar hazard may be caused by placing a sewer on the remnants of old building structures (e.g. walls in unchecked deposits).

The fundamental parameter taken into consideration when selecting structures of ductile sewers (in the considered case, sewers made of thermoplastics) is peripheral rigidity. This parameter is determined at a temperature of 20°C. When sewage with higher temperature is dumped into the sewer, peripheral rigidity decreases. In such cases additional analysis is necessary to evaluate the influence of increased sewage temperature on deflection and stability of the sewer. It should be added that uncontrolled rises in sewage temperatures are observed more and more often, due to the growing number of household devices dumping warm sewage (washing machines, dishwashers). This also causes intensification of biological processes, and thus increase in the risk of biological corrosion.

Another important issue is corrosion of gaskets used in sewer pipe joints. Fats and oil-derived products can be hazardous for elastomers (EPDM, SBR) commonly used in production of seals.

In areas affected by mining damage a hazard for all sewage conduits is posed by terrain deformations, which may cause the loss of leaktightness on sewer pipe joints (on bulges of subsiding troughs), crushing (on concavities of subsiding troughs) or other mechanical damage.

In areas with substantial slopes a hazard for sewers, especially in rain drain systems and combined sewage systems, can be posed by dragged substances causing abrasion of inverts. Abrasion is particularly hazardous for concrete sewers, because abrasion resistance of thermoplastics, vitrified clay and even hardening plastics is substantially greater. Abrasion resistance of concrete depends on its quality, especially including its strength class, water/cement ratio and type of aggregate used. At present usually products made of C35/45 concrete are used, with water/cement ratio below 0.45 and aggregate with high abrasion resistance – as a result contemporary concrete products are characterized by much greater abrasion resistance.

Cascade wells and cascades on the sewer route are used in sewers sited in areas with substantial height differences. In such sewer sections damage to the surface of brick and concrete sewers may occur, as a result of cavitation. This damage occurs due to extraction of particles from mineral binding materials, caused by underpressure in bladders of gases forced into the sewage in a cascade. On the surface of a sewer damaged by cavitation characteristic "pockmarked" cavities are noticeable.

Damage to sewers, especially within joints and cracks, can also be caused by tree roots. Damage to sewers made of thermoplastics caused by rats is also observed. Rats also create corridor systems in the ground around sewers, causing unfavourable changes in foundation conditions.

Other frequent reasons for sewer damage are defects in workmanship, which include, among other things:

- neglectful preparation of the base under the pipes (application of improper filling material and its negligent compacting),
- incorrect water drainage from an excavation (periodic interruptions of water pumping, water pumping directly from excavation made in very fine-grained ground),
- using protections of excavation walls incompatible with the design and removing them inconsistently with the design,
- making the excavation geometry incompatible with the design,
- leaving large stones or remnants of old structures on the excavation bottom,
- incorrect assembly of gaskets, including application of improper lubricant,
- failure to use stub pipes at chambers,
- using products based on cement incompatible with requirements for sulphate-resistant cement, especially products based on fast-setting cement, in sewers exposed to sulphate corrosion.

3 SPECIFICITY OF TESTING UNDEGROUND INFRASTRUCTURE

Diagnostics of underground infrastructure, particularly sewers, differs substantially from other areas of construction expert activity. It has its own specificity resulting mainly from the character of tested objects and conditions occurring inside. These conditions are, in principle, always *"unfavourable"* for the expert and are associated with relatively high testing costs. The most important impediments include first of all:

- lack of access to tested objects from outside, causing the necessity of performing the majority of tests inside, which usually poses a substantial impediment due to the small dimensions of structures,
- practically no possibility of disconnecting a sewer section being tested from the network for a longer period; this is often associated with the necessity of working inside sewers, where sewage flows.

Performing tests inside sewers, where large amounts of deposits and sewage occur, requires special protective clothing, appropriate protection for testing equipment and proper suspensions and fixing of all additional accessories, including electrical wiring. Power supply is indispensable for lighting of test stands and operation of various electric devices. Due to electric shock hazard, in extreme cases special protection is indispensable. Life hazard also results from the presence of harmful gases, particularly hydrogen sulphide as well as deposits which can be its source. Creating safe testing conditions requires professional ventilation of ducts, constant monitoring of gas concentrations and ensuring the possibility of quick and safe evacuation.

Also of great significance is discomfort for the expert, associated with the specific unpleasant smell and necessity of working in the presence of deposits which soil the equipment and clothing.

4 ASSESSMENT CRITERIA FOR TECHNICAL CONDITION OF TRUNK SEWERS

The assessment of sewers' technical condition is based on three basic criteria: strength, ecological and hydraulic.

The strength criterion with reference to rigid conduits (made of concrete, bricks, vitrified clay) comes down to determining the current load-carrying capacity of a sewer (taking into consideration current technical condition and measured strength parameters of structural materials) and comparing this load-carrying capacity with that resulting from actual loads, taking the required safety factor into consideration.

In the case of sewers made from plastics (ductile sewers) measurements of diameter deformations are carried out and compared with admissible values. Moreover, for ductile sewers the value of stresses in the sewer wall caused by the acting load is determined (taking into consideration geological influences) and these tensions are compared with admissible values. For full determination of safety status of ductile sewers verification of the buckling safety factor is necessary.

The ecological criterion with reference to all sewers comes down to checking their leaktightness, as maintaining the leaktightness is necessary for prevention of sewage exfiltration (polluting surrounding ground and ground water) as well as infiltration of ground water (uncontrolled self-consolidation and desiccation of soil).

The hydraulic criterion for all sewers comes down to checking actual flow capacity (taking into consideration current technical condition) and comparing it with the required capacity. It should be noted that in the case of sewers, when there is no scheduled control of their condition, quick detection of damage is often difficult. This is because the gravitational mechanism of sewage transportation causes that consequences of sewer breakdown, even with unmet strength criterion and leaks, are not detected immediately. Such sewers very often still carry the sewage effectively, and, as a result of exfiltration or infiltration, cause hazards to the surrounding environment.

5 TESTING TECHNICAL CONDITION OF TRUNK SEWERS

5.1 Introduction

The main task of technical inspection is to detect possible damage and determine reasons for their occurrence. Check-up tests may be associated with information about network damage, or suspicion that the damage has occurred as a result of observed external forces with unforeseeable intensity or very common observed damage to other structures or terrain in the vicinity of the network. Inspection is then random and implies diagnostics and repair of a diagnosed section of the network. Inspections may also be carried out periodically for the needs of a programmed strategy for system exploitation. Periodic inspections generate costs, but on the whole they may significantly reduce operating costs associated with maintaining operating efficiency of the network – Fig. 1 (Madryas et al. 2010).

The graph shown in Figure 1 indicates that in terms of renovation costs there exists a certain optimum frequency of inspecting the condition of sewers. Localizing damage may be carried out by persons present directly inside sewers (large-diameter sewers only) or by means of remotely controlled equipment (large-diameter sewers and small-diameter sewers). As already mentioned, sewers (as opposed to the majority of technical devices) may perform their functions even when

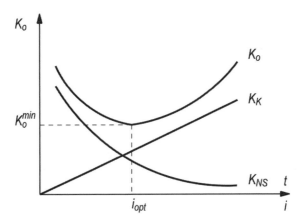

Figure 1.   Relationship between system renovation costs and frequency of inspections where: K_O – renovation costs, K_K – inspection costs, K_{NS} – operating costs of ineffective system, i – frequency of inspections.

several elements are damaged. Due to this fact the inspection must cover the entire network section assumed as potentially damaged and cannot be completed after identification of the first damage.

If the defect and its location (or rather defective area) are found as a result of sewer tests, then diagnostics is started. Its purpose is to determine, as quickly as possible, the causes of the damage and the level of its significance. Tests of this type are most often carried out at the level of technical opinions by construction experts. The tests require application of appropriate diagnostic procedures and specialist testing equipment. Particularly recommended here is considering the possibility of using non-destructive testing methods, allowing quick non-invasive evaluation of technical parameters, having essential influence on correct assessment of the actual technical condition of the tested object. Tests carried out directly on site often also require supplementary laboratory tests in order to determine the parameters of structural materials and the degree of structure degradation caused by chemical and biological corrosion. Requirements for tests of technical condition of sewers are specified in several European standards.

5.2   Diagnostics of concrete arge-diameter sewers

Generally speaking, evaluation of the technical condition of concrete large-diameter sewers comprises the following elements:

- visual inspection comprising both an inventory of existing damage and specification of the indispensable scope of diagnostic tests, aimed at finding the causes of resultant damage, determination of its actual influence on operating values of the tested object and evaluation of its actual technical condition,
- measurements of the object's actual geometry, including determination of its actual grade line and static diagram, as well as estimation of existing loads,
- location and identification of reinforcements in concrete cross-sections along with evaluation of actual thickness of concrete cover layer,
- assessment of compressive strength of concrete through tests of cored specimens and non-destructive methods along with estimation of its compressive strength class,
- chemical tests comprising, first of all, determination of harmful salt content in tested brick and concrete cross-sections, and in the case of concrete structures additional determination of actual pH value, associated with neutralization degree of concrete cladding, as well as evaluation of chemical and biological corrosion hazard, especially taking sulphate-derived corrosion into consideration,
- determination of concrete absorbability on the basis of laboratory tests of specimens taken from the existing structure,

- making indispensable uncoveries to enable random verification of information obtained through various non-destructive measurements as well as estimation of the scope and degree of advancement of corrosion processes,
- evaluation of the tensile strength of concrete by the "pull-off" method in order to establish whether for repair of a given fragment of the structure modern surface repair methods can be applied,
- entirely non-destructive evaluation of thickness for concrete cross-sections accessible from one side by means of the "Impact-Echo" method as well as various defectoscopic measurements made by the "Impact-Echo" and "Impulse Response" methods.

Besides tests mentioned above, very important for correct evaluation of hazards and their possible influence on safe and defect-free operation of a trunk sewer is also determination of chemical composition of the sewage and specialist tests of ground space surrounding the trunk sewer.

5.2.1 *Visual inspection*

The basic form of visual inspection of a sanitary trunk sewer is the inventory of damage, which comprises counting them, description and photographic documentation. Examples of types of occurring damage include: material cavities in sewer walls, exposed reinforcement, internal corrosion, cracks and fractures, and mechanical wear (abrasion). If necessary, visual inspection is supplemented by measurements localizing the damage and specifying its geometric features, e.g. length and width of crack dilation as well as endoscopic examination, by means of modern optical techniques (borescope devices).

The set of damage types which may occur in a sewer is limited, but within each type features associated with it may exceed to various degrees the admissible range. The uniform system for notation of damage adopted in the countries of the European Union has been defined in the second part of European standard EN 13508. Besides the damage coding system the standard also contains a catalogue including 55 photos of damage, described according to the proposed notation.

Visual evaluation is also aimed at recognizing the occurrence of deposits and dripstones in the sewer, infiltration of ground water and all obstacles disturbing the flow of sewage. Moreover, on this occasion possible deformations and deviations of sewer position should be identified and an inventory of equipment and connections should be made

5.2.2 *Sewer geometry measurements and leaktightness control*

Geometrical measurements concern mainly evaluation of the tested trunk sewer in the context of the hydraulic criterion, associated with the analysis of sewage flow conditions in the sewer. Disturbances in sewage flow may be caused by various factors, among which special attention should be paid to inappropriate cross-section (too small or too large cross-section, possibly inappropriate shape), incorrect slopes of conduits in the sewerage system, occurrence of excessive local or line flow resistances and changes in the parameters of carried medium in relation to design conditions.

Geometrical parameters mentioned above are mainly determined by means of geodetic inventory methods, using modern laser rangefinders and three-dimensional scanning devices.

Sewer tests in terms of the ecological criterion refer however mainly to their leaktightness. Phenomena of sewage exfiltration to the ground and infiltration of ground water into the sewer are considered. At present the assessment of leaktightness for sewers in operation is carried out mainly by the air pressure test and, to a lesser degree, by the hydraulic test.

5.2.3 *Localizing and identifying reinforcements in concrete cros-sections*

One of the basic elements in modern diagnostics of concrete trunk sewers is the localization and identification of reinforcements in concrete cross-section. Such tests can be carried out with various methods, which in the great majority are non-destructive. Here, among others, X-ray and gamma-ray radiography, and "Georadar" (Ground Penetrating Radar) should be mentioned. At present, however, the most popular research method used for localizing and identifying reinforcement rods is the electromagnetic method, which makes use of phenomena accompanying disturbances in an electromagnetic field, caused by introduction of ferromagnetic materials, for example reinforcement rods, into the field. Modern devices of this type, (e.g. "Cover-Master", CM-9 model) besides the capability of localizing steel rods at depths up to 90 mm, measuring concrete cladding thickness as well as estimating reinforcement rod diameters immediately during measurement, also have a number of additional functions, among which the following deserve special attention:

- *"Max-Pip"* function, operating so that the device gives out a sound only when its head is located exactly above a steel bar,
- *"Under Cover"* function, operating so that the device gives out a sound when its head is located above a rod placed under cladding with insufficient thickness.

Despite very high effectiveness and reliability of electromagnetic measurements, they should be verified each time empirically, by local reinforcement uncoveries. These uncoveries also enable one to obtain a series of data, which cannot be acquired directly during measurements. These data comprise:

- type of reinforcement steel (smooth or ribbed), type of ribbing,
- actual spacing between reinforcement rods in case they are very concentrated,
- whether double steel bars occur.

Due to an existing impediment consisting in one-sided access to the inspected structure, it is also important to establish whether within the concrete cross-section there are two reinforcement layers, and if so, how they are located with respect to each other. For this purpose most often a special type of rod head is used, inserted into a hole drilled specially for this purpose.

5.2.4 *Assessment of "in-situ" concrete compressive strength*

The main test performed for all types of concrete structures is the determination of the compressive strength of concrete. Due to conditions existing in trunk sewers the series of research methods commonly used in tests of buildings and transport infrastructure cannot be applied. The group of such methods includes both the ultrasonic method and the rebound measurements (Schmidt's hammer). In practice we only have available laboratory strength tests of cored specimens and a quasi non-destructive "pull-out" method – more specifically, its variant called the "CAPO-Test".

Tests of cored specimens, according to currently European standards (EN 12504-1 and EN 13791), should be carried out on samples cut out from the structure, with a diameter close to 100 mm. Such an approach results from the fact that the ratio between the maximum grain of the aggregate to the bore hole diameter should not be greater than 1:3, which in practice means that in the case of aggregate with grain sizes up to 32 mm the preferred diameter is approximately 100 mm. In practice diameters of samples cut out from structures are very variable, which results mostly from the fact that crown drills available on the market are not standardized. Their dimensions range most often between 95 and 105 mm. It should be emphasized that in the case of trunk sewers, parameters of the coring machine should be matched to conditions existing inside such objects. Its dimensions and weight should be as small as possible. The coring machine intended for this purpose must also have wires protected against moisture, as well as an automatic safety switch to guarantee its immediate shut-down in case of danger.

In accordance with regulations specified in the standards mentioned above, a rule is adopted that testing a bore-hole with a nominal diameter of approximately 100 mm gives as a result the strength value corresponding to the strength of a cube sample with a side of 150 mm, made and matured in the same conditions. In practice, due to density of reinforcements, or too small thickness of the element tested, we often deal with cases in which there is no possibility of cutting bore-holes with the recommended diameter of approximately 100 mm. In such situations the solution used most often is to take bore-hole samples with diameter of approximately 80 mm, which at maximum size of aggregate grain of e.g. 20 mm entirely meets the requirements described above, associated with the ratio of aggregate grain to the diameter of the sample. In such a case, when working out the results of destructive compression tests of concrete, the following reasoning can be adopted:

- according to the publication "Concrete Structures" (Bukowski 1972), the following assumption can be made:

$$f_{cylinder}(h=\phi=160 \text{ mm}) \approx 0.85 \, f_{cylinder}(h=\phi=80 \text{ mm})$$

− moreover, according to PN-88/B-06250 [7]:

$$f_{cube}(a=150 \text{ mm}) = 1.15 \, f_{cylinder}(h=\phi=160 \text{ mm})$$

that is

$$f_{cube}(a=150 \text{ mm}) \approx 1.15 \times 0.85 \, f_{cylinder}(h=\phi=80 \text{ mm})$$

− which as a consequence leads to the relationship:

$$f_{cube}(a=150 \text{ mm}) \approx 0.98 \, f_{cylinder}(h=\phi=80 \text{ mm})$$

To ensure correct execution of concrete strength tests on samples cut out from cored specimens, proper preparation of the ends of those samples is of key importance, in order to ensure parallelism of surfaces to which the load will be applied. For this purpose, after cutting the cored specimens into pieces (individual samples), grinding them or applying so-called "*caps*" is recommended.

Unlike testing cored specimens, the "pull-out" method enables assessment of the compressive strength of concrete directly on site. This is an unquestionable advantage of this method, which, in connection with the fact that concrete damage occurring during the measurements is very little and practically does not require repair, often has decisive significance when selecting this testing method. The "pull-out" method is also irreplaceable in case of great density of reinforcements or small thickness of concrete cross-section (smaller than 80 mm), which often makes the cutting-out of test samples from an existing structure impossible. Testing principle of the "pull-out" technique is shown on the Fig. 2.

The CAPO-Test was developed in Denmark (Krenchel & Schah 1985, Krenchel & Bickley 1987) as a variation of the "pull-out" technique for determining on site compressive strength of existing concrete structures. The principle behind the test method is that the force required to pull an insert out of concrete can be correlated with the concrete's compressive strength. To prepare a test, an 18 mm diameter hole is drilled outside reinforcement disturbance followed by recessing a 25 mm diameter groove at a depth of 25 mm. A special steel ring is inserted through the hole in the groove and expanded by means of a special expansion tool until it fits in the inside diameter of the groove (Fig. 2). Finally, the ring is pulled out against a counter pressure 55 mm in diameter. A load is applied through a manually operated hydraulic pull-machine. The pull machine from one side applies the pull-out force to the insert, and on the other side presses the concrete surface via the counter-pressure ring. The ring, thanks to appropriate proportion of its dimensions in relation to the depth of the insert, forces a complex stress state, which in the end results in destruction characterized by strict correlation between the recorded force pulling out the insert and the compressive strength of concrete.

This correlation has a general character, i.e. with the exception of concretes based on light-weight aggregates and concretes with grain exceeding 32 mm, in principle it is independent of the influence of material and technological parameters, such as water/cement ratio, type of cement, maturing conditions, concrete age, content of possible additives, such as silica, dusts, or various types of fibres. This relationship is most often described by the two formulas (Moczko 2004):

$f_c = 1.41 \, P - 2.82$ – for concrete with strength up to 50 MPa
$f_c = 1.59 \, P - 9.52$ – for concrete with strength over 50 MPa

where:
f_c – cube compressive strength of concrete in MPa
P – value of pull-out force in kN

Recently in the professional literature a new proposal has been presented to replace the above curves with one correlative curve, being their mathematical generalization (Fig. 3). Analysis carried out by the authors has shown that both proposals are completely reliable, and occurring differences are insignificant and do not have an important influence on the accuracy of test results.

The existence of an exceptionally good correlation between the value of the recorded pull-out force and compressive strength of concrete, determined on standard laboratory samples (e.g. cubes with 150 mm sides) is explained by the fact that local destruction within the insert extraction zone

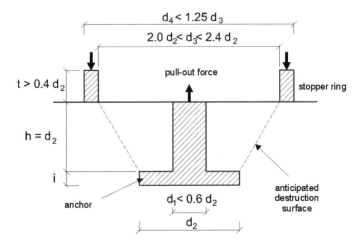

Figure 2. Testing principle of the "pull-out" technique.

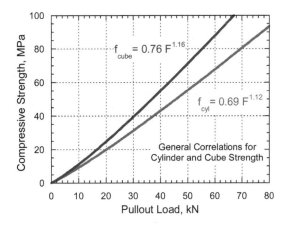

Figure 3. General correlation curves for the "pull-out" method.

is dominated by compression of the concrete between the insert head and the counter-pressure, pressed to the concrete surface by the hydraulic pull machine.

The reliability of evaluating concrete compressive strength by means of this method is reflected by several standards: American standard ASTM C-900-87 (*Standard test method for pullout strength of hardened concrete*) and European standard EN 12504-3 (*Testing concrete in structures – Part 3: Determination of pull-out force*), also adopted by the Polish standardization system in the standard PN-EN 12504-3: 2005.

Determination of strength class on the basis of strength test results obtained "in-situ" is carried out according to the current European standard EN 13791:2007 (*Assessment of in-situ compressive strength in structures and precast concrete components*). At the same time it should be noted that the mentioned standard, in comparison with the standardized rules for assessing the quality of concrete formulated in the new European concrete standard EN 206-1:2002 (*Concrete – Part 1: Specification, performance, production and conformity*), defines slightly different conformity criteria for concrete strength evaluated on the basis of strength test results in an existing structure. The rules for defining the characteristic value of the compressive strength of concrete in a structure come down to two cases. The "A" case is applicable to the situation when at least 15 strength test results are available. The "B" case, however, concerns the situation when from 3 to 14 test results are available. As in practice the possibility of cutting 15 samples from an existing structure or

Table 1. The "k" variable for small number of test results

Number of test results	"k" variable value
From 10–14	5
From 7–9	6
From 3–6	7

performing 15 measurements with the CAPO-Test method is usually not possible, this study is limited to a discussion of case "B", for which the characteristic value of the compressive strength of concrete, determined in a given measuring location, is the lower of the two values:

$$f_{ck,\, is} = f_{m(n),\, is} - k \qquad \text{or} \qquad f_{ck,\, is} = f_{is,lowest} + 4$$

The "k" variable depends on the number of test results and is defined in accordance with Table 1.

The compressive strength of concrete is determined according to the table published in the standard, on the basis of the determined characteristic value of compressive strength of concrete in a structure. However, it should be clearly emphasized that in this case we are dealing with essential reduction in minimum characteristic values of compressive strength of concrete, required for individual strength classes of concrete, compatible with the new European concrete standard (EN 206-1:2002). The European standard (EN 13791:2007) introduces the corrective factor 0.85, which takes into consideration the common conviction that the concrete strength in a structure is generally lower than the strength determined for standardized samples taken from the same batch of concrete. This fact is often attributed to the drilling process, which undoubtedly involves the risk of slight damage to the core material, and partly to the fact that the curing conditions for concrete on a building site are usually worse than those in a laboratory.

5.2.5 Chemical tests

Chemical tests mainly concern the measurement of content of harmful salts in wall and concrete cross-sections. They usually involve determining the content of sulphates, chlorides, and nitrates in samples taken from the existing structure of a trunk sewer. Test samples are most often drilled or knocked off from the surface layer up to a depth of approximately 10–15 mm. In justified cases, for concrete structures samples are taken from various depths, in order to obtain the distribution profile of a given salt content in the tested cross-section. Such tests are usually performed for the following three depths:

- material layer placed directly at the surface (from 0 to approximately 5 mm),
- material layer placed at a depth of approximately 5 mm to half of the concrete cladding on the reinforcement (usually up to a depth of approximately 15 mm),
- layer placed at a depth from half of the cladding thickness up to the reinforcement (usually from approximately 15 up to approximately 30 mm).

Chemical tests may be performed in a chemical laboratory, or by using measurement kits for "in-situ" tests. From among the known kits of this type, "Aquamerck-Test", which in fact is a portable chemical mini-laboratory, has been recognized as most popular. For a concrete structure it is commonly assumed that the limit value of ionic concentration of SO_4^{-2} amounts to 0.50% in relation to the weight of concrete, and ionic concentration of Cl^- is at the level of 0.40% of cement weight. In the case of nitrates the admissible value of their concentration is assumed as 0.15% in relation to the weight of concrete.

Besides the estimation of harmful salt content in the tested structure, an extremely important test is the determination of the actual pH value, associated with neutralization degree of concrete cover layer. This measurement is most often carried out on the same samples on which other chemical tests are performed. Sometimes the phenolphthalein test or the thymolphthalein test is used for this purpose; however, they are approximate tests, useful only for evaluation of pH value in a very narrow scope. Significantly more reliable in this respect is the "Rainbow-Test", which is a

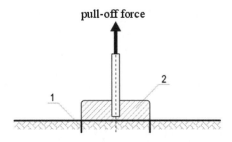

pull-off force

Figure 4. "Pull-off" – the principle of the testing method where: 1 – notch, 2 – steel disc glued to the surface.

composition of chemical reagents, enabling identification of the distribution of pH values along the depth of the tested cross-section. The measurement itself involves spraying onto the tested surface a solution with a specially selected composition of chemical reagents, identifying individual pH values in the range from 5 to 13. The pH value equal to 11, commonly recognized as a limit value, below which the natural ability of concrete to passivate the reinforcement is reduced, corresponds to concrete colouring to violet. Colour transition from violet to green (pH = 9) indicates decrease of pH below the value assumed as a limit signifying potential corrosion hazard to the reinforcement. Colour change to yellow means the pH value is approximately 7, while red colour corresponds to a pH value of approximately 5.

5.2.6 *Determination of concrete absorbability*

The new European concrete standard (EN 206-1: 2002) does not require determination of the absorbability of concrete. However, due to its practical usefulness for evaluation of natural resistance of concrete structure to water absorption, this test in the case of sewers is commonly used in expert practice. Then the recommendations of Polish standard (PN-88/B-06250) are used. In the case of concrete, the admissible absorbability value is 5%.

5.2.7 *Assessment of tensile strength of concrete by "pull-off" technique*

The "pull-off" test was oryginally developed to measure the in-situ tensile strength of concrete by applying a direct tensile force (Long & Murray 1984, Bungey & Madandoust 1992). The method is also used to evaluate the effect of surface preparation procedures on the tensile strength of the substrate before applying a repair material or overlay and for measuring bonding of surface repairs as well (Cleland 1993).

The principle of the "pull-off" method, generally speaking, is measurement of the force required to pull off a metal disc with a known surface area, glued to the tested surface (Fig. 4). Most frequently discs with diameter of 50 mm are used. A centric notch 10–15 mm deep is made around the disc. The recorded value of the pull-off force, divided by the surface area onto which the load is transferred, gives as a result the value of the tensile strength of concrete, also called the bond strength.

Among many devices which make such measurements possible, two solutions have found the widest application in engineering practice: the Swiss device called "DYNA" (Fig. 5) and the Danish device "BOND-Test".

According to technical guidelines for this type of test, a given fragment of a structure may be repaired with the aid of modern surface repair techniques, provided that the following conditions are met:

- average compressive strength of concrete should not be less than 25 MPa,
- average value of bond strength, determined in a given measurement location for all measurements, should not be lower than 1.5 MPa,
- minimum value of a single measurement should not be less than 1.0 MPa.

Figure 5. DYNA – one of the most popular "pull-off" testing systems.

Figure 6. Sewer inspection system with self-driven camera.

5.3 *Diagnostics of small-diameter sewers*

Due to the very small cross-sections of small-diameter sewers, their diagnostics consists mainly in visualization of the sewer interior by means of a video camera moving inside. This type of survey consists in camera passage through the sewer, from which the picture is transmitted to a display monitor located outside on the terrain surface. Traditionally the picture from the camera is recorded on an appropriate carrier (video cassette or DVD disc). Also current comments of the inspector performing the inspection are recorded. In principle it is assumed that a sewer should be cleaned before the inspection. A disadvantage of this testing method is the subjectivity of obtained results, associated with the character of the inspection, as psychophysical features and qualifications of the person performing the observation of the sewer interior are of crucial significance. The concept of inspection with a CCTV camera is shown in Fig. 6.

At present in visual inspections besides traditional cameras more and more often scanning systems are used. Generally speaking, scanning consists in making a series of digital photos covering the entire sewer volume during a passage of the scanning device through the sewer. Photos (scans) are made at equal, short time intervals, at constant device speed (e.g. one scan for each 5 cm of length). High light intensity and short flash duration as well as high-quality lenses ensure very good resolution of images obtained. Passage of the device through a sewer, associated with scanning, is carried out with a speed over twice as high as in the case of inspection with a traditional

160

camera (e.g. over 35 cm/s compared to the recommended 15 cm/s for inspection with a traditional camera). The scanned sewer image is recorded in digital form and processed with the aid of special software, so that already during scanning it is possible to obtain an image of the sewer interior as taken from a straight-looking camera, or (in some sets) from a camera equipped with a two-axis rotary head – called a 'spatial image'. The possibility of two-dimensional expansion of the sewer image has essential importance, as it enables a quick overview of its technical condition and easy determination of dimensions and location of possible damage. Principal notation and classification of damage takes place on the basis of images recorded digitally, which enables transfer of these activities to a specialized entity performing appropriate analysis.

It should be emphasized that in comparison with inspections carried out by means of traditional equipment, the main advantages of using scanning devices are:

- considerable reduction in tasks and responsibilities of the camera operator – he does not make a description and initial evaluation of the sewer on site, or detailed visual inspection of selected damage; in fact his main task is correct introduction of the scanner into the sewer and supervision over its travel,
- as the operator of the scanning camera does not perform the above-mentioned operations, time spent by the device inside the sewer is definitely shorter and depends only on the specific features of device operation (automatic scanning time, trolley speed) and any obstacles restricting its movements inside the sewer,
- detailed description of damage and evaluation of technical condition of the inspected sewer is made by specialists in laboratory conditions.

Description of the damage found by means of a CCTV camera is made, similarly as in the case of large-diameter sewers, using the notation system published in European standard EN 13508.

5.4 *Testing the ground around trunk sewers*

Tests of ground space around the sewers most often concern the following issues:

- exact locations of sewer routes,
- identification of soil zones with different densities, including location of loose soil zones and detection of caverns situated in the vicinity of the sewer,
- identification of ground zones with different moisture, including zones created by sewer leaks,
- seeking areas of possible silting and pollution spreading from the sewer as well as areas with chemical aggressiveness of external origin (not associated with the sewer),
- identification of chemical aggressiveness in the ground surrounding the sewer being tested by way of detailed laboratory tests of soil samples taken on site.

Traditionally, ground tests involve probing, drillings and uncoveries (test excavations). Moreover research equipment is available, partly originating from so-called 'geophysical' methods. These types of methods include, among others, geoelectric, seismic, gravimetric, magnetic and geothermal methods, as well as methods based on nuclear geophysics. From among diagnostic techniques mentioned above, special attention should be paid to the ground-penetrating radar (GPR) method, which uses reflection and refraction of electromagnetic waves during their propagation in a medium with variable electrical and magnetic properties. This method is also ranked in the group of geoelectric methods and adopted in Poland under the name "Ground-Penetrating Radar" tests as a method for identification of ground structure.

Operation of ground-penetrating radar consists in pulse transmission of high frequency electromagnetic waves from transmitting antenna into the ground medium and recording the waves reflected from objects (media) with different values of dielectric constant. By means of this method ground zones with various properties can be located (caverns, underground conduits, remnants of building structures, etc.) As the ground dampness causes significant reduction in the amplitude of the reflected signal, interpretation of test results for the detection of sewer leaks is possible, especially including exfiltration. Localizing a leak consists mainly in observing abnormal indications of sewer foundation depths, whose apparent lowering is caused by ground irrigation around the leak.

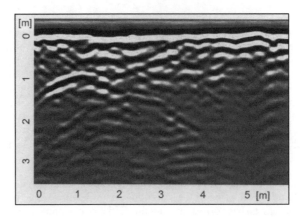

Figure 7. Example profile (wave graph) obtained with impulse GPR and 500 MHz antena.

In traditional presentation, in order to obtain a sectional image of the ground, the measurement (transmission and reception of pulses) is repeated many times along a defined measurement line. A recorded image of reflected waves gives a two-dimensional XZ section, called a wave graph (Fig. 7), where X is the travelled distance, and Z is the wave travel time converted into distance. A combination of many parallel wave graphs allows one to obtain a three-dimensional ground space model with objects located inside.

Depending on antenna operating frequency (frequencies of transmitted signals range from several dozen MHz up to as high as 2 GHz), detection of objects with various sizes is possible. However, errors occurring during tests, resulting mainly from different speeds of antenna movement along the ground, at large density of underground objects may cause difficulties in interpreting a wave graph. A limitation of this method is caused by occurrence of low-resistance ground, such as clay or silt, which strongly suppresses electromagnetic waves. Mostly the high-frequency waves, responsible for detection of small objects, are suppressed; hence high resolution of this method is possible in surface layers of the ground.

In modern solutions sets of connected antennas are used, installed in rows, perpendicularly to the direction of measurements – so-called 'multi-antenna systems'. Simultaneously measurements with various frequencies are carried out here; thus high resolution and greater penetration depths are obtained. Also quicker coverage along the Y axis occurs. There are also solutions in which the measurement method is not based on recording reflections of a single signal with defined frequency, but is based on generating a modulated signal which covers a wide frequency spectrum – broad-band signal (Majewski 2008). Thus efficiency of the method, image resolution, testing speed and the possibility of achieving 3D images are increased.

Final processing of test results is made by means of appropriate software, facilitating the creation of legible maps of underground space, e.g. in the CAD or Microstation system.

6 SUMMARY

Tests of sewers in underground network infrastructure in cities are exceptionally onerous and definitely differ from tests of other building structures; thus they require experts specialized in this field. Tests of sewers are mainly aimed at determining their capability to function properly, i.e. to meet the criteria concerning strength, flow capacity (hydraulic criterion) and environmental protection (ecological criterion). In the case of sewers, the majority of tests involve verification of the strength criterion. The authors of this study have performed such tests for many years. When testing sewers in this respect it is necessary to determine actual loads acting on them (taking into consideration actual parameters of ground and sewer structural materials) and define the safety factor in terms of load-carrying capacity (rigid sewers) and stability (ductile sewers). In Poland the required values of safety factors are most often assumed on the basis of ATV guidelines. According to these guidelines, on the basis of obtained test results and static-strength check-up calculations, sewers are qualified to one of three technical statuses. Qualifying a sewer to a particular status is the basis for making a decision on the scope and method of sewer structure renovation. Hence, carrying out repairs without inspection of the technical condition, an expert opinion and a design is unacceptable. It may lead to incorrect decisions, resulting in a threat to safety (when adopted solutions insufficiently increase the load-carrying capacity of a sewer) or unjustified increase in renovation costs (when adopted solutions excessively increase the load-carrying capacity of a sewer being repaired).

REFERENCES

American standard. 1987. ASTM C 900-87 – Standard test method for pullout strength of hardened concrete. American Society for Testing and Materials, Philadelphia, PA. 19103, USA.

ATV-DVWK – A 127P. 2000. Static-strength calculations for conduits and sewers. Warsaw: Publishing House Seidel-Przywecki Ltd.

Bukowski, W. 1972. *Concrete Structures*, vol. I, part 1 – Concrete Technology. Warsaw: Arkady.

Bungey, J.H. Madandoust, R. 1992. Factors influencing pull-off tests on concrete. *Magazine of Concrete Research* vol. 44, No 158: 21–30.

Cleland, D.J. 1993. In-situ methods for assessing the quality of concrete repairs. *Proc. ICE, Structures & Buildings*, vol. 99, 68–70.

European standard. 2003. EN 206-1: 2002 – *Concrete – Part 1: Requirements, properties, production and conformity*

European standard. 2001. EN 12390-3: 2001 – *Testing hardened concrete. Compressive strength of test specimens.*

European standard. 2001. EN 12504-1: 2001 – *Testing concrete in structures – Cored specimens. Taking, examining and testing in compression.*

European standard. 2005. EN 12504-3: 2005 – *Testing concrete in structures – Determination of pull-out force.*

European standard. 2001. EN 752:2001 – *Drain and sewer systems outside buildings.*

European standard. 2006. EN 13508:2006 – *Conditions of drain and sewer systems outside buildings.*

European standard. 2003. EN 12889:2003 – *Trenchless construction and testing of drains and sewers.*

European standard. 2002. EN 1610:2002 – *Construction and testing of drains and sewers.*

European standard. EN 1917 – *Concrete manholes and inspection chambers, unreinforced, steel fibre and reinforced.*

European standard. 2005. EN ISO 22476:2005 – *Geotechnical investigation and testing – Field testing*

European standard. – EN 598: *Ductile iron pipes, fittings, accessories and their joints for sewerage applications.*

European standard. EN 1916 – *Concrete pipes and fittings, unreinforced, steel fibre and reinforced*

European standard. EN 295 – *Vitrified clay pipes and fittings and pipe joints for drains and sewers.*

Krenchel, H. Shah, S.P. 1985. Fracture analysis of the pullout test. *Materials and Structures*. RILEM, vol. 18, No 108.

Krenchel, H. Bickley, J.A. 1987. Pull-out testing of concrete. *Historical background and scientific level today*. The Nordic Concrete Federation, No. 6.

Long, A.E. Murray, A.McC. 1984. The pull-off partially destructive test for concrete. *Spec. Publ. SP 82-17*. Detroit: American Concrete Institute: 327–350.

Madryas, C. Przybyła, B. Wysoki, L. 2010. *Testing sewers*. Wrocław: Lower Silesia Educational Publishing House.

Majewski, J. 2008. 3D Georadar vs pulse georadars. *Trenchless Engineering* 2/2008.

Moczko, A. 2004. Testing cored specimens in the light of current standard regulations – Part 2 – strength tests and interpretation of test results. *Construction, Technologies & Architecture* vol. 26 (2): 32–35.

Polish standard. 1988. PN-88/B-06250 – *Plain Concrete.*

Underground Infrastructure of Urban Areas 2 – Madryas, Nienartowicz & Szot (eds)
© 2012 Taylor & Francis Group, London, ISBN 978-0-415-68394-4

Selected exploitation problems of underground municipal utilities

H. Michalak
Warsaw University of Technology, Warsaw, Poland

ABSTRACT: This paper will cover the problems of the impact of damages to elements of the municipal infrastructure (district heating mains, water pipelines and the lining of collecting channels) on the technical condition and durability of existing buildings and those under construction in the direct neighbourhood. The examples will provide evidence of the necessity to check the technical condition of infrastructure components, maintenance and possible repair. Evidence of the necessity for the impact analysis of a new building undertaking on the existing technical infrastructure and development will be provided. The paper describes the problems of designing and constructing garages erected in dense urban conditions, in deep trenches with shielding made of monolithic diaphragm walls, propped, according to the principles of the roof method or rigid bracing. Studies on the impact (effects) of deep-embedded new buildings upon ground deformation in the nearby are have been carried out.

1 INTRODUCTION

The proper functioning of an underground infrastructure has a significant impact on the technical condition and usability of the nearby building projects. In some specific situations, a failure or disaster affecting the infrastructure may cause a failure or disaster in the buildings themselves. Failures of both the infrastructure and building projects may occur in the construction period and later on, even after many years of exploitation.

According to data for the year 2008 (Kwietniewski, 2011), in Poland, the mains water supply is more than 262 thousand km long and the sewerage system – approx. 95 thousand km. In the material structure of the mains water supply, pipes of grey cast iron, steel, PVC and PE prevail, whereas in the sewerage system – those of stoneware, concrete, reinforced concrete and PVC. In the large cities in Poland, the share of water-pipes which have been in use for over 50 years can be estimated at about 50% of the total network length, and those 25–50 years old at about 30–45%. The durability of steel pipes is estimated at around 50 years and those of grey cast iron at 75-80 years. It should be mentioned here that in towns and cities, sewers installed at the end of 19th and the beginning of 20th century are still functioning, which means that they have been in use for 100 years and more (Kwietniewski, 2011).

In this study, the problems of interactions between the urban underground infrastructure (collecting channels, heat distribution networks and mains water supply) and building projects situated in the direct vicinity, as well as the mutual impact of damage occurring to their technical condition and durability are considered.

2 HAZARDS RESULTING FROM THE OCCURRENCE OF EMERGENCY CONDITIONS IN THE UNDERGROUND INFRASTRUCTURE OR BUILDING PROJECTS

The most fundamental and frequent technical dangers related to the construction of new building projects in towns usually result from inadequate recognition of ground and water conditions of the ground base or from the improper interpretation of the same. The wrong recognition of the ground base and the assumption of overestimated values of the endurance parameters of the ground base result in underrating the value of earth pressure to the underground part of building projects, e.g. on the casing. In consequence, in the the design, insufficient endurance parameters, improper casing geometry, or insufficient anchoring of the casing beneath the bottom plate, etc. are adopted. The wrong recognition of the location of aquifers results in the possible occurrence of a local emergency condition of the excavation of the ground base or a loss of stability of the bottom and walls of the excavation. The improperly adopted technology for the construction of the underground part of a building project may lead to excessive horizontal dislocations of the excavation and the ground base behind it, a change in the condition (loosening or plasticizing) of the ground base behind the excavation, etc. Such situations may adversely affect the neighbouring infrastructure and they frequently lead to a disturbance in operation or to the occurrence of emergency conditions.

In respect to the underground infrastructure, two types of situations occur most frequently – damage to the casings of collecting channels or conduits. Damage to collecting channel casings may result in the penetration of ground water with ground particles into the inside of the channel and may lead to the loosening of the ground base behind the casing and, in consequence, to dislocation (settlement) of the ground surface together with the buildings erected. Damage to underground conduits usually causes a leakage of utilities (water, sewage) into the ground base and ground degradation with a range depending on the leakage scale. The effect is usually the subsidence of the ground base and building projects situated in the vicinity.

3 CASES OF THE IMPACT OF UNDERGROUND INFRASTRUCTURE DAMAGE ON NEIGHBOURING BUILDINGS

Two cases of the failure of a collective channel casing and a heat distribution network conduit and of the failure impact on neighbouring facilities (railway tracks, a housing building structure) will be presented, as well as the proposed methods of repair and removal of the dangers occurred.

3.1 *A collective channel*

Two tunnels with an inner diameter of 3200 mm with the mains water supply, 1200 mm in diameter (Fig. 1), are situated under the railway tracks. The tunnels were made with the mining method in the years 1973–1975. Their lengths, including the chambers situated at the ends, are, respectively, around 403 and 270 m. The provisional structure supporting the excavation is made of reinforced concrete frames spaced in 1.0-metre intervals. Each frame consists of a sill piece which is 200 mm wide and 160 mm high, two side elements, 220 mm high and 120 mm wide and a cross-bar with the same cross-section as the side elements. The reinforced concrete elements of the frames have frontal plates which make it possible to connect the elements by means of four bolts, 180 mm in diameter in each connection.

The excavation between the frames was supported by means of reinforced concrete staves driven in during the tunnelling. After completion of the excavation which was supported with a provisional casing, a permanent concrete casing was built. The concreting was done with sections about 2.5 m long, upon setting the formwork.

The thickness of the concrete casing between the provisional frames is about 400 mm. In the tunnel design, the application of class C12/15 concrete was provided for. The ground overlay ranges in case of the first tunnel from 3.70 to 11.80 m and in case of the second one – from 4.90 to 6.70 m.

At both ends of the tunnel are chambers with a reinforced concrete structure covered with reinforced concrete prefabricated slabs. The dimensions of the chamber enable the disassembly of the mains water supply and its reassembly with pipe sections 6.0 m long.

The mains water supply in both tunnels are supported on reinforced concrete frame supports. The shortest (horizontal) distance of the mains from the outside tunnel surface is 400 mm.

The technical condition of both tunnels was appraised in 2006, upon the previous cleaning of its inner surface with pressurised water. In the course of the cleaning, the coating resulting from flooding the tunnel by ground water, as well as larger efflorescences of calcium carbonate were removed from the concrete surface. The deposit gathered in the bottom part of the tunnel was removed.

Upon cleaning the tunnel, samples (boreholes) were collected for testing the concrete compression strength and sclerometric tests and stripping strength tests of the concrete were carried out. Also an inventory of tunnel damage (cracks) was made.

Figure 1. View of the inside of the tunnel.

Figure 2. Damage to the tunnels – circumferential cracks spaced at 2.45-metre intervals on working section joints.

Two main types of cracks (of a width of 0.3 to 5 mm) were found: circumferential cracks, spaced rather regularly in 2.45-metre intervals on the whole length of the tunnel (Fig. 2), i.e. on joints of concrete sections (working gaps), and longitudinal ones in the tunnel course (Fig. 3).

Ground water penetrated the tunnel through a considerable part of those cracks. The water inflow inside the tunnel was estimated at around 4 dm^3/(h · m) in the first tunnel and around 15 dm^3/(h · m) in the second tunnel.

In the tunnel, earlier seals of circumferential and longitudinal cracks were visible. However, the seals applied were not fully effective.

On the basis of the tests of concrete borehole samples from the tunnel casings it was found that the concrete compression strength is differentiated, but generally higher than in the design. It was assessed that the concrete inhomogeneity resulted from the manual technology of concrete mix laying, with a differentiated condensation degree, and the preparation of the mix at the construction site. It was assumed for the purpose of strength verification calculations that the concrete met the requirements for class C16/20 concrete.

Similar results as regards concrete strength were obtained in sclerometric tests.

The stripping strength tests with the "pull-off" method were made along the tunnels in selected places every 20 metres. It was found that the strength ranged from 0.8 to 3.5 MPa. The scatter of results proved to be high, which confirms the inhomogeneity of the concrete in the tunnels, similarly to the concrete compression strength tests made on the samples collected from the construction.

The verifying statistical calculations were carried out with the assumption (as confirmed in tests) that ground water is stabilised at approx. 0.5 m below the outer vault of the first tunnel and approx. 2.0 m above that vault in the second tunnel. Also the assumption was made, as confirmed

Figure 3.   Damage to the tunnels – longitudinal cracks in the key course.

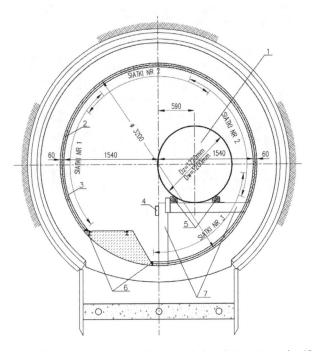

Figure 4.   Cross-section of the tunnel with the reinforcement designed: 1 – water mains 1200 mm in diameter, 2 – reinforcement fabrics, 3 – reinforcement of the gallery with a gunite layer, 4 – reinforced concrete board, 5 – oaken block, 6 – permanently plastic putty, 7 – prefabricated supporting structure spaced at 2.0-metre intervals (Szulborski et al. 2010).

by soundings results, that a vault in the ground had formed which reduced the tunnel load. The fact that no subsidence of railway tracks was found on the ground surface was acknowledged as confirmation of the existence of the vault.

Calculations of bending moments and longitudinal forces in the tunnel cross section were carried out in the same way as in the case of the structure built with the tunnel method, while using the formulas given in this work, as well as with the finite element method (Szulborski et al. 2007).

It was found on the basis of verifying calculation results and tunnel cracks recorded, that in order to meet the critical requirements for the usability condition, all the cracks which had occurred in the tunnel casing must be sealed (liquidated). A general opinion was expressed that tunnels exploited

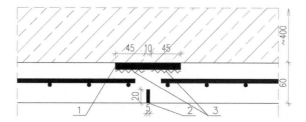

Figure 5. A detail of the gunite layer dilatation: 1 – dilatation rubber band, 2 – epoxy resin strewed with quartz sand, 3 – permanently plastic putty (Szulborski et al. 2010).

for more than 30 years in very difficult ground and weather conditions required major repairs, including the sealing and strengthening of the casing.

In the tunnel sealing and strengthening design, a proposal to conduct the works in three stages was included (Fig. 4).

The first stage provided for (Szulborski et al. 2010):

- pumping out water from the tunnels, then protecting the existing mains water supply 1200 mm in diameter, e.g. with polyethylene film 0.5 mm thick to protect its insulation coating against destruction,
- cleaning the inner surface of the tunnel, e.g. by means of hydro-blasting,
- removing, after the cleaning, the waste generated during the operation.

The second stage provided for sealing the cracks which had occurred in the tunnel casing. It was assumed that the sealing of the cracks should include:

- hammering out the crack and boring holes 14 mm in diameter alternately on both sides of the crack at an angle of 45° to the injected element surface, in the crack direction,
- mounting injection packers in the holes bored and stopping the cracks with rapid-hardening material; the distance between the packers and between the crack axis and a packer should be half the thickness of the element injected, however not more than 200 mm,
- making crack injections with appropriate sealing preparations.

In the third stage, prior to proceeding to gunite preparation, mounting the reinforcement delivered to the placement site in the form of welded reinforcement fabrics 2.0 m wide was planned, mounted on the inner circumference of the tunnels. The fabrics were designed made of rods 6 mm in diameter: circumferential fabrics spaced in 60-mm intervals and longitudinal ones spaced in 100-mm intervals, of class A-IIIN steel.

It was recommended that the fabrics should be mounted with a 25-mm distance maintained between the fabric reinforcement rods and the tunnel surface. A junction between the fabrics and the existing tunnel structure was designed made of rods 6 mm in diameter inserted at a depth of 100 mm. Around 10 such rods were to be applied on 1 m², placed along the circumference, 400 mm apart and along the tunnels – 240 mm apart. Fabric joints on the circumference were planed to be reinforced with lap joints 400 mm wide.

Gunite 60 mm thick was designed, made in two stages, in layers approx. 30 mm thick, with the use of off-the shelf gunite mix of specialist manufactures. Gunite of class C30/37 was adopted. Adding mix additions in order to reduce the contraction of the gunite prepared was recommended.

Dilatation of the gunite layer was provided for in 16-metre intervals (Fig. 5), i.e. every 8 widths of reinforcement fabrics. In view of the slight thickness of the layer, instead of typical dilatation inserts, bands of modified rubber were designed, 100 mm thick, glued to the concrete surface of the tunnel with epoxy resin. It was planned that the inner surface of type band sides would also be covered with epoxy resin to the width of about 45 mm, then strewed with quartz sand to ensure interaction between the band and the gunite. According to the design, on the middle surface of the band, 10 mm in width, a tape. e.g. of plastic was to be placed and removed prior to gunite application.

Figure 6.   Reinforcement fabrics mounted.

Figure 7.   Dilatation band of modified rubber.

Reinforced concrete chambers are situated at both ends of the tunnels. Cracks on the chamber walls were to be sealed with the use of the same solution as described above and applied for the tunnels.

In January 2008, the tunnel user ordered the repair and reinforcement of the tunnels. Works were performed first in the first tunnel and then, after pumping out water, in the second.

The works started with the careful washing of the tunnel walls with water under high pressure and the removal of all impurities. Then the leakages of ground water penetrating through the casing were sealed. First the leakage place (crack) was hammered, then holes were bored and injection packers were mounted. Cracks were stopped with rapid-hardening material. In the case of water conducting cracks, first a two-stage resin injection was carried out, in the case of dry or wet cracks, the injection was one-stage. On sealing the cracks, reinforcement fabrics were mounted (Fig. 6) and gunite layers were laid.

Off-the-shelf concrete mix was applied for the gunite fabrication (as applied in two stages with layers approx. 30 mm thick in each stage) and the gunite surface was left without troweling. The gunite layer was dilated in approx. 16.0-metre intervals, with the application of modified rubber bands, 100 mm wide (Fig. 7). To maintain the regular shape of the crack in gunite, after its application, a cut was made in the fresh gunite in the dilatation point (a gap 5 mm wide and 20 mm deep). Once the gunite was set, the gap was filled with permanently plastic putty.

The view of the tunnel inside during gunite application is presented in Fig. 8.

Repair works were conducted with the uninterrupted flow of water in the mains water supply and with uninterrupted railway traffic over the reinforced tunnels. They were completed in June 2008. The designed method and technology of the repair proved to be effective (Szulborski et al. 2007).

Figure 8. Gunite application.

The reason for the tunnel damage (cracks) and the water penetration inside the tunnels was the application of concrete structure for the casing (instead of a reinforced concrete one) and the manual technology of the concrete mix laying, as well as the lack of dilatation and dilatation inserts at single concrete section joints. An additional factor that furthered the occurrence of circumferential cracks was the phenomenon of concrete contraction and the transmission of vibration caused by the railway traffic on the tracks laid on the ground surface of the gallery structure.

Provision for the sealing and reinforcement of the tunnel structure were acknowledged to be indispensable for eliminating water leakages and ensuring parameters of the structure carrying capacity and usability. The method of the repair work execution proposed in the design was accepted by the the user of the tunnels.

The tunnel examples discussed demonstrate that in the case of structure design, structure exploitation conditions must be taken into consideration and solutions must be applied that are characterised by proper durability and reliability.

3.2 A dwelling of a traditional design

A dwelling of a traditional design was put to use in 1950. It has four storeys above-ground and a complete cellar. The load-bearing system is wall-based, longitudinal, with two-aisles. The building was erected directly on bricked strip foundations set 1.75 to 2.25 m below the ground.

The strip foundations were built of brick or crushed-stone concrete. The foundations of longitudinal load-bearing walls consist of old bricked strips of ceramic brick, full of earlier structures, built at a depth of 1.72 to 2.25 m below the ground, in fine sands with a compaction degree of $I_D = 0.55$.

The inter-storey ceilings are the Klein type, made without a horizontal coping and supported on steel beams, with a ceramic or crush-brick concrete slab. The ceiling under the flat roof has a light ceramic slab.

The building has been provided with vertical and horizontal damp-proofing insulation of the cellar walls. In the exploitation period, the building was equipped with hot and cold water, central heating, gas, electric and telephone installations.

In 2004, by reason of the sudden enlargement of the cracks in the walls and ceilings (Fig. 9), the building was completely emptied, the utilities were disconnected and provisional safeguards were introduced. The nature of the damage, its concentration in a single broken line starting at the ground floor indicated the cause, namely the uneven settlements of the ground base.

It follows from the analysis of the ground and water conditions in the area of the building that the ground base consists of sandy formations (fine and dusty sands) which are medium condensed directly in the foundation zone and condensed deeper on. Under half of the building, at a depth of around 3 m below the ground level, organic formations occur in the form of a layer with a thickness increasing from 0.1 m (in the hole situated near the middle of the building length, i.e. in the region

Figure 9.   Damage to the building.

of the building "break") to 1.0 m at the gable wall. Hence, the building foundations should be adjusted to the conditions resulting from the ground base structure and they should not have been built on the fragments of an earlier pre-war building and partially on crushed-brick concrete strips.

As for the trees growing at the said building, it should be mentioned that these are trees several score years old, with a developed root system, situated at a distance of about. 6 m from the outer walls of the building. The trees (poplars, maples and abeles) are from 1.0 to 0.6 m in diameter. The pavement around the abeles is locally uplifted, which indicates that the trees have a developed root system.

The applied design of the building – faulty but characteristic for buildings of the period after the second world war – is characterised by a low resistance of the structure to additional loads, including those caused by uneven subsidence of the ground base.

Taking into account the chronology of the phenomena considered, it has been assessed that the original cause of the damage to the building was the foundation and building structure design which was faulty from the point of view of the existing hydrological conditions. An example here is the lack of coping in planes of inter-storey ceilings. The adopted design of the building – resulting probably from the lack of a complete recognition of the ground base – did not assure a safe taking over by the building structural elements of additional loads coming from uneven subsidence and, in consequence, additional tensile stresses in the foundations with a bricked, not reinforced structure.

The development of the cracks (their propagation and growing opening width) was caused by underground infrastructure failures in the building vicinity, including those related to the sewerage system in 2002 and the heat distribution system in 2004. The outflowing water caused the loosening and plasticization of the ground base under a part of the building and, in consequence, a rapid increase of uneven subsidence of the ground base in the region. The consequence was the failure of the building covering a part of it. Subsidence and vertical deviations of the walls occurred in the corner area in the vicinity of the sewerage system and heat distribution system routes.

A significant cause of the initial damage to the building was a natural factor, namely the trees with significant annual growth, which were planted too densely and too close to the building (Jeż & Jeż, 2003). It brought about a change in the hydrological conditions of the ground base which under half of the building contains an inter-bedding with variable thickness in the form of organic ground susceptible to subsidence when the water table level changes.

Figure 10.   View of the excavation.

On the basis of results of "in situ" studies and analyses, the technical condition of the walls, ceilings, flat roof and staircases were deemed as unsatisfactory and that of the walls, ceilings and flat roof in the crack occurrence area as almost a failure.

It was indicated that bringing the building back to a usable condition was technically feasible, but required some reinforcing and repairing works which were complicated and expensive. The works should include, fro example: reinforcing the foundation zone of part of the building, e.g. through transferring loads to load-bearing layers of the ground base by means of micro-piles, repairing cracks in the wall structure, e.g. with the use of re-erection and other specialist technologies, as well as repairing ceilings, etc.

4   CASES OF THE IMPACT OF AN EMERGENCY CONDITION OF THE UNDERGROUND
    PART OF A BUILDING TO THE CONDITION OF THE NEIGHBOURING
    UNDERGROUND INFRASTRUCTURE

A public utility building with three underground storeys and a multi-storey overground part has been designed in a dense urban building development as neighbouring with an extended underground infrastructure network (Fig. 10).

Construction of the underground part of the building was planned in an excavation 15.3 m deep, protected with diaphragm walls 0.60 m thick. In static calculations, two phases of the diaphragm wall structure were practically analysed (the ordinate of the diaphragm wall top was adopted as 1.5 m, and that of the excavation bottom as 15.3 m below the surface of the ground):

– phase I: an excavation made to the depth of 9.6 m and further down, a wall support from the excavation side with a cut slope 5.7 m high, width at the base: 11 m and at the top: 7.5 m (slope gradient: approx. 58°),
– phase II: an excavation made to the depth of 15.3 m, a wall support with a horizontal counter-tie on the ordinate 9.30 or 10.40 m below the surface of the ground, supported on the load-bearing structure of the erected building (in the middle zone of the excavation).

The static work scheme for the supportive building around the excavation in phase I was designed in the form of a cantilever with a reach of about. 8–10 m (in the bottom part of the excavation, a cut slope was designed to ensure support for the diaphragm wall). In phase II of the construction, application was provided for of counter-ties and angle braces supporting diaphragm walls as resting on the completed middle part of the base slab and the structure of the underground part of the building with the cantilever, with the aforementioned reach, remaining.

Figure 11. Deformation of the ground adjacent to the cavity wall.

In static calculations for the supportive building around the excavation, calculation values of ground parameters were assumed that were much higher (approx. 2–3 times) than the ones determined on the basis of the initially made analysis of results obtained in geo-technical tests. On the assumption of these increased parameter values, in diaphragm wall calculations, earth pressure forces were obtained approx. 3 times lower than those resulting from taking into account the actual parameters characterising the grounds of the building ground base. On the basis of the static analysis, the largest horizontal dislocation (in the excavation direction) of the upper edge (crown) of the wall was obtained $u = 122.61$ mm. The dislocation proves that the applied method of wall support was erroneous. It follows from experience that such movements of the supportive building work around the excavation also result in a significant (ca. $\Delta s = 0.5u = 0.5 \cdot 122.6 = 61.3$ mm) deformation of the ground adjacent to the diaphragm wall and, in consequence, also in damage to buildings and technical infrastructure in the vicinity.

Measurements of the horizontal dislocations of the upper edges of the diaphragm wall, in the phase corresponding to digging the excavation to the full depth, indicated that the dislocations significantly increased, up to as much as 208 mm. Such dislocations resulted in an active pressure release, as well as in ground deformations behind the excavation and lead to damages in the water supply system in the street adjacent to the erected building. The water flowing out of the installation worsened the strength properties of the ground base and, in consequence, disaster of the supportive building work around the excavation – a breakage at the depth of 14 m below the surface of the ground and an overturn of the diaphragm wall (Fig. 11). Steel counter-ties and fragments of the ceiling and pole structure of the underground part of the erected building as well as the surface of the street and installations in the landslip area. The disaster developed gradually – it lasted about half an hour, which enabled the evacuation of construction site workers from the danger zone.

In the light of the analyses conducted, it was found that the primary (original) cause of the disaster was the adoption of a changed design of much overstated parameters characterising the base ground, and in particular, a very high value for the ground cohesion; as a result significantly decreased (approx. 3 times) earth pressure forces were obtained, and therefore also the internal forces in the diaphragm wall and in the strut structure. There is a very high probability that due to excessive deformation of the diaphragm wall and then of the ground behind the wall from the eastern side, a water conduit was damaged. That fact might have additionally adversely affected the ground condition and worsened its strength parameters and may therefore have had an effect on releasing a greater earth pressure on the diaphragm wall and the straining structure.

Immediately after the disaster, the nearby utility lines were put out of operation and at the same time, protective works were commenced (reinforcing the buttresses, applying additional flying shores supporting diaphragm walls and straining tubes between the north and south one walls).

According to recommendations, the ground base under the foundations of the building situated in the landslip were reinforced with jet-grouting columns approx. 1.00 m in diameter to the depth of 9.0 m below the bottom of the foundation of the building (11.0-m below the surface of the earth). The depth of the injections resulted out of the necessity to "cut' with columns of the supposed slide curve which could have occurred at the depth of about 9.0 m, i.e. at the border between clays and loams, as well as the necessity to transfer the loads from the building to the loam layer. In case the foundation reinforcement should prove to be insufficient, a system of steel tie-beams was designed to brace the whole building in several planes.

The new excavation securing wall was designed as a palisade of jet-grouting columns reinforced with steel profiles. For the whole period of the works, precise geodesic measurements of the deformations of diaphragm walls and the subsidence of the surrounding buildings were carried out. The construction of the underground part and the rest of the building has been completed (Szulborski & Pyrak, 1998).

5 SUMMARY

In view of the frequently inadequate technical condition of the much exploited underground infrastructure of towns and cities, as well as the frequent construction of deeply founded buildings with many underground storeys, close cooperation is necessary between representatives of the installation sector, designers and geo-technicians in designing and preparing such investment projects. The presented situations indicate the need for taking into consideration the problems of interactions of between the urban underground infrastructure and buildings situated in the direct vicinity, as well as of the mutual impact of the damage on their technical condition and durability. The safe usage of building projects depends frequently to a large extent on the technical condition of the underground infrastructure situated in the direct vicinity.

REFERENCES

Jeż, J. & Jeż, T. 2003. Niespokojne wzgórze w Poznaniu. *Conference papers XXI Konferencja Naukowo-Techniczna „Awarie Budowlane 2003", Międzyzdroje, may 2003.*

Kania, M., Niedzielski, A. & Duda, A. 2009. Katastrofa kolektora sanitarnego spowodowana osuwiskiem podczas robót ziemnych. *Conference papers XXIV Konferencja Naukowo-Techniczna „Awarie Budowlane 2009", Międzyzdroje, may 2009.*

Kotlicki, W. & Wysokiński, L. 2002. Ochrona zabudowy w sąsiedztwie głębokich wykopów. *ITB, praca nr 376/2002. Warszawa.*

Kwietniewski, M. 2011. Awaryjność infrastruktury wodociągowej i kanalizacyjnej w świetle badań eksploatacyjnych. *Conference papers XXV Konferencja Naukowo-Techniczna „Awarie Budowlane 2011", Międzyzdroje, may 2011.*

Lessaer, S. 1979. Miejskie tunele, przejścia podziemne i kolektory. Wydawnictwa Komunikacji i Łączności, Warszawa 1979.

Madryas, C., Kolonko, A., Szot, A. & Wysocki, L. 2006. Mikrotunelowanie. *Dolnośląskie Wydawnictwo Edukacyjne, Wrocław.*

Michalak, H. 2006. Kształtowanie konstrukcyjno-przestrzenne garaży podziemnych na terenach silnie zurbanizowanych. *Prace naukowe – seria architektura, vol. 2. Oficyna Wydawnicza Politechniki Warszawskiej, Warszawa.*

Michalak, H. 2009. Garaże wielostanowiskowe. Projektowanie i realizacja. Arkady, Warszawa.

Michalak, H., Pęski, S., Pyrak, S. & Szulborski, K. 1998. O diagnostyce zabudowy usytuowanej w sąsiedztwie wykopów głębokich. „Inżynieria i Budownictwo", nr 6/1998.

Michalak, H., Pęski, S., Pyrak, S. & Szulborski, K. 1998. O wpływie wykonywania wykopów głębokich na zabudowę sąsiednią. *„Inżynieria i Budownictwo", nr 1/1998.*

Przystański, J. 1990. Posadowienie budowli na gruntach ekspansywnych. *Instrukcja ITB nr 296, Warszawa.*

Szulborski, K. & Pyrak, S. 1998. O katastrofie obudowy wykopu głębokiego pod budynek przy ul. Puławskiej w Warszawie. *„Inżynieria i Budownictwo", nr 12/1998.*

Szulborski, K., Michalak, H., Pęski, S., Pyrak, S. & Przybysz, P. 2007. Degradacja, uszczelnienie i wzmocnienie betonowej obudowy podziemnych tuneli zbiorczych. *Conference papers XXIII konferencja naukowo-techniczna „Awarie budowlane", Szczecin – Międzyzdroje, may 2007.*

Szulborski, K., Michalak, H., Pęski, S., Przybysz, P. & Pyrak, S. 2010. Uszczelnienie i wzmocnienie betonowej obudowy podziemnych tuneli zbiorczych. *„Inżynieria i Budownictwo", nr 4/2010.*

Tarnawski, M. 2011. Geotechniczne przyczyny awarii budowlanych. *Wydawca Przedsiębiorstwo Produkcyjno-Handlowe ZAPOL Dmochowski, Sobczyk Spółka Jawna, Szczecin.*

Underground Infrastructure of Urban Areas 2 – Madryas, Nienartowicz & Szot (eds)
© 2012 Taylor & Francis Group, London, ISBN 978-0-415-68394-4

Monitoring of structures adjacent to deep excavations

Z. Muszyński, J. Rybak & A. Szot
Wroclaw University of Technology, Wroclaw, Poland

ABSTRACT: Geodetic survey constitutes the main source of information on the behavior of the soil and the structures in the vicinity of deep excavations. That survey makes it possible to determine the absolute values of vertical and horizontal displacement and deformation of a structure under scrutiny against an external system of reference. That system of reference is defined by the network of reference points which should maintain their fixed position in time. The article presents modern techniques and technologies providing valuable information about displacements which occur in the course of geotechnical works (excavations).

1 INTRODUCTION

The structure under observation is represented by a group of control points stabilized at the key places on the structure. The position of the control points changes together with the alteration of the position or shape of the structure under survey. The control points, together with the reference points, constitute a control-measurement network, which is subject to systematic geodetic survey. The first measurement in the network is called the initial (zero) survey, and the subsequent measurements are called periodic (cyclic) surveys. The measurement data are further corrected in relation to the impact of the environment. Their quality is also checked, the supernumerary observations are averaged and the outliers are eliminated. Subsequently, the fixed reference points are identified. The detection of the potential displacement of those reference points is crucial for the proper calculation of the coordinates of the control points. Those coordinates are calculated by means of the least square method in relation to the reference points which have been determined as fixed. Once the coordinates of the control points in a current survey are known, they are compared with the coordinates calculated on the basis of the initial survey and the previous cyclic survey. The values of short- and long-term displacement are thus determined. The next stage entails geodetic description of the structure's deformation (Prószynski & Kwasniak 2006). The calculated vectors are reduced by the impact of the structure's translation and rotation. While determining the structure's displacement parameters, it is checked if the structure behaves like a rigid body or if it is subject to permanent deformation. An overview below presents the fundamental surveying methods used for measuring the displacement of structures and the soil in the vicinity of deep excavations.

2 GEODETIC METHODS OF MEASURING DISPLACEMENT AND DEFORMATION

2.1 *Methods of vertical displacement determination*

2.1.1 *High-precision geometric levelling*

The primary method of surveying vertical displacement of structures and the area surrounding a deep excavation is high-precision geometric levelling. In the method of geometric levelling, the axis of sighting of the levelling instrument indicates in space a geometric level which is locally perpendicular to the direction of the gravity force. By means of placing the levelling staffs vertically at the points under survey, it is possible to read the distance of those points from the level of the geometric levelling instrument. For a measured pair of points, the difference of those distances depicts the difference in those points' altitude, i.e. the so-called height difference. By connecting the subsequent measurement stations, it is possible to determine the altitude of control points on the basis of the altitude of the reference benchmarks. High-precision geometric levelling uses the instruments and techniques that ensure the accuracy of $0.1 \div 0.3$ mm. Nowadays, two types of high precision levelling instruments are used: the optical and the digital ones. The most popular optical levels made by Carl Zeiss Inc. are the models: Ni 007 (with the precision of ± 0.7 mm per 1 km of double high-precision levelling), Ni 005 (± 0.5 mm/1 km), and Ni 002 (± 0.2 mm/1 km). The readouts are carried out on the levelling staff with an invar band. Digital (electronic) levels perform the readouts automatically via the analysis of the image of the invar bar coded rod (a staff into which an invar bar code scale is built). The sample digital levels are the following models: DNA 03 manufactured by Leica Geosystems (± 0.3 mm/1 km), DiNi 0.3 – by Trimble (± 0.3 mm/1 km), and DL-101C – by Topcon (± 0.4 mm/1 km).

In designing a control-measurement network it is crucial to assess properly the zone of the deep excavation impact. The greater the distance from the construction site is, the greater are the chances that the altitude of the control points will be determined correctly. At the same time, when the length of the level circuit increases, the precision of the determination of the control points' altitude declines. The value of the expected vertical displacement of the surface of ground in the vicinity of a deep excavation depends on manifold factors. The key factors are: local soil conditions, lowering of the water table, the construction technology and the schedule of works related to the deepening of the excavation, the construction of the underground and ground-based part of the structure. The work (Michalak 2007) presents an overview of the indicators describing the extent of an excavation's impact, the value of the expected vertical displacement of the ground and the horizontal displacement of the excavation support, depending on different soil types. The suggested zone of impact is in the range of 1.5 to 4 times the excavation depth, however, the largest vertical displacement occurs in the zone of 0.5 to 0.75-tuple depth. In the aforementioned study it was proved that the vertical displacement of the ground directly behind the excavation support constitutes 50-70% of the value of the expected horizontal displacement of the excavation. Usually one should expect with the displacement of about 0.25% of the excavation depth and the settlement of about 0.20% of the excavation depth (Rychlewski 2006). That fact is significant for further prognosis of the behavior of structures neighboring the construction site.

In selection of the location of reference benchmarks it is vital to take into account the geotechnical conditions: the soil type and the information of the water table alterations. That refers to, in particular, long-term investigation of vertical displacement. The study (Wolski & Kwiecien 2006) provides the examples of the situations when persistent drought periods or heavy rainfall have contributed to the loss of stability of the reference points. Lowering of the water table caused by the construction works or leaking sewage system which washes out soil particles may cause even the distant buildings to settle and thus they cannot serve as the location of the reference points. It is also crucial to select properly the time intervals at which periodic measurements are taken.

2.1.2 *High-precision trigonometric levelling*

Another method of determining the vertical displacement of control points is high-precision trigonometric levelling. The method makes it possible to measure the points located in inaccessible places. The points under measurement are indicated by special targets. Those control points are observed from special stations (concrete pillars enabling the forced centering of the instrument), at the distance of no more than 100 m. The measurement of vertical angles is taken synchronously from a couple of stations, using electronic total stations with the accuracy of the vertical angle measurement of 2cc. The readouts are repeated in several measurement series. The distance from the surveyed check point is measured with the help of electronic distance meter (EDM) or it is determined on the basis of the intersection. The obtained accuracy of vertical displacement ranges between $\pm 0.5 \div 0.8$ mm for a distance below 50 m, and $\pm 0.8 \div 1.5$ mm – for a distance up to 100 m (Brys 1996). The decreasing precision of measurement in the case of longer sight lines is caused, in the first place, by the phenomenon of refraction.

2.1.3 *High-precision hydrostatic levelling*

The third method of determining vertical displacement is high-precision hydrostatic levelling. Unlike the two methods described above, it consists in the determination of relative height difference using communicated vessels (Żurowski & Kuralowicz 1999). Hydrostatic level consists of two glass cylinders connected by means of a rubber hose ($20 \div 40$-meter long) and filled with water (Brys 1996). The height difference of the measured benchmarks cannot exceed 30 cm. The

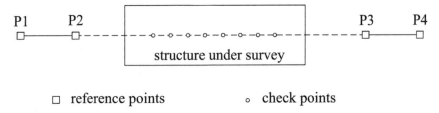

P1   P2   P3   P4

structure under survey

□ reference points   ○ check points

Figure 1.   Alignment method scheme

measurement of the water level in the cylinder is taken by means of lowering a special spike fastened to the micrometer scale. When the spike touches the water surface, an electric bulb goes on. The readouts are made for the two cylinders simultaneously, repeated four times and averaged to the value of 0.01 mm. Then, the whole measurement is iterated after the cylinders' places have been exchanged. The most modern hydrostatic levels use silicone oil instead of water, whereas the level of the liquid is established by three independent methods, i.e. using ultrasonic, dynamometric and microwave sensors. It is possible to automate the whole measurement process and to transmit remotely the measurement data. Hydrostatic levelling is at present the most precise method of relative displacement determination in the engineering geodesy (Brys 1996). It may be used in hardly accessible places where no other methods are effective. In order to obtain the absolute values of the displacement, one end of the measured section should be related, by means of geometric or trigonometric levelling, to a reference benchmark.

2.2   Methods of horizontal displacement determination

2.2.1   Alignment method
One of the simplest ways of determining the horizontal displacement of an excavation support is the alignment method, also known as constant straight line method. By means of that method it is possible to determine horizontal displacement in the direction that is perpendicular to a vertical reference plane. The system of reference is established by four points joined with a straight line, two at each side of the structure under survey, outside the zone of deformation (Fig. 1).

Whenever possible, the location of the points is selected in such a way so that the control points are situated along the reference plane, at a short distance (up to several centimeters) and at a similar height. The instrument is placed on point P2, via forced centering, and the target – on point P3. The surveying is carried out by means of special stable theodolites with telescopes with large magnification, called alignment telescopes, or high-precision electronic total stations. The measurements are taken in two positions of the telescope, sometimes in several measurement series. Mutual fixity of the points of reference is controlled by analyzing their horizontal displacement and by checking the horizontal angles in relation to the established additional reference points.

Depending on the method of horizontal displacement measurement at the control points, there are the following varieties of the method in question (Brys 1996):

– Geometric-alignment method, in which the vertical reference plane is determined by the target axis of a theodolite or an alignment telescope placed in point P2 and oriented to point P3. In each of the control points a special target is placed, which is shifted by means of a micrometric screw to fit into the target line (Fig. 2). The values of horizontal displacement in the crosswise direction to the reference line are calculated by means of a comparison of the readouts from the micrometric screw that come from different survey periods. Sometimes immobile targets are used and the value of the transverse displacement is read out from an alignment telescope with an optical micrometer.
– Trigonometric-alignment method, in which a high-precision total station is sighted on an immobile target (situated in the control point) and the parallactic angle is measured.
– Electromechanical alignment method in which the reference line is established by means of a metal string stretched between points P2 and P3. In each of the control points, a special device is fixed, equipped with a container filled with liquid. In the container there is a float attached to the metal string. The displacement of the control point results in the change of the position of the

a)   b)

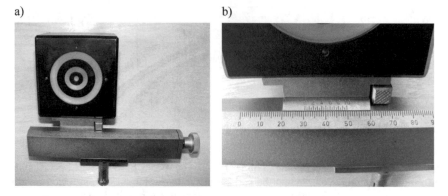

Figure 2. Mobile target (a) with a scale and a nonius (b)

float in the container, which is recorded by electric sensors and may be transmitted telemetrically to the surveying station.
– Laser-alignment method, in which a laser beam stands for the reference line. The target placed at the control point may be manually introduced into the laser beam by means of a micrometric screw and then the displacement value is read out of the micrometer. Another option is an immobile target equipped with a photodetector, which records the position of the laser spot in relation to the center of the target.

2.2.2 *Trigonometric method*

Trigonometric method consists in taking manifold measurements, at specific time intervals, of horizontal angles and distances, aiming at the control points from the fixed reference points located outside the zone of impact of the deep excavation. The surveying is carried out with the help of total stations with high accuracy of angle and distance measurement. The ordinates of the control points are calculated by means of manifold computations of angular, angle-linear and linear intersections. The differences of the coordinates of the points under investigation, calculated from the particular measurement cycles, make it possible to calculate the displacement of those points.

The trigonometric method enables the measurement of deep excavation displacement at manifold levels of the control points. It must be, however, remembered that the initial measurement of the subsequent lower levels is postponed in relation to the measurement of the highest control points' level. That is caused by the progress of the construction works. In order to determine the displacement of the lower parts of the excavation support before uncovering them in the earth and fixing the control points, it is recommended to combine geodetic (absolute) methods with the relative methods. The primary technique in that case is the inclinometric survey. When the displacements of the inclinometer tubes are measured using geodetic methods, relative inclinometric survey of the displacement may take place within the assumed system of reference.

2.3 *Structural monitoring*

In recent years, the investigation of displacement has employed with a growing frequency the so called structural monitoring. That is a modern system combining different absolute measurements of geodetic survey with the relative measurement of displacement, as well as with various types of sensors recording the behavior of the structure under scrutiny. Structural monitoring makes it possible to record automatically the state of the structure, cyclically at predefined time intervals, for example every 10 minutes. Thus collected data are gathered, corrected in relation to the influence of the survey environment and calculated in the appropriate manner. The obtained values of displacement are compared with the permissible values, visualized and sent remotely to relevant supervisory bodies. The system ensures a round-the-clock control of the structure under scrutiny in the real time, together with the possibility of automatic alarm signaling. The main component of the system are the motorized electronic total stations, which offer the measurement of the angles with the mean error that does not exceed ±1". For standard distances smaller than 1000 m, the mean error of the distance measurement ranges from ±0.5 mm to ±2 mm. These total stations have the function of the automatic target detection; the target is indicated by means of a geodetic mirror and located of at the distance of no more than 3000 m. That function consists in scanning the area around the mirror, further analysis of the strength of the reflected signal and the automatic aiming of the telescope (by means of servo-motors) at the point of the strongest reflection (the mirror's center), and then taking the appropriate measurement. The total stations are usually placed at the fixed reference benchmarks, in special glazed containers equipped with air conditioning and the source of power supply. Those containers protect the instrument from the changing weather conditions and enable its continuous operation (Karsznia 2008). In order to ensure the control of the stability of the total stations position, it is subject to angular-linear observation in relation to other points of reference. The second component of the structural monitoring system are the satellite GNSS receivers operating in different modes: the static or the real-time kinematic (RTK) mode. The receivers which operate in the static mode are placed at the reference points in order to control the stability of their position. Those points may be, at the same time, equipped with geodetic mirrors and thus they may be related to the stations of the total stations (Karsznia et al. 2010). If the stability of the tacheometric stations cannot be ensured, they are integrated with the GNSS receivers in order to record their position (Van Cranenbroeck & Brown 2004). The receivers operating in the RTK mode are also installed at the key control points, where they enable the recording of the dynamic changes of the structure even at the intervals smaller than one second (Wan Aziz et al. 2005). The remaining measurement devices used in structural monitoring are all sorts of sensors that record significant phenomena from the point of view of the structure's safety, e.g. standard inclinometers, high-precision electronic inclinometers, piezometers, meteorological stations that register the temperature, humidity and atmospheric pressure, etc. The system also requires the appliances and software that ensure the undisturbed transmission, gathering, processing, analysis and visualization of data.

2.4 *Ground-based laser scanning*

The ground-based laser scanning, used in the survey of displacement and deformation, has recently become an increasingly frequent and rapidly evolving technology which makes it possible to survey a great number of points in an instant. For example, the laser scanner ScanStation C10 (Fig. 4a), manufactured by Leica Geosystems, is able to survey up to 50,000 points per second. A horizontally rotating prism sends the impulses of a laser beam which is reflected from the structure under scrutiny. The scanner rotates simultaneously at a predetermined leap in the horizontal plane. For each point under survey the reflectorless measurement of the distance (up to 300 m) is taken, and the horizontal and vertical angles are recorded. The accuracy of the measurement of a section of no more than 50 m amounts to 4 mm. The maximum visual field of the scanner equals 360° horizontally and 270° vertically. The accuracy of the angle measurement amounts to 60 microradians. The accuracy of the determination of the points located at the distance of no more than 50 m equals 6 mm. The density (the number) of the points under survey is defined by way of determining the horizontal and vertical dimensions of the mesh of rectangles which the points under survey constitute on the predefined surface. The minimum distance between the points is smaller than 1 mm. The example of a scanned fragment of a structure with maximum resolution is shown in Figure 4b. The scanner is equipped with a high-resolution digital camera with an automatic zoom, which takes the photos of the structures under survey and thus enables the archiving of their condition and their realistic visualization. In the case of large structures, the separate scanner stations are connected with one another by way of scanning the linking points indicated by special targets. The linking points are the reference points, if the range enables that. In the case of the deep excavations with the reference points situated at a considerable distance, a control-measurement network may be set up. The main components of that network, surveyed by means of standard geodetic methods, are: the planned scanner station, the linking points and the reference points. Example of using the laser scanner to survey displacements is described in the article (Gordon et al. 2001). Examples of other models of laser scanners is shown in Figure 3.

Despite its relatively small accuracy of determining the position of a single point, the laser scanning has a great advantage. In a short time we are able to obtain the measurement of the position of millions of control points which form a fully metric, three dimensional model of

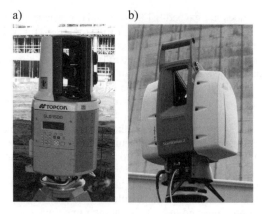

Figure 3.   Laser scanners: GLS-1500 (a) manufactured by Topcon and ScanStation 2 (b) manufactured by Leica Geosystems.

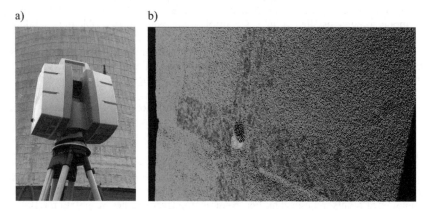

Figure 4.   Laser scanner ScanStation C10 manufactured by Leica Geosystems (a) and example of a scanned fragment of cooling tower with photopoint (b).

a deep excavation, its support and the neighboring structures. The model additionally includes photographic documentation. The technology is remote and does not require contact with the structure under survey. It makes it possible to register the condition of the whole structure and not only of some selected characteristic points. The laser scanner may also be integrated with the satellite GNSS receiver.

3   CONCLUSION AND FINAL REMARKS

Although the monitoring programs may be in practice diversified, complex and perfectly adjusted to even very demanding construction technologies, the difficulty lies in the assessment of the proportions between the real needs, the proper monitoring scope and the financial resources of the investor, who, in the end, pays for such investigation. It is however worth emphasizing that the costs of such examination are much lower than the potential damages that may be thus avoided. Passive behavior on the part of the investors and/or contractors leads to downtime and additional cost of technology changes.

ACKNOWLEDGMENTS

The authors would like to thank Mr. Waldemar Kubisz from Leica Geosystems for providing the scanner ScanStation C10 to be tested.

REFERENCES

Bryś, H. 1996. *Geodezyjne pomiary odksztalcen i przemieszczen zapór wodnych.* Kraków: Wydawnictwo Politechniki Krakowskiej.

Gordon, S., Lichti, D. & Stewart M. 2001. Application of a high-resolution, ground-based laser scanner for deformation measurements. *The 10th FIG International Symposium on Deformation Measurements 19–22 March 2001.* Orange, California, USA. 23–32.

Karsznia, K. 2008. Geodezyjny i geotechniczny monitoring obiektów inżynierskich w ujęciu dynamicznym. Wykrywanie slabych punktów. *Nowoczesne Budownictwo Inzynieryjne* <http://www.nbi.com.pl/assets/NBI-pdf/2008/4_19_2008/pdf/23_wykrywanie_slabych_punktow.pdf>

Karsznia, K., Czarnecki, L. & Stawowy, L. 2010. System ciągłego monitoringu przemieszczeń i deformacji wyrobisk górniczych w PGE KWB Bełchatów S.A. Aspekt funkcjonalny i dokładnościowy. *Górnictwo i Geoinżynieria* 34(4): 279–288.

Michalak, H. 2007. Wplyw realizacji obiektów głęboko posadowionych na przemieszczenia podłoża gruntowego. *Inżynier Budownictwa,* <http://www.inzynierbudownictwa.pl/drukuj,159>

Prószynski, W. & Kwaśniak, M. 2006. *Podstawy geodezyjnego wyznaczania przemieszczeń. Pojęcia i elementy metodyki.* Warszawa: Wydawnictwo Politechniki Warszawskiej.

Rychlewski, P. 2006. Głębokie wykopy w zabudowie miejskiej. *Geoinżynieria drogi mosty tunele.* 10(03): 44–52.

Van Cranenbroeck, J. & Brown, N. 2004. Networking Motorized Total Stations and GPS Receivers for Deformation Measurements. *FIG Working Week May 22–27, 2004.* Athens, Greece.

Wan Aziz, W.A., Zulkarnaini M.A. & Shu K.K. 2005. The Deformation Study of High Building Using RTK-GPS: A First Experience in Malaysia. *FIG Working Week 2005 and GSDI-8, April 16–21, 2005.* Cairo, Egypt.

Wolski, B. & Kwiecień, A. 2006. Monitoring uszkodzonych konstrukcji budowli zabytkowych. *Inżynieryjne problemy odnowy staromiejskich zespołów zabytkowych; VII Konferencja Naukowo-Techniczna „REW-INZ'2006" 31 maja – 2 czerwca 2006.* Kraków. 189–198.

Żurowski, A. & Kurałowicz, Z. 1999. Zastosowanie geodezyjnych metod badawczych w budownictwie. *Inzynieria i Budownictwo,* 55(12): 686–690.

Underground Infrastructure of Urban Areas 2 – Madryas, Nienartowicz & Szot (eds)
© 2012 Taylor & Francis Group, London, ISBN 978-0-415-68394-4

Rehabilitation of sewer channels. Investigations of load capacity of channels renovated with GRP liners

B. Nienartowicz
Wrocław University of Technology, Wrocław, Poland

ABSTRACT: In the paper, the problem associated with dimensioning of linings applied for pipeline renovation was elaborated. As it is indicated by the results of numerous investigations, the linings are very frequently overdimensioned due to simplifications applied in the assumed structure operation models. In the paper, laboratory investigations were described in detail, the objective of which is the verification of the assumptions accepted for creation of the guidelines applied presently for dimensioning of this type of structures.

1 INTRODUCTION

Sewerage networks are a very important component of the underground infrastructure of towns. To ensure the comfort of use of the networks, satisfactory for the town residents, it is necessary to perform regularly the control of their technical condition and hydraulic properties. This gives also a possibility of an economically effective network management. Making regular local repairs and routine cleaning of the network contributes to extending of its life. Unfortunately, in many (Polish) towns, the actions are often neglected. Many times, the sewerage conduits feature with significantly decreased hydraulic efficiency due to deposits gathered in them. Decidedly too often, one can hear about cases that, due to intensive precipitations, rain ducts are not capable of receiving precipitation water. This results in flooding streets, premises and cellars of buildings. Lately, the European climate has been changing and, more and more often, high-intensity precipitations happen that not more than a dozen or so years ago were happened once for every few years. Therefore, so important is to maintain the sewerage networks at a high hydraulic efficiency. However, to correct operation of the waste water discharge systems, other aspects contribute also; besides those referring to the hydraulics, such as the constructional requirements, the requirements referring to the environment protection or the requirements ensuring safe operation (Kolonko et al. 2011). Therefore, it is very important that the network maintenance is not limited only to making repairs of failures endangering collapsing of the structure. Performing complex and, at the same time, regular supervision of networks and carrying out current maintenance actions contributes to extension of their life and enables effective planning of costs connected with their maintenance.

The majority of (Polish) pipelines that need renovation now was built several dozen years ago as reinforced concrete or bricked facilities. When years passed, their technical condition deteriorated considerably due to the progressing corrosion as well as numerous structural damages caused, among other things, by the increased usable load connected with the constantly developing transportation management. Presently, a significant part of them requires complete reconstruction. The damages occurring in the pipelines are of a significant influence not only on functioning of the network itself but also on the state of the ground/water environment. It is since leakages that come into existence in the system may lead to occurrence of ex-filtration of pollution into the natural environment or to infiltration of ground water into the network depending on the ground water level. Both phenomena are equally hazardous and result in occurrences of both ecological as well as economic losses.

Neglecting to perform the actions connected with maintenance of the pipeline network leads ultimately to the necessity of the complete renovation of the pipelines. Due to occurred damages, the supporting structure of a conduit loses the capability of transferring loads and, to avoid collapse

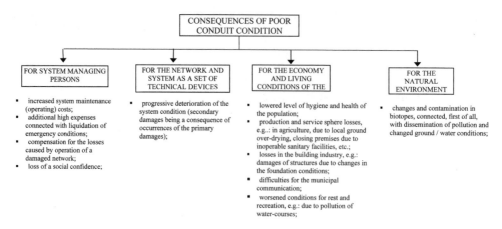

Figure 1. Consequences of poor condition of sewerage conduits (Przybyła 1999).

of the structure, it becomes necessary to reinforce it or to replace it completely. The consequences of the poor condition of the conduits are presented in Figure 1.

At present, non-excavation technologies are commonly applied and even though they often remain a solution more expensive than the excavation methods but taking into consideration the benefits they bring, makes them more and more popular solution, especially in the centres of big towns. The economic, social, ecological, technical and, also, legal considerations for the decision of application of such technology are not without significance. In the cases of necessity to renovate the conduits located at difficult ground conditions, at large depths or in a close neighbourhood of other underground infrastructure networks, non-excavation renovation may prove to be, however, more economically profitable. Considering, also, the social aspect, the choice of such solution seems to be reasonable, since, nowadays, the limitation to the possible minimum of the necessary difficulties in the pedestrian and automobile traffic is highly desirable. Carrying out work in excavations is, also, of ecological impacts: the necessity of lowering the ground water table, the hazard or ground water pollution or the necessity of temporary demolition of the green areas are undesired interference into the natural environment (Kolonko et al. 2011). The renovations carried out by non-excavation methods are limited to placing lining reinforcing the conduit structure within the existing pipeline. In accordance with (PN-EN 13689:2004), six basic lining sorts are distinguished: of continuous pipes, of tightly fitted pipes, of pipes hardened on the spot, of segment pipes, of inserted sleeves and of pipes coiled helically. The decision of selection of the specific technology must take into consideration the following criteria (Kolonko et al. 2011):

- The technical criteria (technical condition of the facility, structural material, foundation depth)
- The economic criteria (foundation depth, ground/water conditions, costs associated with by-pass road organisation)
- The ecological criteria (lowered ground water level, protection against pollution of water, protection of greenery)
- The social criteria (difficulties in pedestrian and automobile traffic, organisation of by-pass roads)
- The operational criteria (various sorts of forecasts of impact on the use of the facilities being renovated).

The analysis of the above mentioned criteria and the knowledge of the renovation technology are the basis for selection of the adequate renovation method

2 LINING STRUCTURE DIMENSIONING

After making the choice of the renovation method it is necessary to realise the structural design. The scope of the stage includes, among other things, determination of the geometric characteristics of the lining, determination of the loads acting on the lining, as well as verification of the strength conditions. The ultimate result of the designing process is the determination of the lining thickness. At present, many types of guidelines for designing this type of a structure are used. On their basis, computer programs are created, used by engineers around the world. The computational algorithms ATV-DVWK-M127P (Germany), ASTM F 1216 (USA), WRc SRM (United Kingdom), RERAU (France) are only some of those applied currently. In order to make the computations, it is also possible to take advantage of computer programs making use of the finite element method. Such a solution is applied mainly when designing non-standard cross-section profiles. The method enables very precise matching of actual operating conditions of the structure and assuming that an experienced engineer makes use of it, it provides the most precise results.

Though each of the generally available guidelines is correct, the selection of the computation method influences the final result. The differences in the assumptions accepted in individual guidelines are of a significant impact upon the result of computations in the form of the lining thickness. Admittedly, acceptance of an increased lining thickness for realisation is of no adverse impact on the structure safety, nevertheless, the consequences of this are perceptible both in the economic scope of the entire realisation (increased lining manufacturing costs, its transportation and mounting costs) and in the scope of the workmanship (difficulties connected with the increased weight of the lining).

Depending on the computation method selected the divergences in the result, as shown by the results of the analyses done a few years ago (Thèpot 2004), are considerable and may reach, at extreme cases, even as much as several dozen percent.

The divergences follow, among other things, from the differences within the assumptions accepted for creation of the above mentioned computational algorithms. The determination of the technical condition of the existing duct, the determination of the load values acting on the lining, the way of supporting of the lining in the conduit, taking into consideration of matching of the mounted lining and the existing conduit, the impact of the conduit ovalisation on operation of the whole structure, taking into consideration the possibility of occurrence of local deformation or the way of ring space filling are barely some selected problems that are defined in a various way in individual guidelines for designing. Presently, in many research centres, it is undertaken to take research investigations and analyses in order to verify the assumptions accepted in the guidelines used, in relation to the actual operation of the structure located in the soil. This is aimed towards eagerness to improvement of the guidelines used presently, so that the investment projects conducted within this scope might be optimised both in the domain of the economy and in the widely understood ecology as well.

3 LABORATORY INVESTIGATIONS

During the last years, in various research centres, investigations of operation of pipelines subject to renovations have been carried out. In the Hanover University, there were carried out the laboratory investigations of the pipeline samples renovated with tightly fitted linings, at the scale 1:1 (Doll 2010). Their purpose was to determine the actual load carrying capability of the renovated pipeline being in the II technical condition. In 2008, in the laboratory facility of the Wrocław University of Technology, there were carried out the investigations of the load carrying capacity of the pipelines being in the II technical condition, renovated with the Trolining lining, and the investigations of the load carrying capacity of the lining itself, of the circular and egg-shaped profile. The investigations were carried out on the pipeline samples at the scale 1:1 located in the soil environment. Their purpose was to determine the actual load carrying capacity of the pipelines and to state the influence of the thickness of the lining applied on the increase in the load carrying capacity of the renovated structure (Abel 2010). In 2010, numerical analyses were carried out also, the purpose of which was a wider recognition of the static aspects of inter-working of the existing pipe, the inject and the free lining occurring in the case of pipeline renovation by the relining method (Doll 2011). The results of the investigations mentioned above and of other investigations carried out around the world indicate that the guidelines used presently include significant simplifications what is of a direct impact on the designing result. It proves univocally that the undertaken actions are well-grounded and that there is a necessity of their continuation. It is because the knowledge acquired during conducting experiments at the scale 1:1 as well as in the progress of the MES analyses enables

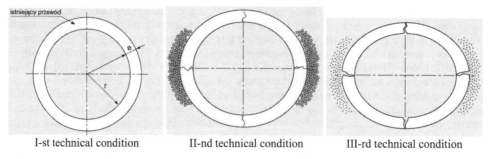

| I-st technical condition | II-nd technical condition | III-rd technical condition |

Figure 2. Technical condition of the existing duct, in accordance with (ATV-DVWK-M127P – 2. 2000).

a credible validation of the models of work of the structures subject to renovation, assumed in the guidelines. Therefore, it was decided to carry out, in the Wrocław University of Technology laboratory, the investigations on samples of pipelines renovated with a use of a free lining, by the method of relining, at the scale 1:1. Their main assumption is to investigate the phenomenon of the impacts of the inter-working of three layers of the system being created during renovation by the relining method: the existing pipe, the inject and the lining introduced on the load carrying capacity of the renovated structure. Based on the auxiliary material (ATV-DVWK-M127P – 2. 2000) to the guideline (ATV-DVWK A127P 2000), during developing the investigation plan, for the realisation samples were accepted of the pipeline exiting as a concrete structure of the internal diameter 600m, existing in the III technical condition. In accordance with (ATV-DVWK-M127P – 2. 2000), there are three technical conditions of the existing pipeline distinguished, different with regard to the damages occurring in the plant, possible deformations of the cross section and the load values assumed for the computations of the lining serving for the renovation. The technical conditions are presented in Figure 2.

The III-rd technical condition assumes creation of longitudinal conduit cracks at the four characteristic spots: the top, the bottom and the abutments of the conduit, as well as the occurrence of the cross-section ovalisation. If, during investigation of a damaged conduit, the ovalisation is visible but there is no data on its accurate dimension, the guideline (ATV-DVWK-M127P – 2. 2000) requires to assume, for the computations, the value equal to 3%. It is assumed also that the lining being mounted in the pipeline classified in the III-rd technical condition should be capable of bearing all loads acting onto the pipeline. The necessity to meet this condition is a consequence of assuming the structure operation model defined because the guideline (ATV-DVWK-M127P – 2. 2000) assumes that the free lining does not inter-work with the existing pipeline in which it is placed. At no degree, it is taken into consideration also the inter-working of the inserted lining and the inject filling the ring space. However, the statement is justified that, in the case of renovation with free linings, the impact of the inject presence and of the existing duct on the load carrying capacity of the lining may be significant. Therefore it was recognised well-grounded to undertake actions aimed at detailed analysing of the problem.

The plan of the investigations being conducted presently in the Wrocław University of Technology covers preparation of 15 pipeline samples and subject them to a load after inserting them into a soil environment. It was foreseen to investigate III sample types:

Type I – the samples consisting of a concrete pipe in the III-rd technical condition, mounted inside the lining of polyester resins reinforced with fibreglass, with the space between these components being filled with the inject,

Type II – the samples consisting of the same components as the Type I samples but with a foil layer inserted additionally between the inject and lining layers,

Type III – the samples of pipes of polyester resin pipes reinforced with fibreglass (GRP).

For preparation of the samples, the GRP linings of the HOBAS company were applied. Each of the samples was made in 3 identical copies. The pipeline samples prepared for laying in a soil environment are presented in Figure 3.

In conformity with the assumed III-rd technical condition of the pipeline existing on each of the concrete pipe samples, 4 longitudinal cuts were made, matching the longitudinal conduit cracks

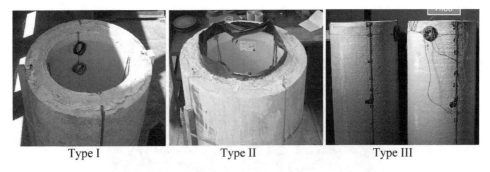

| Type I | Type II | Type III |

Figure 3.   Pipeline samples prepared for the investigations.

Figure 4.   General view of the research stand.

as well as the initial cross section deformation of the value of 3% of the internal pipe diameter was forced. At the contact spots of individual components, along the cuts done, rubber spacers enabling articulated operation of the structure were installed. The investigations are aimed at proving and quantitative determination of the inter-work between the existing conduit, the inject and the lining. In addition, they will enable also state precisely the influence of inter-work at the interface of the layers the inject – the lining on the load carrying capacity of the complete structure. After carrying out the investigations, it will be also possible to find out in what way the quality of the applied inject influences the operation of the renovated structure. It is because the samples of the I-st and II-nd types have been made in the two identical series for various inject sorts.

To carry out the laboratory investigations, the authoring laboratory stand of Mr. Tomasz Abel, PhD. Eng., is used. The investigations (Abel 2010) carried out by him in 2008 and cited above referred to load carrying capacity of concrete pipelines renovated with the use of the Trolining lining. The research stand designed and made for the purpose of the said experiments provides a possibility of matching the actual operating conditions of a renovated structure by locating it in a soil environment. The stand is of a box shape, the supporting structure of which is constituted of C140 rolled profiles, while the walls and bottom are created of tight boarding. The general view of the stand is presented in Figure 4. The dimensions: 150 cm × 150 cm × 160 cm enable convenient placing of the sample on a pre-prepared sand sub-crust and, next, preparation of a soil fill and backfill. Figure 5 presents a sample of Type II during laying of the pre-prepared substrate.

Prior to carrying out the investigation, on each of the internal lining samples, two cross section lines were determined, in which the strain gauges were installed, at the top and the bottom of the conduit from the inner side and on the abutments on the external side. The load is transferred uniformly along the full sample length, via a steel beam consisting of three HEB200 profiles. Under the beam, a spacer of elastomer rubber of 7 mm thickness is applied in order to eliminate the

Figure 5.   A sample of Type II during laying on a sand sub-crust.

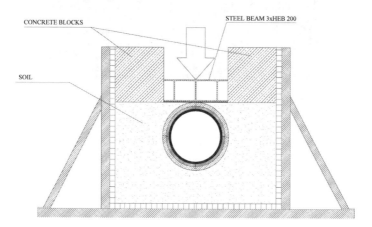

Figure 6.   The scheme of the stand.

possibility of occurrence of stress concentrations connected with possible local surface unevenness of the concrete pipe. The scheme of the research stand prepared for conduction of investigations is presented in Figure 6.

The first investigation was carried out on a sample of Type II. Inside the pipe, two pairs of displacement sensors were installed, recording the change of the diameter in the horizontal and vertical direction. In additions, sensors were located on the upper surface of the concrete blocks in order to record their displacement connected with displacement of the soil. The sample was subjected to load. The desirable effect was to achieve the decreased vertical diameter value of the inner lining of 10% (5 cm). As per the guidelines (ATV-DVWK A127P 2000), the pipelines, the vertical deflection of which exceeded this limit, are not suitable for further use. To maintain the deflection assumed, it was necessary to apply a force of the value 670 kN. The weight of additionally loading concrete blocks is chosen so that the pipeline operating conditions are imitated for the pipeline laid in the soil and poured with a layer of sand fill of 60 cm thickness. None of the standards applied presently and defining the load values from the automobile fleet

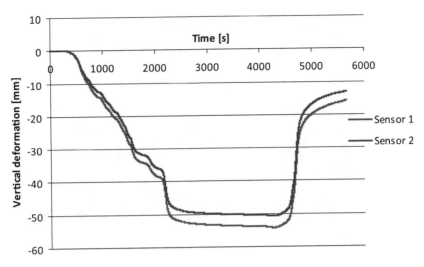

Figure 7.　The plot presenting the change of the vertical deflection value versus time.

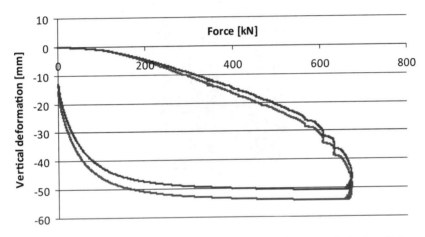

Figure 8.　The plot presenting the change of the vertical deflection value versus the load applied.

expects the occurrence of such high load in normal operating conditions. The results of the trial conducted are shown in Figures 7 and 8.

The stand on which the investigation was carried out was built, as mentioned already, with a thought of carrying out investigations of pipelines renovated by the Trolining method. In that case, the internal diameter of the concrete pipe amounted to 500 mm and the peripheral hardness of the Trolining lining itself was decidedly lower that that of the GRP lining that is subject to investigations presently. These differences in the initial assumptions were of an influence on that, during the first investigation, the research stand frame underwent deformation. The pressure of the ground against the walls of the stand were so high that it resulted in their displacements. Figure 9 presents a sample of Type II during conducting the investigations; Figure 10 – the view of deformed structural components of the stand, already after completion of loading and emptying the box.

After completion of the investigation of the first sample, it was decided on the expansion and reinforcement of the supporting structure of the stand. Analyses were carried out with the use the finite element method and, on the basis of their results, the decision was undertaken on widening the supporting frame of 0.5 m towards each side and on reinforcing all walls of the stand, as well as on mounting additional pulling components. The deformed side components were removed and, in their place, a new structure was designed, composed of a higher number of the supporting components,

Figure 9.   A sample of Type II during investigating.

Figure 10.   Deformed structural components of the stand.

complete with additional stiffening. The scheme of the stand prior and post the reconstruction is depicted in Figure 11.

   After completing the reconstruction, further investigations were commenced in conformity with a predefined plan. It was decided to apply an additional pair of sensors recording the conduit diameter change in the horizontal and vertical directions. Presently, samples of Type III are subject to the investigations. The general view of the research stand, after the reconstruction, during the investigations, is presented in Figure 12. The result of investigations of an exemplary sample is depicted in Figure 13.

4   RESUME

The forces that should be extorted to achieve the lining deflection of 10% of its internal diameter (5 cm) are several times higher than the loads admissible by the standards applied presently. The GRP lining accepted for conduction of the present investigations corresponds, with respect to the strength characteristics, to the linings applied commonly for renovation of pipelines of the similar dimensions and technical condition. A complement of the investigations carried out will be the analyses of operation of the structures investigated, that will be carried out with the use of the finite element method, after completion of the investigations. The initial conclusions that may be drawn from the experiments conducted presently support the thesis of a considerable overdimensioning of

a)

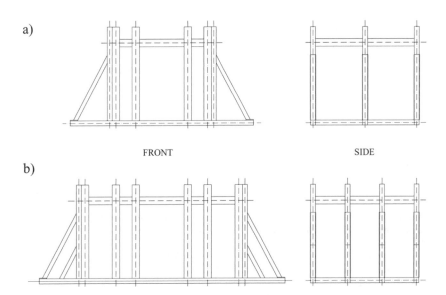

FRONT   SIDE

b)

Figure 11.   The diagram of the research stand a) prior to the reconstruction b) post the reconstruction.

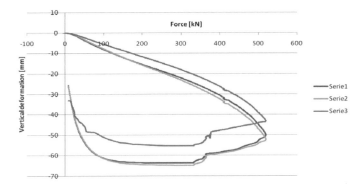

Figure 12.   General view of the stand after the reconstruction, during the investigations.

Figure 13.   The plot presenting the vertical deflection value change versus the load applied.

193

the linings used for renovation of pipelines by the relining method, the reason of which is, mainly, the assumption of the oversimplified models of the structure operation. Therefore, it is justified to undertake all actions leading to creation of computational models that map the actual operation of the structure better. The Wrocław University of Technology responds to the need conducting numerous investigations aimed at widening the state of knowledge concerning the actual character of operation of the structures being subject to renovation.

REFERENCES

Abel T., 2010 *Badania nośności kanałów rekonstruowanych wykładzinami ściśle pasowanymi na przykładzie technologii Trolining,* Praca doktorska, Instytut Inżynierii Lądowej Politechniki Wrocławskiej, 2010
ATV-DVWK A127P 2000 *Obliczenia statyczno-wytrzymałościowe kanałów i przewodów kanalizacyjnych,* Edition 3, Warszawa 2000
ATV-DVWK-M127P – 2. 2000 *Obliczenia statyczno-wytrzymałościowe dla rehabilitacji przewodów kanalizacyjnych przez wprowadzenie linerów lub metodą montażową,* Uzupełnienie do Wytycznej ATV-DVWK A127P, January 2000
Doll H., 2010 *Dimensionierung von Kunststoff-linern Close-Fit-Verfahren,* Mitteilungen des Instituts für Grundbau, Bodenmechanik und Energiewasserbau, Universität Hannover, Heft 59, 2001
Doll H., 2011 *Sanierung schadhafter Abwasser kanaele verdämmten Kunststoffrohren, statishe Aspekte*
Kolonko A., Kujawski W., Przybyła B., Roszkowski A., Rybarski S. 2011. *Podstawy bezwykopowej rehabilitacji technicznej przewodów wodociągowych i kanalizacyjnych na terenach zurbanizowanych,* Standard Izby Gospodarczej "Wodociągi Polskie", Bydgoszcz
PN-EN 13689:2004 *Zalecenia dotyczące klasyfikacji i projektowania systemów przewodów rurowych z tworzyw sztucznych stosowanych do renowacji*
Przybyła B. 1999 *Ocena i kształtowanie konstrukcji przewodów kanalizacyjnych w ujęciu teorii niezawodności,* praca doktorska, Instytut Inżynierii Lądowej Politechniki Wrocławskiej, Wrocław
Thèpot O., 2004 International Comparison of Methods for the Design of Sewer Linings, *3R International* (43) Heft 8-9/2004, pp. 520–526

Non conventional polyethylene pipe installation techniques: A trigger point for PE100RC development

C. Penu
Total Petrochemicals Research Feluy, Belgium

ABSTRACT: For more than half a century, polyethylene pipes have been used for pressure applications thanks to their excellent properties such as lightweight, flexibility, resistance to corrosion, weldability, durability . . . Today polyethylene is the preferred material for potable water networks in Europe and more than 95% of local gas distribution networks in Germany are made of polyethylene pipes.

PE100RC is the latest generation of polyethylene pressure pipe resin. This latter is particularly well suited to non conventional pipe installation techniques which present some clear advantages in terms of cost, time saving, environmental friendliness . . . when compared to the convention open trench sand bed technique. The development of PE100RC material included the elaboration of new test methods and the publication of local standards which is essential to spread confidence among involved parties.

1 INTRODUCTION

Today, more than ever, the lack of space in urban as well as in rural areas has pushed energy providers and utility companies to build their networks underground. Space is not the only reason why piping networks belong to the underground area, aesthetics play also an important role, a nice example is all the critics that were made to the Pompidou museum in Paris (see fig. 1).

Even if most of these networks are invisible, it is not really the case during their installation. Every single person living in modern society has been confronted at least once with the pleasure of being stuck in a traffic jam due to long lasting workloads for the installation of these networks. Most of today's pipe network installation is still made with the conventional installation technique which consists of digging a trench all along the network location.

For a decade, however, the apparition of a new generation of HDPE materials, namely PE100RC, has allowed an important development of the non conventional installation techniques which present several important advantages (cost and time saving, environmental friendliness, traffic disruption . . .) when compared to the conventional one. After a short introduction on polyethylene pressure pipe history/evolution, the advantages of PE100RC when combined with non conventional installation techniques will be presented.

2 KEY DEVELOPMENTS IN POLYETHYLENE PRESSURE PIPES

Polyethylene presents some clear advantages when compared to other plastic and non plastic materials:

- lightweight
- no corrosion
- excellent resistance to impact
- high flexibility
- excellent weldability for more reliable joints
- long term durability
- . . .

Figure 1.   Pompidou museum in Paris: an example of modern art building using pipes as decoration.

In mid 1950s, just after the discovery by Ziegler and Natta of the catalyst that was given their names, the first high density polyethylene (HDPE) commercial grade for pressure pipe application made with this catalyst was introduced on the market. At that time HDPE pressure pipes had to compete with more traditional materials such as cast and ductile iron, concrete or PVC. This first polyethylene generation for pressure pipes presented however a relatively limited resistance to hydrostatic pressure (PE63–see insert here below) and also a relatively limited resistance to slow crack growth.

In the 1970s the second generation HDPE resins for pressure pipe were introduced on the market. These latter presented a higher resistance to hydrostatic pressure (PE80) and also a higher resistance to slow crack growth mainly due to the addition of a comonomer to the polymeric chain. This comonomer increased the presence of the so called "tie molecules". These latter, in addition to physical entanglements, provided a higher resistance to slow crack growth by creating a physical link between crystalline zones. Some of these grades were medium density polyethylene (MDPE) and are still appreciated today for their high flexibility especially for small diameter pipes up to 180 mm in diameter.

At the beginning of the 1990s, the third generation appeared on the market after a significant change in the production process. The polymerization of ethylene and comonomer was done in at least two different reactors coupled in serie. This allowed the production of bi or multimodal materials with a controlled distribution of the comonomer. Thanks to this new molecular design, a new class of pressure resistance was achieved leading to the well known PE100 class. With their tailored molecular design, PE100 materials reached very high levels of slow crack growth and impact resistance.

3   MARKET REQUIREMENTS FOR THE DEVELOPMENT OF PE100RC

Most of the developments of new materials described here above were mainly, but not only, driven by an increase in the resistance of the pipe to hydrostatic pressure. Such an increase allowed HDPE pipes to enter new segments or to consume less material for a given segment. For example, for a

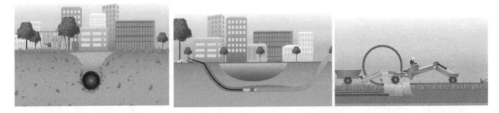

Figure 2. Open trench sand bed free – Horizontal directional drilling – Ploughing and milling.

given pipe diameter the use of PE100 would require 33% less raw material than a PE63 in order to obtain the same level of resistance to pressure (Le Roux et al. 1998).

The incentives in developing a PE100RC material were not exactly the same. As its name mentions, a PE100RC is still a PE100, consequently it has the same resistance to pressure as a PE100. The development of such a material was not driven by a higher resistance to pressure, which is indeed a critical property but rather by a higher resistance to slow crack growth propagation which is also a critical property. If a crack is initiated by whatever source at the inner or outer surface of the pipe, with time, it will propagate all over the thickness of the pipe leading then to pipe failure.

The resistance of material to slow crack growth propagation is then an important parameter and has been continuously improved. Moreover, in order to avoid such crack apparition on pipes, the conventional installation technique requires installing the new HDPE pipe on a dedicated trench and surrounding it with a sand bed. This conventional installation technique still represents almost 80% of the today's installations in Germany, the rest being considered as non conventional installation techniques. However, it's important to mention that non conventional installations techniques are being more and more used and have reached up to 50% market share in big cities such Berlin in order to reduce traffic and business disruption.

> The level of resistance to hydrostatic pressure allows the correct design of polyethylene pressure pipes. Thus according to ISO 9080 and to ISO 12162, a PE100 material has a Minimum Required Strength (MRS) of 10 MPa at 20°C after 50 years. A PE80 is defined by an MRS of 8 MPa, a PE63 by an MRS of 6.3 MPa and so on. Knowing the MRS value, one can choose the correct pipe, diameter and thickness, for a given network.

4 NON CONVENTIONAL INSTALLATION TECHNIQUES

A non conventional installation technique, by opposition to the open trench sand bed technique, consists in a HDPE pipe installation technique which requires neither open trench, neither sand bed nor none of them. Those non conventional installation techniques can be divided into 2 categories:

- Non conventional installation technique for new networks installation (fig. 2)
  – Open trench sand bed free
  – Horizontal directional drilling
  – Ploughing and milling
- Non conventional installation technique for rehabilitation of existing networks (fig. 3)
  – Pipe bursting
  – Sliplining
  – Swagelining, rolldown and U-lining

Some of these techniques have been used for a long time. For example early mentions of the ploughing technique can be found in 1970 (Jackson 1970) or of the relining technique in 1979 (Ashman 1979). At that time those techniques were employed only in some specific cases where the open trench sand bed technique could not be used; for example networks under rivers, motorways or railways or for rehabilitation of an existing network. Other techniques such as pipe bursting are more recent.

Figure 3. Pipe bursting – Sliplining – relining.

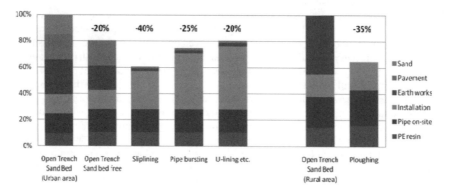

Figure 4.   Cost advantages of non conventional installation techniques when compared the conventional open trench sand bed technique.

Non conventional installation techniques, with the exception of the open trench sand bed free technique, present the particularity of requiring specific installation equipment and qualified workers. This could be one of the reasons why those techniques have not gained a higher interest and a higher presence on the market, although they present some important advantages when compared to the conventional installation technique:

- Significant reduction of the total cost of installation
- Significant reduction of the installation time
- Lower impact on environment mainly due to the absence of sand (transports . . .)
- Reduced disturbance/annoyance for inhabitants
- . . .

As an example, a recent market study (Haubruge et al. 2011) showed that the cost reduction of non conventional installations vs. the conventional one can be as high as 40%, as shown by figure 4.

Despite the above mentioned advantages brought by these techniques, they present a major challenge which is a higher "aggressiveness" towards the pipe. The presence of rocks (open trench sand bed free, ploughing), the deformation of the pipe (sliplining, relining) or the presence of the old pipe sharp edges (pipe bursting) can produce notches or scratches at the outer or inner surface of the pipe. These latter can be considered as the starting point of a crack which can propagate and lead to the failure of the pipe. Consequently, this aspect had to be taken into consideration, in order to further introduce these non conventional installation techniques in the market. This led to the development of PE100RC materials 10 years ago.

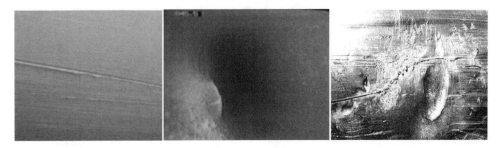

Figure 5.   Practice relevant defects: scratches and point loading.

5   PE100RC FROM DEVELOPMENT TO STANDARDIZATION

As mentioned above, a PE100RC material is a PE100 with outstanding resistance to slow crack growth propagation. This outstanding resistance to slow crack growth was needed for non conventional installation techniques. The idea was to have a pipe that preserves its theoretical lifetime even if some cracks or scratches were created at its inner and/or outer surface by the installation technique (see figure 5). This idea, although quite simple, revealed several questions:

• What is the minimum resistance to slow crack growth needed for non conventional installation techniques?
• Are all the notches/scratches similar? For example, a rock continuously pressing a pipe or a notch produced during a pipe bursting installation. In other words, how to measure the slow crack growth resistance? Are the existing slow crack growth tests OK or should one develop new testing techniques?
• How to relate laboratory test results to real cases?

The problem is certainly not trivial, and it has required methodology, exhaustive testing and the setting-up of several correlations. The first interrogations started in 1995 in Germany and were refined over the next five years when boundary conditions were agreed between market experts, which can basically be summarized as:

• PE 100 pipe material are exclusively considered because representative of current market situation
• 100 years minimum lifetime expectation at maximum design pressure
• Maximum outside defaults size is a sharp notch of 20% of the wall thickness
• Maximum point loading intensity is a permanent punctual strain well above yield point (>9% local strain).

The heart of the process has obviously been the laboratory evaluations and the establishment of correlations between the different practice relevant parameters. This work occurred in the period 1998–2007 and resulted in 2009 by the publication of a voluntary standard DIN (German Institute for Normalization): PAS 1075 (PAS = Publicly Available Standard).

PAS 1075 document proposes, for the first time, a definition of the minimum level of performance, which guaranties an operation of >100 years of PE pipes installed under the worst local stress concentration conditions. A material meeting those performance requirements is entitled the acronym PE 100RC.

Beside a specific test for the estimation of thermal stability of the material on a period of 100 years, PAS1075 specifies mainly 3 tests evaluating the slow crack growth resistance and their corresponding requirements. These tests are the following:

• Notch pipe test (ISO 13479)
   This test illustrates the resistance of pipe materials to notch propagation under high constant load conditions. The test consists in measuring the failure time of a pipe, under pressure, with a machined notch at its outer surface. This method has been selected for its high availability and

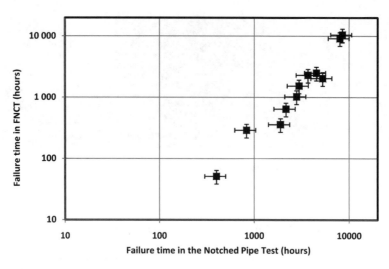

Figure 6.   Observed relationship between failure time in Notched Pipe Test (9.2 bar, 80°C) and FNCT (2% Arkopal N100, 80°C, 4.0 MPa) performed on PE 100 and PE 80 HD 110 mm SDR 11 pipes.

spread on the market and because it illustrates the resistance to the propagation of a notch made at the outer surface of a pipe by a sharp rock (or any other sharp surface such a tool . . .). Other relevant test methods exist such as the cone test (ISO 13480) and others are into development.

- Full Notch Creep Test (ISO 16770)

  This test is very useful for development purposes and quality control of raw materials or pipes. It requires only a small amount of material and samples can be obtained either directly from pellets (used to make plates) or from pipes. The test is a creep tensile test on notched bar in presence of water containing a surfactant at a temperature of 80°C. The presence of the surfactant accelerates the propagation of the crack by a factor up to 27 when compared to standard water (Beech et al. 2004). Moreover, a correlation between the Notch Pipe Test (NPT) and the Full Notch Creep Test (FNCT) has been made on PE80 and PE100 materials as it can be seen on figure 6.

- Point loading test

  The point loading test was developed by Hessel Ingenieur Technik (Hessel 2001). The point loading conditions are standardized to produce a constant local strain well above yield point, and a solution of surfactant is used to accelerate the stress cracking process. This test simulates what could happen in a real installation such as the trench installation without sand bed. Due to the fact that the excavated ground is reused to fill back the trench once the pipe is installed, a rock could get in close contact with the pipe and because of the above ground make a continuously constant pressure on the pipe. Since the first approach, in 1999, test conditions have been confirmed and a correlation between full notch creep test and the point loading test could be obtained as showed by figure 7.

As for the thermal stability, different activation or pseudo activation energies could be determined for each slow crack growth test. As all the above mentioned tests were performed at higher temperatures than the standard ambient temperature, the lifetime of the pipes in presence of notches or point loading could be estimated. Generally it was agreed that an extrapolation factor of 100 is conservative and suitable for the thermal stability or the point loading performed in Arkopal N100 between 80°C and 20°C.

Based on these results the minimum requirements for PE100RC as specified by PAS 1075 are the following.

Point loading, FNCT and NPT are different tests developed to measure the slow crack growth resistance of a material. Although they are well correlated one to another, each test presents some particularities like, for example, a close simulation of real cases for point loading and notch pipe

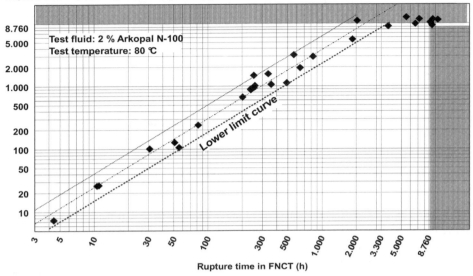

Figure 7. Correlation between FNCT and Point loading test (PAS 1075).

Table 1. Requirements for PE100RC materials as specified by PAS 1075

Method	Standard	Value
FNCT (80°C; 4 N/mm², 2% Arkopal N-100) on moulded specimens	ISO 16770	>8760 h
Point Load test on 110 SDR11 pipes (80°C; 4 N/mm², 2% Arkopal N-100)	PAS 1075	>8760 h
Notched Pipe Test on 110 SDR11 pipes (80°C; 9.2 bar; water)	ISO 13479	>8760 h
Thermal stability	PAS 1075	>100 y at 20°C

test or an easy/rapid method of raw material/pipe Quality Control or Product Verification Testing for the FNCT.

6 PERSPECTIVES AND FUTURE PE100RC DEVELOPMENT

The minimum requirements prescribed by PAS1075 are based on laboratory tests. The tests conditions were chosen high enough (high temperature, presence of surfactant ...) so that worst conditions during real non conventional installations could be taken into consideration. Of course, like for the conventional installation technique, state of the art rules have to be followed during non conventional installation techniques. When installed correctly, PE100RC materials should, then, give enough confidence to guarantee the same network lifetime for a non conventional installation made with PE100RC pipes as a conventional sand bed installation made with PE100.

The confidence brought by this new generation of materials is confirmed by its consumption evolution as it can be seen on figure 8.

Fast, economical and non-disturbing pipe installation practices are growing rapidly. And the association of PE 100RC to those techniques will result in growth for the whole plastic pipe business. Pipes made of PE 100RC are more and more prescribed in Germany, Poland, Austria and neighboring countries, to an extent that it represents today about 20% of the PE pressure pipe market in that region.

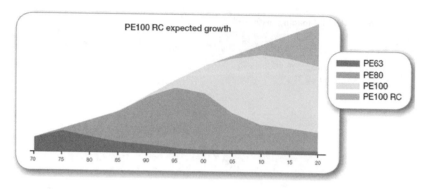

Figure 8.   HDPE market evolution since 1970 and growth estimation up to 2020.

To support this development, further recognition into standards will help. Currently the PE 100RC is defined in DIN PAS 1075 and in ÖVGW PW 405/1. It is part of the technical prescriptions of Gas and Water companies and will be recognized by Gas and Water associations like DVGW in Germany. National standardization committees are working on this topic in several others European countries such as Italy or Belgium and that could open the way to a harmonized European Norm.

REFERENCES

Ashman, R. 1979, Present and projected markets for HDPE pipe in Scandianvia, *Plastic Pipe Conference 4*, Brighton.
Beech, S., Clutton, E. 2004, Interpretation of the Results of the Full Notch Creep Test and a Comparison with the Notched Pipe Test, *Plastic Pipe Conference 12*, Baveno.
Haubruge, H., Laurent, E. 2011, Non-Conventional Pipe Laying Techniques: Quantitative Modeling of Total Cost of Installation, *Inzynieria Bezwykopowa*, 2, 36–39.
Hessel, J. 2001, Minimum service-life of buried polyethylene pipes without sand-embedding, *3R International*, 40, 4–12.
Jackson, R. T. 1970, Moleplough laying technique for plastic pipes, *Plastic Pipe Conference 1*, Southampton.
Le Roux, D., Ahlstrand, L.-E., Espersen, H. 1998, PE100 Opens new horizons for plastic pipes, *Plastic Pipe Conference 10*, Gothenbourg.
PAS 1075, 2009, Polyethylene Piping for Alternative Installation Methods, *DIN Publicly Available Specification*.

Underground Infrastructure of Urban Areas 2 – Madryas, Nienartowicz & Szot (eds)
© 2012 Taylor & Francis Group, London, ISBN 978-0-415-68394-4

Research over water loses and preventing of them due to usage of monitoring system in chosen water distribution companies

F.G. Piechurski
Institute of Water and Sewerage Engineering, Silesian University of Technology, Gliwice, Poland

ABSTRACT: Two water network system were taken into consideration. In both of them a research was made before and after installation of monitoring system. A water loses were analyzed in both cases. Obtained results clearly show that right after installation of water monitoring system we can observe a major decrease of water loses. In mean time amount of located water network failures is highly increased which is a direct result of introduce water network monitoring system. Research was made on the water networks made of different materials and for networks of various age.

1   INTRODUCTION

Water loses are common phenomenon in every water Network. There is no possibility to predict the time or the place of the failure. In many cases we are not able to realize that the failure already exists as long as it results in major water loses. In every case the only method of reducing water loses is fast detection before major loses appear. Detection and removing even small water leakages has a plenty of advantages. First of all it is financial profit due to reduction of water loses. Second is protection against greater failures. Disadvantage is cost of introducing of efficient monitoring system. Another important future is environment protection due to smaller power consumption by system pumps, smaller amount of chemicals used for water treatment, better propagation of sweet water resources.

2   SOURCE OF WATER

Supplying customers with water in the region of Upper Silesia is realized mainly through central water network owned by Górnośląskie Przedsiębiorstwo Wodociągowe in Katowice. Other water distribution companies buy a water from central water network and then distribute it via own local networks to their customers.

Such kind of company is analyzed PWiK in area A. Company has 62 points of supply. Water consumption in various points differs and in summary it was 24 575 m³/d in 2009.

A source of supply for analyzed PWiK area B determine underground waters that are taken from own 51 deep wells. All of them are connected with 10 compensatory tanks and 15 pump rooms. Productivity is $Q = 119\ 280\ m^3/d$ and water consumption was in 2007 $Q = 42\ 815\ m^3/d$ which is about 36% of possible productivity.

Pay attention on difference between water source of PWiK A that buys a water and PWiK B that uses its own water. In first case measurement of water is made in measurement chambers, in second case in water source.

3   WATER NETWORK SYSTEM

The oldest water network ducts in PWiK A come from the end of XIX century and together with ducts from the beginning of XX century they stand for 38% of Total Network length. As presented

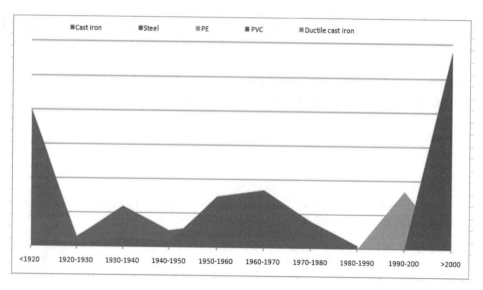

Figure 1.   Age of Network with respect to the material. Source PWIK A.

on fig 1. Water net is made of cast iron. Only a small amount of it made of steel. Since the second world war was finished we can observe we can observe increase of usage of steel instead of cast iron.

Steel was most common material used for water networks in 70's and 80's. From that time period till now about 13% of water network is still in use. In a last few years traditional materials were substituted with plastic. At the beginning PVC pipes were used. Now most common are polyethylene pipes.

Length of ducts is almost 329.2 km considering 243.4 km of distribution network and 85.8 km of domestic connections

Most significant part is represented with pipes with diameter 100 mm or less. It is equal to 62.32% of overall length. Ducts over 100 mm and up to 200 mm represents 25.88% of overall length. Ducts with diameter over 200 mm represents only 11.8%. PWiK B supply with a water area about 1000 km^2. Overall length of the network is 2201.80 km. Distribution network is 138.60 km long, splitting network is 1285.59 km long and domestic connections are 777.61 km long. Mainly water network is made of cast iron (44%). Second most important material is PVC (39%). 7% of the network is made of steel. The same amount (7%) is represented with PE. The company is all the time upgrading the network but still about 3% of network is made of asbestos and cement (fig. 2).

Water network is made of pipes with diameter range from Dn80 up to DN 1000.

4   WATER NETWORK FAILURE CHARACTERISTICS

One of the most important coefficients used for evaluation of water network condition is damage intensity coefficient. It is said that value shouldn't exceed 0.5 number of damages/km year. Taking into consideration types of network one should try to achieve following parameters: distribution network 0.3. splitting network 0.5. and domestic connections 1.

Despite using different methods of preventing damages in PWiK A a water network failure coefficient is high. In 2005–2009 there was average 1.77 damage interventions a day. A great influence on number of removed water network failures has a fact that detection of water failures was greatly improved. This directly results with minimizing of water losses. In PWiK A in 2009 this coefficient was 1.6 dam./km year for splitting network and 1.3 dam./km year for domestic connections. Those coefficients are much above allowed values. Comparing his coefficient with type of material used in network we can observe that for network built of PVC, PE, ductile iron this coefficient is fulfilled. For steel and for cast iron the coefficient is highly exceeded.

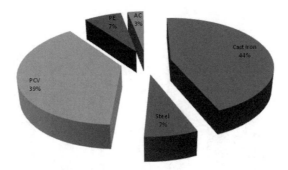

Figure 2. Materials used in water network in PWiK B.

Table 1. Damage intensity coefficient depending on type of used material in 2009 in PWiK A.

No.	Material	Number of leakages [pcs.]	Network length [km]	Damage intensity coefficient
1	PVC	6	22.4	0.3
2	PE	7	131.5	0.1
3	steel	535	133.6	4.0
4	cast iron	62	30.3	2.0
5	ductile cast iron	1	7.9	0.1
6	no data	12	–	–
7	different material	–	3.5	–
Total:		623	329.2	

In practice value of this coefficient is compared in different zones. Zones with the highest value of coefficient are qualified to be replaced. Taking a look at this coefficient values we can observe that water network greatly exceeds allowed level of coefficients

In the water network of the PWiK B we can see 8 different zones of increased pressure. All procedures are based on data received from monitoring system. System is successfully expanded on new areas of water network. First of all an analysis was made to determine changes of basic parameters in time. On fig. 3. One can observe damage intensity coefficients λ for proper years for different type of network. The highest value of the coefficient is noted for domestic connections and is about 0.42–0.51 dam./km year. For distribution network this parameter was significantly lower, about 0.09–0.41 dam./km year.

Average value for all types of materials are as follows:
   distribution network   0.25 dam./km year
   splitting network   0.36 dam./km year
   domestic connections   0.47 dam./km year

Summarizing received data with coefficients that we can find in literature we can notice that number of water network failures is lower than average received from data from several water distribution companies. Average for 2006 is 0.46 dam./km year for distribution network and 0.70 dam./km year for domestic connections. Analyzed water network has a higher coefficient value than country average only for distribution network for which λ = 0.16 dam./km year.

5 WATER NETWORK MONITORING

In PWiK A GIS system is used for locating water network failures on vector map of installation. Worker responsible for removing the leakage is ordered to prepare a protocol. Protocol consist

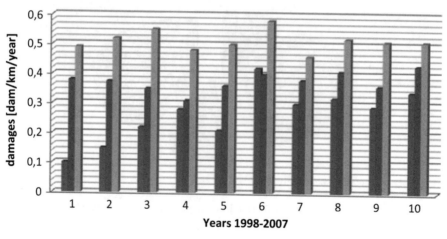

damages [dam/km/year]

Years 1998-2007

■ Distribution network   ■ Splitting network   ■ Domestic connections

Figure 3. Damage intensity coefficient in 1998–2007 in PWiK B.

Table 2. Damage intensity coefficient in 2006 i 2007 in PWiK B.

L.P.	Duct type	Damage intensity coefficient 2006 [dam./km* year]	Damage intensity coefficient 2007 [dam./km* year]
1	distribution network	0.25	0.34
2	splitting network	0.37	0.44
3	connections	0.51	0.54
Mean		0.41	0.40

important data about dimension of the pipe, type of the material, cause of the leak, method and place of the removal of the leak. That procedure gives a plenty of possibilities. However plenty of reports has a fields with „no data" information. This can have a great influence on evaluation of water network failures. It can be seen clearly that PWiK has to improve quality of those protocols.

To determine the optimal value of failure coefficient a discussion need to be done. Parameters that need to be taken into consideration are: spatial distribution of failures, diameter of failure ducts, costs of repairs and water loses. Average values of coefficients give a clear view over the condition of the network and shows how close the water network parameters are close to optimal values.

In cases where a coefficient value exceeds 1 dam./km year and no seismic activity is noticed one should find the source of problems and develop the method of improving.

In PWiK B to obtain constant and failure free cooperation between pump rooms, tanks and sources a water monitoring system is used. For most sources, technical and electrical parameters such as productivity, level, time of working are remote controlled. Workers have remote access to all devices and can change any parameter if needed. Monitoring is realized via radio frequency. It gives a possibility to continuously observe all the parameters of chosen point, register parameters and display alarms if any value is exceeded over it's normal operating condition. Present status of all instruments in given point is demonstrated in graphical way.

6   WATER LOSES

One of the accurate coefficients used for comparison of condition and exploitation method of water network is unit water loses factor. It determines water loses with or without own needs with respect to length of the network without domestic connections (m^3/h km or m^3/d km). Factor calculated this

Table 3. Unit water loses factor in PWiK A in 2005–2009.

Year	Water loses [m³/h]	Water loses with own needs [%] S	Water network length [km]	JWST m³/h km
2005	283.5	29.7%	257.5	1.10
2006	315.6	32.2%	258.8	1.22
2007	308.7	32.1%	260.7	1.18
2008	233.4	23.6%	241.6	0.97
2009	160.9	16.6%	243.4	0.66

Table 4. Total length of the network, percentage coefficient of water loses and load of analyzed network.

Year	Total network length [km]	Coefficient network load [m³/d × km]	Percentage water loses coefficient [%]
2002	1744.50	32.78	18.00
2003	1646.30	34.58	18.30
2004	1664.60	32.48	18.06
2005	1683.35	31.46	19.27
2006	1702.05	30.68	18.81
2007	1729.15	28.64	18.57

way allows to analyze different structures of water supply system and different length of the ducts. Allowed value of this coefficient is in range 0.2 m³/h km–0.3 m³/h km (Dohnalik & Jędrzejewski 2004).

Taking into consideration allowed value of this factor equal to 0.2 m³/h km a company should reach the value of loses about 48.7 m³/h. Changing this amount into precentage value we obtain 4.77%. In German guidelines from 1986 there is an information about allowed unit water loses depending on the type of ground. For sand coefficient is equal to 1.2–1.6 m³/d km, for gravel 2.4–6.0 m³/d km, for rocks 4.81–4.4 m³/d km. Taking into account above parameters in PWiK A in 2009 this coefficient was 15.9 m³/d km. German factor for gravel allows 6.0 m³/d km. The length of the network is 243.4 km, so allowed water loses are 1460.4 m³/d. Taking into the calculation this value and changing it to percentage form we obtain 5.94%.

Tab. 4 presents length of the network, percentage coefficient of water loses and load of analyzed network in 2002–2007 in PWiK B.

In last few years amount of water loses is in the range of 18%–19.7%. The greatest loses were noticed in 1999. in 2002–2005 we can observe a small increase of loses however in last two years loses are decreasing. A direct influence on this process can have a fact that diagnostics of water network was greatly improved.

7 INFRASTRUCTURAL LEAKAGE INDEX

Infrastructural leakage index is at hem moment the only real loses index used all over the world. It is so because it takes under the consideration actual condition of water network. It shows multiplicity of real loses with respect to minimal water loses level. ILI is much better factor than used before for comparison between different water network systems. In tab. 5, we can observe value of ILI for PWiK A.

Taking into consideration the best practices in the world, and comparing them to values obtained in PWiK A we can observe that they differ significantly. Analyzing the cause of such difference

Table 5.   Value of ILI for PWiK A.

	Area A	Area B	Total
2005	9.52	7.64	8.6
2006	10.38	7.09	8.76
2007	10.22	6.41	8.36
2008	5.93	5.07	5.5
2009	5.77	3.85	4.8

Table 6.   Real loses, inevitable loses and ILI for PWiK B.

Year	Amount of real loses [m^3/year]	Amount of inevitable loses [m^3/year]	Technical [dm^3/connection/day]	ILI [–]
2000	3516.20	1146.42	0.25	3.07
2001	2494.20	1162.63	0.25	3.01
2002	3821.40	1178.38	0.27	3.24
2003	3803.10	1118.71	0.28	3.40
2004	3563.90	1133.10	0.26	3.15
2005	3723.60	1148.38	0.27	3.24
2006	3585.00	1163.77	0.25	3.08
2007	3355.20	1185.89	0.23	2.83

one should pay attention that area of water network is located in the region of mining activity. We can clearly see that for regions with mining activity the coefficient is much higher. There appears a doubt about judgement over the condition of network with respect to (Speruda & Radecki 2003). Interesting idea would be comparing the value of coefficient in different companies.

Tab. 6. Introduces the real and inevitable water loses in 2000–2007 in PWiK B. Value of ILI for tested network was in 2007 2.83. Condition of the network is assumed that it need to be replaced. In every water network it should be introduced: policy of quality of repairs, single leak detection and active leak control together with pressure management system.

From obtained data we can see that considered water network has real loses coefficient below average of the country (Speruda & Radecki 2003), ILI comparing to international standards is below average. For each water network there is a level of water loses below there is no sense from economical point of view to minimize them any further This are so called inevitable loses. Especially small water leakages (below 0.5 0.5 m^3/h) from economical point of view are hard to find.

8   ACTIVE LEAKAGE CONTROL

Trying to minimize the water loses or in general terms or in chosen region, systematic method should be considered. First of all standard analysis has to be performed. All needed coefficients need to be determined. On the base of above factors we can determine economical level of water leakages. Determining above factors is invaluable help for planning detection, conservation or investment. Taking into account costs, Active leakage control is the cheapest process of minimizing water loses. In regions equipped with monitoring system most effective procedure will be time of repairs. Planing regulation of the network in wider aspect one should pay attention on damaging influence of high pressure and its rapid changes. Optimal solution should be based on active pressure control for regions with varying height. Effective water supply would be based on communication of reducers with important points in water network equipped with pressure control system. In that case reducers would not work in constant settings but they would adjust pressure to actual needs. In this case we

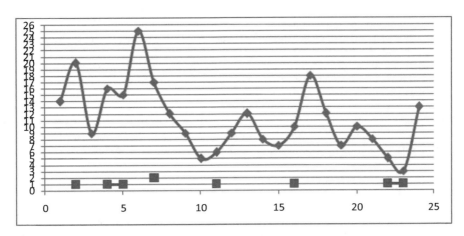

Figure 4. Amount of removed leakages with usage of ALC in 2008–2010r.

Table 7. Found failures in PWiK B in 2007.

No.	Quarter/ year	Length of checked network (meters)	Range	Amount of found leakages	Type of leakage		
						Not called	
					called	visible	invisible
1	I	11 250	Area A	14	11	1	2
		32 400	Area B	22	0	3	19
2	II	87 005	Area A	53	33	1	19
		25 800	Area B	13	3	0	10
		9 700	Area C	0	0	0	0
3	III	78 130	Area A	63	28	6	29
		20 350	Area B	7	3	1	3
4	IV	81 315	Area A	51	31		20
		3 500	Area B	14	6		8
		100	Area C	1	1		
	Total	349 550		238	116	12	110

can easily minimize overpressure during the night time. Due to high costs while reducing water loses the condition of ducts in at the top priority. Rehabilitation is necessary process but usually not performed to whole network. Only the most failure parts are done with it.

In 2008 in PWiK A monitoring system was expanded to 17 of 62 water supply points. Additionally acoustic leakage detection system was expanded to 30 sensors. A detailed analysis of night and daily water flows. That research allows fast detection of regions with possibility of leakage. After recognition and precise location of the leakage an information is passed to exploitation department which is responsible for it's removing.

In 2008 amount of found and removed leakages with ALC was 137 and it was 28% of all leakages removed by the company. In 2009 amount was 112 and it was 18% of all leakages. In 2010 so far there were 21 noticed leakages. It should be pointed out that leakages found with this method were leakages that were invisible from the surface what was additional disadvantage in it's location. Except leakages counted above, there was a part of calls that were not leakages. Total number of such false calls was 9 and they are marked red on the fig. 4. Comparing correct calls with those missed we can say that accuracy of calls was 96.7%

Except above actions in PWiK A there is done analysis with a usage of acoustic detection system. There are 3 methods of analysis First is reaction for increase of water consumption in analyzed region. Second method is based on assumption that whole area of the network should be checked at least once a year. Third method is waiting for failure, by placing a sensor in most probable places for failure (Speruda 2006).

In PWiK B in 2007 made ALC in 3 separate areas. Effect of this action are presented in tab. 7. Especially important is a number of invisible leakages compared to visible ones. Elimination of such leakages is very important to minimize water loses in water network. In 2007 a number of called failures 116 is comparable with found invisible failures that wax 110.

Summarizing analysis of water loses and failures in PWiK A and PWiK B we can say that chosen way of leakage detection is giving positive results. Basing network diagnostics on ALC and water monitoring system in proper zones caused continuous decreasing of water loses.

9 SUMMARY

Water loses in water distribution system in Poland and all over the world are big and highly exceeds allowed values. Minimizing the water loses should be one of the main tasks of water distribution companies because it leads to lower costs of water.

High water loses in Poland are caused by long term neglect of modernization, repairs and expansion in water networks. This fact leads to water network degradation. Another fact is improper performance in planed economy time.

Introducing planed movement leading to minimizing water loses both PWiK A and PWiK B the same as other companies in country receive expected results. However to achieving proper results requires time and money. One should pay attention that minimizing water loses is depended on many factors. Minimizing water loses in polish companies is one of the newest branches of activity, with additional handicap with a years of neglect form ages of planed economy.

REFERENCES

Dohnalik, P & Jędrzejewski, Z. 2004. *Efektywna eksploatacja wodociągów ograniczanie strat wody*. LemTech. Kraków
GIS. 2007. *Modelowanie i monitoring w zarządzaniu systemami wodociągowymi i kanalizacyjnymi*. PZIiTS. Warszawa
Speruda, S. 2006. *Straty wody w Polskich sieciach wodociągowych*. Akademia strat wody WaterKey.
Speruda, S. & Radecki, R. 2003. *Ekonomiczny poziom wycieków*. Warszawa: Wydawnictwo Translator

Underground Infrastructure of Urban Areas 2 – Madryas, Nienartowicz & Szot (eds)
© 2012 Taylor & Francis Group, London, ISBN 978-0-415-68394-4

Identifying sewer damage condition for estimating network reliability indexes

B. Przybyła
Technical University of Wroclaw, Wroclaw, Poland

ABSTRACT: The study contains considerations connected with assessment of technical condition of sewers, in association with possibility of using this assessment for estimating network reliability indexes. Consideration here is given most of all to effects resulting from difficulties in finding sewer damages and operation (performance of their function) of sewers in damaged condition. Classification of damaged condition has been proposed, which, in the author's opinion, is suitable for reliability calculations, as well as outline of method for counting damage conditions which makes their number independent of renovation policy used by network operator.

1 INTRODUCTION

The task of sewerage system is to carry away required amount of sewage from defined area – usually residential unit or industrial plant. In order to perform its function, the network must by appropriately designed and built, and then correctly operated. One of the basic tasks for companies responsible for operation of networks is to ensure their appropriate reliability, which traditionally is defined in the following way (Kłoss-Trębaczkiewicz et al. 1993): "Reliability of sewerage system denotes its capability to carry away the planned amount of sewage from given area to defined location (treatment plant), at specific conditions for the existence and operation of the network, during assumed operating period".

Assessment of reliability of existing networks is an essential issue, which constitutes the basis for starting appropriate actions ensuring correct functioning of the network in the future, and making conclusions as for the principles applied in designing, constructing and operating the network.

From the point of view of reliability theory, sewerage systems consist of serial systems, creating together a hierarchical structure, which as a system is characterized by multistateability. Determination of reliability is usually carried out by way of operating tests and using various analytical methods described in the literature of this subject, e.g.: (Kłoss-Trębaczkiewicz et al. 1993, Kwietniewski & Rak 2010, Madryas 1993, Wieczysty 1990). Sewers creating the system are reparable objects, which in case of damage are repaired or replaced with new ones. A sewer is therefore a technical object subject to renovation, which is called a transition from unsuitability (damage) condition to suitability condition (correct operation). Figure no. 1 shows a graphical illustration of operation process with non-zero renovation, adopted for sewers. As the duration of renovation is not negligibly short, it is included in reliability calculations carried out for sewers.

In quantitative characteristics of technical objects so called reliability indexes (inter alia) are used, which are defined, for example, in the Polish standard PN 77/N-04005. In Poland reliability indexes are successfully used in research on reliability of water and gas conduits, at least from the beginning of the 1970s. Also attempts to use them in examinations of sewers have been made.

Taking into account operating model of such sewers, indexes essential for their description are described (inter alia) in (Przybyła 1999). The following among them are deemed particularly essential:

– probability of operation (serviceability) of a sewer $R(t)$,
– average operating time between damages T_p,
– parameter of damage stream $\omega(t)$,

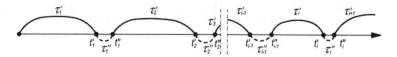

Figure 1.   Graphical illustration of a process with non-zero renovation (Kłoss-Trębaczkiewicz et al. 1993): $\tau_1', \tau_2', \ldots$ – operating (serviceability) periods, $\tau_1'', \tau_2'', \ldots$ – non-operating periods (including technical rehabilitations), $t_1, t_2', \ldots$ – temporary unserviceability periods, $t_1'', t_2'', \ldots$ – periods of introducing an object to operation.

– average duration of renovations T_o,
– intensity of renovation $\mu(t)$,
– probability of renovation $P_o(t)$.

These indexes have been defined in the subject example literature mentioned above. Also formulas useful in practical estimation of those indexes are quoted there. And thus, for example, average operating period and average renovation time are estimated as (Madryas 1993):

$$\overline{T}_p \cong \frac{1}{n}\sum_{i=1}^{n} \tau_i', \ \tau_i' = t_{p,i} \, , \qquad (1)$$

$$\overline{T}_a \cong \frac{1}{n}\sum_{i=1}^{n} \tau_i'', \ \tau_i'' = t_{a,i} \, , \qquad (2)$$

where: $t_{p,i}$ = duration of i-th operating period, $t_{a,i}$ = duration of i-th renovation, n = number of renovations (operating periods) in the examined period.

The parameter of damage stream determining the likelihood of object damage in the time interval $(t, t + \Delta t)$, regardless of whether at the "t" moment the object was operative or not, in relation to linear objects whose length does not change may be estimated by using simplified formula (Kłoss-Trębaczkiewicz et al. 1993):

$$\overline{\omega}(t) = \frac{m(t, t + \Delta t)}{L \cdot \Delta t} \, , \qquad (3)$$

where: $m(t, t + \Delta t)$ = number of damages of all examined sewers in Δt time interval, L = total length of examined sewers, Δt = time interval in which observation is carried out.

Taking into account standard assumptions for the stream of damages to which sewers are subject – a single stream, stationary and without memory acc. to (Kłoss-Trębaczkiewicz et al. 1993), intensity of damages designated as $\lambda(t)$ and determined for non-renewable objects or in case of sewers applicable to time intervals between damages is constant and equivalent to the parameter of damage stream $\omega(t)$:

$$\lambda(t) = \lambda, \ \omega(t) = \omega \ \text{and} \ \lambda = \omega \qquad (4)$$

Moreover an assumption is made, that in the period of normal operation, reliability of sewers is characterized by expotential distribution. With such assumptions the sewer reliability function and probability density of operating time are described by the following formulas:

$$R(t) = e^{-\lambda t} \, , \qquad (5)$$

$$f(t) = \lambda e^{-\lambda t} \, , \qquad (6)$$

Reliability indexes such as T_p, $\omega(t)$ and the likelihood of sewer operation $R(t)$ are called "failure (or non-failure) frequency measures" of sewers.

2 THE PROBLEM OF IDENTIFYING SEWER DAMAGE CONDITION FOR THE NEEDS OF QUANTITATIVE ANALYSIS

In order to estimate failure frequency measure for a sewer, knowledge of the number of its defects within examined time interval is required. Due to the specificity of sewer operation, identifying and classifying sewer condition as the one in which defect has occurred is, in the author's opinion, the main problem which determines the correctness of calculations or real comparability of results obtained by researchers independent of each other.

Sewers usually carry sewage gravitationally. In such systems situation in which the flow of sewage is interrupted occurs very rarely. Even damages causing essential destruction of construction need not cause noticeable disturbances in the flow of sewage, although a sewer operating in such condition causes adverse consequences for the environment and the sewerage system, as well as epidemiological and sanitary hazard for the population. Consequences of sewer operation in damage condition are usually shifted over time, and negligence in identifying this status may intensify over time. Difficulties in detecting defects distinguish the sewerage systems from other networks of urban underground infrastructure – those ones which carry fluids under pressure and in which defects are found soon after their occurrence.

At present, the basic tools used in diagnostics of sewers are video inspection sets (cctv), which transmit the sewer interior picture for analysis carried out by a specialized technician. Inside the sewers various changes are observed, for example: cracks and fractures, displacements of connections of individual pipes, deformations of transverse section, deposits and swellings limiting the flow capacity of sewers, established roots, symptoms of corrosion and mechanical wear. These changes have different influence on operation of a sewer, cause various consequences depending also on its location in the system. During inspections usually many of such changes are found in a sewer – these are documented, and then finally the assessment of sewer condition is carried out, which usually classifies the sewer to a certain group. This results in practical recommendations concerning operating actions or connected with technical rehabilitation (renovation) of a sewer. Practice shows that even sewers in the worst condition, classified in accordance with certain method to a group that requires urgent action, are able to effectively carry sewage, thus performing their proper function. On the other hand, occurrence of easily removable obstacle in the flow of sewage may even entirely block this function. Remarkable subject of classifying and assessing sewers condition has been considered in many literature studies. Overview of solutions can be found in e.g. (Bölke 2009, Madryas et al. 2010, Stein 1998).

When considering the notion of a defect in a sewer, as such each change can be treated in relation to condition assumed as ideal and which can be found during video inspection. However, then the problem of actual condition of sewers appears, i.e., it might turn out, that practically most of them are still damaged. This is not the only problem, because a sewer even without those changes may not have the required flow capacity, which may result from incorrect designing or unforeseen development of settlement unit. The sewer is therefore unreliable, does not perform its function and can be classified as damaged if this classification criterion is taken into consideration.

When determining the number of sewer defects indispensable for calculating reliability measures one may assume, that each change observed in a sewer during video inspection is treated as a damage. Also another approach is possible, in which condition of sewer between wells is assessed, and outcome of this assessment will be decisive for claiming that damage has occurred or not. Each sewer consists of many such sections connected in series.

This approach corresponds to the practice adopted for assessing condition of sewers, where initially the condition of sections between wells is analysed. Finally, as already mentioned, a sewer consisting of such sections is classified into a group that also points at the size or extent of damage (these notions, as ambiguous, each time require detailed elaboration).

For example, according to the popular ATV M 149 guidelines of the German DWA organization, sewers are classified under one of five groups described as follows:

    class 0 – urgent (emergency) actions are required,
    class 1 – actions within short period of time are required,
    class 2 – actions within medium period of time are required,
    class 3 – actions within longer period of time are required,
    class 4 – no need to take actions.

Here the classifying is carried out by assigning appropriate number of so called "penalty points" to individual changes found in a sewer, moreover this value is modified by considering various factors, such as current hydraulic conditions for the flow of sewage, type of sewage, location of a sewer in the ground in terms of specific factors, e.g.: protection of water-bearing grounds.

In Polish literature on this subject, the notion of damage condition of a sewer was considered in detail in the works (Madryas 1993, Przybyła 1999), in which also their definitions were suggested. And thus, the term "damage condition" is also applicable to a random event, in which at least one characteristic within a set of measureable or non-measureable characteristics describing the sewer ceases meeting requirements demanded of it (Madryas 1993):

$$\exists_i, C_{mi} \notin \left[\underline{C_{mi}}, \overline{C_{mi}} \right] \vee \exists_j, C_{nj} = 0 \quad i,j = 1,2,3, \dots , \tag{7}$$

In this and in other formulas quoted further, the following mathematic symbols are used: $\exists$ – exists, $\in$, $\notin$ – belongs, does not belong, $\vee$ – or (logical alternative).

Measurable (quantitative) characteristics describe objectively (by means of real numbers) those functions and properties of sewers, which can be measured within a given time interval. Non-measurable (qualitative) characteristics objectively describe those properties and characteristics of sewers, which cannot be unambiguously measured. Measurable characteristics usually include: geometrical characteristics of sewers, their structural and strength-related features, hydraulic parameters and age. Non-measurable characteristics usually include resistance to chemical corrosion, symptoms of technical and economic ageing, durability if a sewer, etc. Non-measurable characteristics listed here have general meaning and have not been given an explicit definition so far. Obviously, this situation in individual cases may change – when commonly accepted definition describing the phenomenon, appropriate measure and measurement method appear.

Formal notation of damage condition, quoted in formula (7), is based on the zero-one system of identifying non-measurable characteristics and does not take into consideration the partial loss of sewer's ability to operate. By using the limits of changes in measurable characteristics C_{mi}^1, $\overline{C_{mi}^1}$ and fuzzy logic for identification of non-measurable characteristics, the formula (2.1) describing full damages (the sewer has entirely lost the ability to perform its tasks) is modified to the following form (Przybyła 1999):

$$\exists_i, C_{mi} \notin \left(C_{mi}^1, \overline{C_{mi}^1} \right) \vee \exists_j, C_{nj} : \chi_{nj} = 1 \quad i,j=1,2,3,\dots \tag{8}$$

For partial damage condition (the sewer has partly lost the ability to perform its tasks) the notation has the following form:

$$\exists_i, C_{mi} \in \left[\left(C_{mi}^1, \underline{C_{mi}} \right) \vee \left(\overline{C_{mi}}, \overline{C_{mi}^1} \right) \right] \vee \exists_j, C_{nj} \cdot \chi_{nj} \in (0,1) \quad i,j=1,2,3,\dots \tag{9}$$

The χ parameter (affiliation degree) used in formulas (8) and (9) determines the truthfulness of the sentence based on fuzzy logic, formulated as follows: „the C_{nj} characteristics does not fulfil the conditions for sewer operation". Assigning of extreme values, i.e. $\chi = 0$ or $\chi = 1$ denotes full knowledge of the C_{nj} characteristic, while assigning a value from the range (0, 1) denotes lack of such certainty as for such knowledge and thereby decides about the truthfulness of the above formulation. Extreme values, therefore, may be interpreted accordingly – $\chi = 0$ denotes the state of correct sewer operation, while $\chi = 1$ denotes occurrence of full damage.

Obviously a similar, fuzzy analysis of the set of measurable characteristics is possible. For example, when analysing deformations of transverse section of a sewer, they can be described by the affiliation degree referred to the statement on exceeding a limit value specified by e.g.: standards or guidelines. When considering the entire set of measurable and non-measurable characteristics describing a sewer it is therefore possible to obtain the χ_{ni} set of affiliation degree and then the X global characteristics presenting the sewer damage condition. This assessment method is an alternative to point-by-point methods characterising the condition of a sewer and similarly here the range of X values can be determined, pointing at the necessity of taking appropriate actions.

To sum up, when willing to count damages occurring in sewers creating a certain system, it is advisable to use an approach based on identifying damage condition for sections of sewers confined

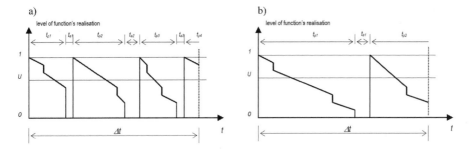

Figure 2. Example diagrams of operation for identical sewers with different renovation intervals (Przy-była 1999); $t_{p,i}, t_{a,i}$ – temporary realisations of random variables of operating time and renovation period, Δt – observation period.

within wells. Then a problematic necessity occurs of adopting conventional level of changes in those sections, which will characterize the damage condition rational in the reliability calculations for sewers. This conventional situation may be represented by appropriate number of penalty points, global X affiliation degree etc. An alternative to this approach is, obviously, counting all changes which have occurred in the sewer, and are recorded most of all during visual inspections. Yet, like the basic vice of the first approach is the very conventionality of the damage state, so in this situation it is hardly comparable influence of various observed changes on the reliable performance of functions by the sewer.

Let us then assume, that a certain U level has been adopted, reflecting the beginning of partial damage of a sewer. Therefore this is a clearly conventional boundary between the serviceability and unserviceability of a sewer in accordance with figure 1. In the considered model it is represented, for example, by: the number of penalty points – A or adequate affiliation degree $X = B$, $B \in (0,1)$. Depending on arrangements, the assumed level of damage (U) may, but does not have to classify the sewer for conducting its renovation urgently. In practice such emergency situation does not occur often, and by referring to the assessment method acc. to the ATV M 149, this corresponds to classifying the sewer to the "0" class. An assumption that the damage level U dictates that renovation is carried out within scheduled period of time will be more likely. Despite detection of sewer defect (classifying it to this status) the process of gradual worsening of sewer condition proceeds until the renovation has been made. It is carried out according to the schedule, or due to various reasons it is postponed. If the process of sewer operation is traced over time taking into consideration this aspect it will turn out, that it may cause consecutive problems in determining the number of sewer damage states. This is depicted in figure 2 (Przybyła 1999). In order to illustrate the example, the drawings are made with assumption, that the sewer worsening process is described by the first degree function complemented by step changes associated with random coercions acting within a short time. In both cases (a and b) the first degree curves are the same.

In "a" case, during observation period Δt, partial damage condition U occurred three times. In "b" case renovation was carried out only once, its quality also did not recover the sewer to its original condition. In this case, during Δt the partial damage U occurred twice. Paradoxically, the number of damages in this case is smaller – despite action incompatible with the schedule. Number of observed damage conditions U in those cases is distinctly dependent on the number of renovations and their quality, and this is incompatible with commonly adopted assumptions for the stationary character of the stream of damages. This example shows that reliability calculations carried out on the basis of transitions through the damage condition U counted over a long time interval may produce incorrect results.

3   PROPOSED METHOD FOR CALCULATING THE NUMBER OF SEWER DAMAGES

A method is proposed that enables the number of sewer damage conditions to be calculated for authoritative estimation of its reliability indexes, taking into consideration remarks from chapter 2.

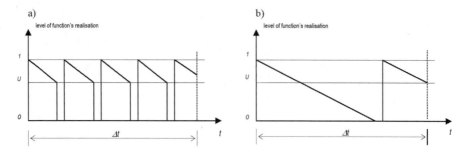

Figure 3.   Diagrams of sewer operation – extreme cases: a) renovation is carried out immediately after the occurrence of U damage condition, b) renovation is carried out only after the occurrence of full damage.

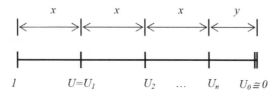

Figure 4.   Levels of function performance expressed by x sections with length of $|x| = |\overline{1U}|$.

Referring to the example from chapter 2, the aim is to obtain identical number of downtime states for both cases "a" and "b".

For simplification let's assume, that function describing the worsening of sewer condition over time will be a descending straight line. Moreover, in the sewer operating model two boundary cases occur, graphically illustrated in figure 3:

- sewer renovation is carried out immediately after finding the damage condition U;
- renovation is carried out only after the occurrence of full damage condition.

Between points 1 and 0, which define respectively the state of no damage and the full damage state (interpreted also differently as full performance of the function by the sewer or no performance of this function), there is a finite number of x intervals with a length of $|x| = |\overline{1U}|$ (section between points 1 and U). The x interval shows the difference between the state of correct sewer operation and a conventional level of partial damage state, authoritative for reliability calculations. Its is proposed that intervals with such length define the levels of sewer worsening condition (U_i) during its use between renovations – figure 4.

As sewer damage states assumed for calculations of its failure frequency measures, it is therefore proposed to adopt levels $U_1, \ldots, U_n$ and U_0 level close to the level of full sewer damage ($U_0 \cong 0$). In extreme cases of renovation the number of damages is calculated within the time interval Δt – figure 5:

- for the "a" case, the number of damages which occurred in Δt – 4,
- for the "b" case, the number of damages which occurred in Δt – 5.

The lack of similarity between results is caused by the difference in total lengths of renovation periods in both cases. In the "a" case four renovations were carried out, while in "b" case one renovation. These are extreme cases and the difference results from total lengths of renovation periods. However, in typical situations this difference will additionally be dictated by the length of x section assumed; the shorter the section is, the smaller difference between the results will be.

Figure 6 shows examples of simulated numbers of damages obtained depending on random instances abruptly changing sewer condition and renovations carried out inconsistently. It was shown in two variants, counting damages resulting from passing only the level $U = U_1$ and the resulting from passing all levels U_i. Purpose of the proposed method becomes visible – the

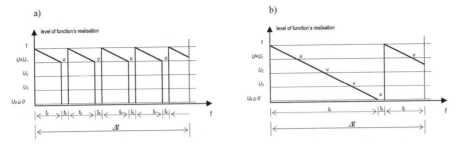

Figure 5.   Sewer damage conditions for calculating non-failure frequency measures: a) renovation is carried out immediately after occurrence of U damage condition, b) renovation carried out only after occurrence of full damage.

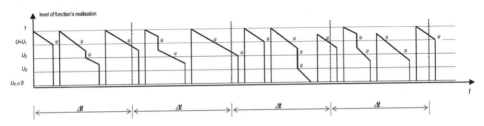

Figure 6.   Damage states obtained for various strategies of technical renovation (rehabilitation) of sewers within identical observation period. Amount of damages in the following time intervals Δt: 4, 3, 6, 4 (by passing all levels $U = U_i$); 3, 2, 3, 2 (by passing the level $U = U_1$).

additional damage levels allows for at least, partial account of deteriorating state of sewer's history in the reliability calculations.

The levels of damage state $U_1, \ldots, U_n$ and U_0 must be presented adequately to the method selected for the assessment of sewer condition. These may be the previously mentioned boundary numbers of penalty points or values of affiliation degrees X_i. Conceivable is also adoption of other method for representing the states of partial damage of sewers.

It also seems that practical number of U_i levels will be small – perhaps three or four levels including U_1 and U_0.

The solution proposed is approximate. Boundary values established for individual methods of assessment, determining the U level, are conventional, therefore the length of x section is relative. At the same time it is difficult to state unambiguously, how much the real nonlinearity of the function which describes the worsening of the damage state over time (or represents descending performance of the function assigned to the sewer) will influence the accuracy of determining the number of damages over time – in view of previously adopted simplifying assumption of its rectilinear variability.

A sine qua non for adopting this method is also the knowledge of changes in sewer condition over time. This assumption is conditioned by systematic examinations of sewers to be carried out by operators of sewerage systems. These inspections are classified as running inspections, repeated in certain time intervals, constant or adapted to sewer condition (usually frequency of inspections is increased with the worsening condition of sewers). Conducting such inspections can presently be deemed a standard in operation of sewerage systems, and assessing the state of inspected sections is a natural consequence of such inspections. In such situation one may admit, that identifying the levels of damage states U_i until renovation is carried out, is unrealistic.

4   SUMMARY

In Polish technical literature the subject of reliability of water and sewer infrastructure has been under consideration as early as from the beginning of the 1970-s. Over the next decades numerous works on this subject were presented, also after 2000. Summary of those extensive achievements including reference and comparison with studies created outside our country is contained in the PAN (Polish Academy of Sciences) compilation (Kwietniewski & Rak 2010). This work also points at significant disproportion between the number of studies concerning water and sewerage systems, to the disadvantage of the latter. At the same time increasing interest in the research on reliability of sewerage systems was underlined, especially in the studies written after 2000.

When looking at the literature of this subject, e.g.: (Kapcia 2006, Królikowski & Królikowska 2008, Leśniewski 2007), one may note, that complexity of assessing sewer condition and finding the occurrence of sewer damages is treated here sometimes too superficially and without reference to real practice in operation of systems. Therefore, this study focuses on this issue, and simultaneously maintains that it is one of the main problems limiting the possibility of direct transfer of solutions already proven in water supply networks to sewerage systems. Methodology proposed further for counting the states treated as sewer damages is a supplementary to the issue expressed earlier and has not undergone practical verification. Thus, it may be judged critically, posing a certain type of challenge for discussions on the issue considered here.

REFERENCES

ATV M 149: Zustandserfassung, -klassifizierung und –bewrtung von Entwasserungssystemen außerhalb von Gebäuden. 1999. Deutsche Vereinigung für Wasserwirtschaft, Abwasser und Abfall e.V. Hennef.
Bölke, K.P. 2009. Kanalinspektion, Zustande erkennen und dokumentieren, Berlin: Springer.
Kapcia, J. 2006. Ocena niezawodności podsystemu sieci kanalizacyjnej dla miasta Nowy Sącz. Konf. nt. Nowe technologie w sieciach i instalacjach wodociągowych i kanalizacyjnych. Wisła.
Kłoss-Trębaczkiewicz, H. & Kwietniewski, M. & Roman, M. 1993. Niezawodność wodociągów i kanalizacji. Arkady: Warszawa.
Królikowska, A. & Królikowski, J. 2008. Dwuparametryczna metoda oceny niezawodności podsystemu sieci kanalizacyjnej za pomocą metody MP+F, Konf. nt. Zaopatrzenie w wodę, jakość i ochrona wód, Gniezno-Poznań. Poznań.
Kwietniewski, M. & Rak, J. 2010. Niezawodność infrastruktury wodociągowej i kanalizacyjnej w Polsce. Studia z zakresu inżynierii, nr 67. Warszawa: Warszawska Drukarnia Naukowa PAN.
Leśniewski, M. 2007. Ocena zawodności kanalizacji deszczowej dla potrzeb analizy ryzyka z wykorzystaniem monitoringu parametrów hydraulicznych. Rozprawa doktorska. Warszawa: Wydawnictwo Politechniki Warszawskiej.
Madryas, C. 1993. Odnowa przewodów kanalizacyjnych. Prace Naukowe Instytutu Inżynierii Lądowej Politechniki Wrocławskiej. Seria Monografie. Wrocław: Wydawnictwo Politechniki Wrocławskiej.
Madryas, C. & Przybyła, B. & Wysocki, L. 2010. Badania i ocena stanu technicznego przewodów kanalizacyjnych. Wrocław: Dolnośląskie Wydawnictwo Edukacyjne.
PN-77/N-04005: Niezawodność w technice. Wskaźniki niezawodności, nazwy, określenia, symbole. Warszawa: PKN.
Przybyła, B. 1999. Ocena i kształtowanie konstrukcji przewodów kanalizacyjnych w ujęciu teorii nieza-wodności. Rozprawa doktorska. Prace Naukowe Instytutu Inżynierii Lądowej Politechniki Wrocławskiej. Wrocław: Wydawnictwo Politechniki Wrocławskiej.
Stein, D. 1998. Instandhaltung von Kanalisationen. Berlin: Ernst & Sohn.
Wieczysty, A. 1990. Niezawodność systemów wodociągowych i kanalizacyjnych. Kraków: Wydawnictwo Politechniki Krakowskiej.

Underground Infrastructure of Urban Areas 2 – Madryas, Nienartowicz & Szot (eds)
© 2012 Taylor & Francis Group, London, ISBN 978-0-415-68394-4

Application of liquid soil in an experimental construction of sewer pipeline

A. Raganowicz
Zweckverband zur Abwasserbeseitigung im Hachinger Tal, Taufkirchen, Germany

ABSTRACT: The article presents an application of the liquid soil in an experimental installation of stoneware sewer pipeline (DN 250). It consisted of three sections with a total length of 65 meters, situated $1.5 \div 2.3$ meters deep. What inspired the test was the necessary installation of a sewer pipeline in the immediate proximity of a heavily damaged residential building. The expert's report on the technical state of the building recommended a non compaction method of installation of a sewer pipeline within a radius of 100 meters.

1 INTRODUCTION

The Zweckverband zur Abwasserbeseitigung im Hachinger Tal (association for sewage in Haching Valley) operates sewer systems of three communities: Oberhaching, Taufkirchen and Unter-haching, which are located near Munich. The association area is drained in a separation sewer system. The sanitary sewer pipeline consists of two main collectors (concrete, diameter: $600/1100 \div 900/1350 \, mm$), public sewer (stoneware, diameter: $200 \div 400 \, mm$) and house connections (stoneware, diameter: $100 \div 200 \, mm$). The waste water from the association territory is discharged into the sewer system of the city of Munich.

During the planning of the sewer in the district of Oberbiberg in the community of Oberhaching it was discovered one of the pipes was planned to be constructed in the immediate proximity of a heavily damaged residential building. The expert's report on the technical state of this residential building recommended a non-conventional procedure of sewer installation. In such cases only a non-vibration construction of sewer pipeline is possible.

The company Readymix (now Cemex) from Aschheim nearby Munich offers two products mixed in plant for backfill trenches. A fine grained self-levelling material called "fuema boden" which can be applied to the backfill in the pipe zone and above up to the level of the bituminous surface a coarse-grained material called "fuema rapid". Related to the fact that this procedure was very unpopular in 2005 in southern Germany the association for sewage in Haching Valley took a decision on an experimental track for the construction of the sewer in the vicinity of the previously damaged house.

2 TECHNICAL DATA OF "FUEMA BODEN" AND "FUEMA RAPID"

"Fuema boden" is a very free-flowing and self-levelling material, which can be applied to non-conventional backfill of pipeline trenches and backfill of any kind of trenches. The product, which is mixed in a plant is composed of a mineral filler (grain size $0 \div 4 \, mm$), plasticizer, stabilizer, foaming agents, air-entraining agents and other additives. The final product is able to fill gaps at the pipe where a conventional compaction machine would simply fail. The liquid soil stays flexible even after the hardening process and the structure of this product is quite close to good conventional soil groups. This makes it easy to later work in the pipe zone, for example for a house connection which has to be added years after the job site has been finished. The liquid property of flexibility after the hardening process is a big advantage for flexible pipe systems, as pipe and soil supplement each other in an optimal way.

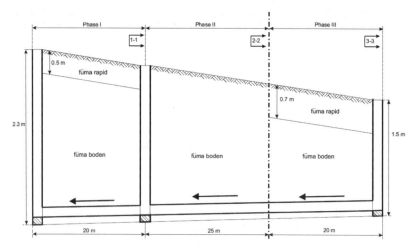

Figure 1.   Installation scheme of two liquid soils: "fuema boden" and "fuema rapid".

The mechanical strength of "fuema boden" is equivalent to the soil groups $3 \div 4$ according to the DIN 18300. After one day the "fuema boden" is walk-in and can then be built on. The other specifications are as a follows (Cemex):

– density: $1.4 \div 1.8$ kg/dm^3,
– e – modulus after 28 days according to DIN 18136: $120 \div 150$ N/mm^2,
– water permeability according to DIN 18130: $10^{-6} \div 10^{-7}$ m/s,
– E_{v2} – value according to DIN 18134: after 3 days > 45 MN/m^2,
    after 7 days > 120 MN/m^2,
    after 28 days > 180 MN/m^2.

The liquid soil "fuema rapid" is a free-flowing, coarse hydraulically stable detrained mineral mix. After 30 minutes of hardening it is possible to walk on the material and after 3 hours it is passable by vehicles. Special additives and a high water content ensure that "fuema rapid" in short-term can be brought into a flowable form and compressed independently. If the excess water can be discharged into existing soil, the material reaches its capacity very quickly. The e – modulus after 28 days according to DIN 18136 is only 60 N/mm^2 and the E_{v2} – value according to DIN 18134 after 2 hours is 20 MN/m^2 but after 28 days it is 200 MN/m^2 (Cemex). The analysis of the physical and mechanical properties of both liquid soils promised a successful construction of the planned sewer pipeline.

3   THE EXPERIMENTAL INSTALLATION OF THE SEWER PIPELINE

The use of liquid soil demands a careful planning of the job site and the work steps on the job site. The construction company has to prepare itself intensively in the handling of this embedding material. The producer of "fuema boden" and "fuema rapid" Cemex offers not only the materials but recommends a complete training package for consultant engineers as well as for the construction company, before the liquid soils are used for the first time.

The engineering office from the city of Landshut in consultation with the association for sewage in Haching Valley planned a test-installation of a sewer stoneware pipeline (L = 65 m), which was divided into two sections and three phases with a length of 20 or 25 m (figure 1).

Phase I is section 1 and phases II and III correspond to section 2. Phase II was backfilled exclusively with the "fuema boden" but the pipe-zone of phase I and III was backfilled equally with "fuema boden" and the deposit above the pipe-zone was filled with "fuema rapid". Three simplified geological profiles (profile 1-1, 2-2, 3-3), which were based on samples of the ground, were defined for the test pipeline and are shown in figure 2.

220

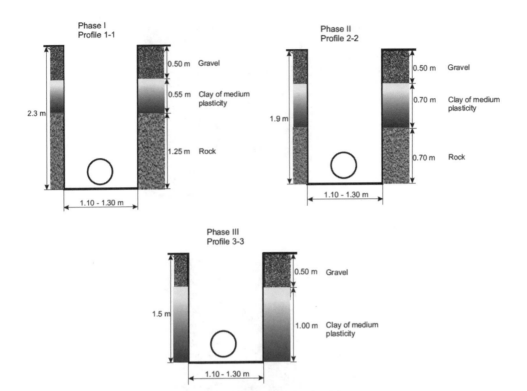

Figure 2.   Geological profiles.

Figure 3.   Sandbags in the area of the pipe connection.

   The construction company from the city of Landshut nearby Munich started to install this experimental stoneware pipeline with a diameter of 250 mm on a base of liquid soil in year 2005. The
depth of excavation was 1.5 m to 2.3 m. The stoneware pipes were positioned directly on small sand
bags and not on the excavation bottom. The exact levelling of the pipes could be done easily by
changing the form of the bags (figure 3). The shaft-pipe-connection was supported by two or three
bags. As an embedding liquid soil has higher specific weight compared to the pipes, the pipe is

Figure 4.   In the first stage of backfilling the excavation pipeline was completely covered without cavities.

Figure 5.   Even distribution of the liquid soil "fuema boden" in the pipe excavation.

under a buoyancy load during the backfill of the trench. Without a special construction method the pipes would swim on the surface of the liquid soil.

The pipes have to be fixed by an application of special banks from a rigid "fuema boden". In order to achieve a stiff consistency of the "fuema boden" one have to add less than $80 \div 100\,l$ water per cubic meter of the composition. A long route (approximately 25 km) between the plant in Aschheim and the pipe trench caused a change in the material, which meant that the banks could not be formed. In this context the pipes were filled with water. The liquid soil "fuema boden" was mixed in the plant in Aschheim and a concretemixingtruck delivered it to the job site. The installation of the "fuema boden" began at the deepest point of the pipe excavation (shaft 1).

A dangerous buoyancy force was prevented by a step by step installation of the liquid soil. Following this special method, the pipes constructed last in the trench were just filled up to the pipe axis, the pipe length already filled in the last step was filled up to the pipe crown and the pipes from first phase were filled up to the ground level (figure 4 and 5).

On this part of the pipes one had to pay serious attention until the liquid soil reached the pipe crown. The last 50 cm up to the ground level were on the next day filled with "fuema rapid". As part of the section 1 (phase I) a total 48 m^3 of "fuema boden" and 25 m^3 of "fuema rapid" were installed.

Figure 6. Fixing of a pipe with lean concrete.

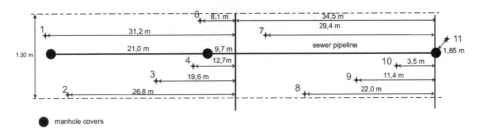

manhole covers

Figure 7. Localization of penetration tests.

An especially important aspect of this procedure is the pulling of the trench support (steel support). In this case the following two factors play a prominent role: the beginning of the solidification process of the liquid soil and the pressure resulting from the level of the fill material.

The filling of the trench in the phase II and III (section 2) was done similarly as in phase I (section 1). On account of the experience from the phase I the procedure was modified and the pipes fixed with concrete. The installation of "fuema boden" within the limits of phase II is shown in figure 6.

4 INVESTIGATION AND CONTROL

Directly after the installation of the liquid soils the condition of the pipe excavation was tested by using a penetration test (LPT 5). The whole experimental sewer pipeline was tested at 11 points in time, after 1, 7 and 28 days (figure 7).

As a result of the investigation it was found that both materials were able to achieve a degree of density R = 0.6 after 28 days. A typical increase of degree of density of two liquid soil zones after 28

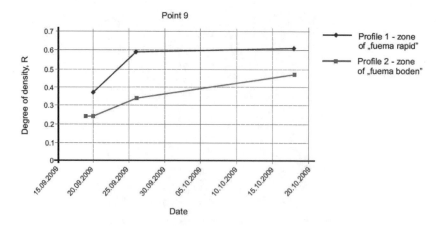

Figure 8.   Increase of the storage density of two liquid soil zones.

Figure 9.   Installation of a house connection.

days, which is taken from the penetration test at point 9, is shown in figure 8. The material "fuema rapid" reached the final degree of density $R = 0.6$ after one week. This effect results from the water permeability of the soil within the "fuema rapid"-area. In the case of "fuema boden" a steady increase in the degree of density was recorded during the hardening process. It was caused by low water permeability of the soil. Immediately after the backfilling of the excavation a CTT-inspection of the pipeline was carried out and no relevant defects were recorded.

5 CONCLUSION

The realisation of the experimental sewer pipeline installation showed that the liquid soils "fuema boden" and "fuema rapid" made a vibration-free construction of sewers in the district of Oberbiberg possible. The material "fuema boden" based on the grain size $0 \div 4$ mm is particularly appropriate for backfill of pipe zones. The liquid consistence of this material ensures that all gaps at the pipe are uniformly filled. This kind of bedding of pipes is very difficult to reach by conventional sewer construction.

"Fuema rapid" is particularly suitable for the backfilling of pipe excavations between the top of pipe zones and the lower edges of the bound ceiling. This material is frost resistant and can very quickly reach the following strength parameters, walked-in after 20 minutes and after 3 hours passable for automobiles.

Two months after the installation of the experimental sewer two house connections were constructed on the base of "fuema boden" and "fuema rapid". Both liquid soils were mechanically detached (figure 9).

The cost analysis of this construction process must take into account local conditions. In this case the analysis was partially distorted by special subsurface conditions. This procedure is economical when the price of the liquid soil is 50 €/m³.

It is advantageous in that the excavation width to the required level can be reduced to 0.85 m for pipes of diameter 250 mm.

The experience from the tests shows that the stoneware pipes must necessarily be supported on the complete pipe crown.

All in the pipe excavation located infrastructure-lines can be backfilled without complications in comparison to conventional installation.

Liquid soil is not the solution for all problems, but it can be of great help for difficult installation conditions. The material can reduce construction costs and can improve the quality of the embedment. It is very important to choose the right type of liquid soil, as there are big differences in the hardening process and the stability values.

The analysis of the results of the conducted pipeline constructions and the scientific research allowed us to formulate conclusions which state that the tested technology is very useful in shockless installation of sewer pipeline in urban areas, having taken into account local ground and water conditions. After comparing the technology costs of the liquid foundation and the conventional one, it turned out that the former is the more beneficial solution in cases, where it is necessary to replace the ground substratum within the area of the planned sewer.

The positive results of the conducted tests determined the decision to install a 300 meters-long stoneware pipeline (DN 250) on the base of "fuema boden" and "fuema rapid" in the immediate proximity of the damaged residential building.

REFERENCES

Cemex. 2007. *Selbstverdichtender Verfüllbaustoff für den Leitungs- und Kanalbau.*
DIN 18134: 2010 *Baugrund – Versuche und Versuchsgeräte – Plattenversuch.*
DIN 18136: 2003 *Baugrund – Untersuchungen von Bodenproben – Einaxialer Druckversuch.*
DIN 18300: 2010 *VOB Vergabe- und Vertragsordnung für Bauleistungen – Teil C: Allgemeine Technische Vertragsbedingungen für Bauleistungen (ATV) – Erdarbeiten.*

Sustainability in pipe rehabilitation

T. Schmidt
Institute of Urban and Pavement Engineering, TU Dresden, Germany

ABSTRACT: The phrase sustainability is often used, but hard to define in detail for projects in pipe rehabilitation. The paper gives some short introduction in the world of sustainability with its three pillars environment, society and economy. An approach is shown how to analyse sustainability. On the first general level of sanitary engineering, individual targets for each one of the pillar can be found. The second and more detailed level is the measurement and evaluation of sustainability in pipe rehabilitation projects. There are two main aspects: the choice of material and the choice of rehab technology. For both aspects criteria for the evaluation of sustainability are found. Although the evaluation of sustainability will be always an individual case-by-case problem, the author tries to give some recommendations for decision makers.

1 INTRODUCTION

Sustainability has become a key word for almost all aspects of social development. Concepts and solutions have to be sustainable in financial, ecological and general point of view. Sustainability is one of the most important criteria for the evaluation of projects and strategies.

But what exactly means sustainability? How is it defined? Which criteria can be used for measuring and evaluating sustainability, especially for urban water management and its projects? This paper shall discuss some of these aspects and gives some conclusions to the topic.

2 SUSTAINABILITY

Sustainability or sustainable development in its today's meaning has become widely known by the definition of the Brundtland commission, named after its chair Gro Harlem Brundtland, the former Norwegian prime minister. This World Commission of Environment and Development presented in 1987 a report called "Our common future" (United Nations General Assembly 1987) with the following definition for sustainability:

> "Sustainable development is development that meets the needs of the present without compromising the ability of future generations to meet their own needs." (United Nations General Assembly 1987)

This statement is usually projected on the three constituent levels society, environment and economy, which are called the three pillars of sustainability (United Nations General Assembly 2005). Sustainability is just reached by equal consideration of ecological, economical and social questions. The pillars are shown in figure 1, the overlapping area of all three circles indicates a sustainable state. Such a sustainable state is reached by measures which lead to economic prosperity, preservation of environmental integrity and social stability.

3 SUSTAINABILITY IN SANITARY ENGINEERING

The definition of sustainability as described above can be generally adapted to sanitary engineering. Table 1 shows a selection of targets which have to be fulfilled for a sustainable operation of sanitary networks, drinking water supply as well as wastewater removal networks. The targets are sorted following the idea of the three pillar model of sustainability.

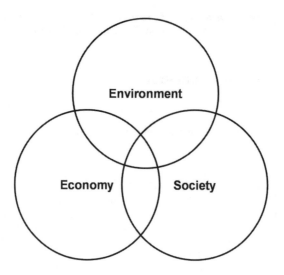

Figure 1.   The three pillars of sustainability.

Table 1.   Targets of sanitary networks (selection)

Environmental targets	Social targets	Economical targets
Water pollution control	Guarantee of sanitation	Financing of assets
Soil pollution control	Preservation of flooding	Continuous fees
Undisturbed hydrological cycle	Fair fees	Yield

The environmental and social targets are fulfilled by having water-tight, stabile and reliable networks. This can be reached independently from the applied building or rehabilitation technologies. Precondition for such a network is – besides a correct building process itself – a careful planning. The planning should be done on several levels, from the long-term network rehabilitation strategy to a detailed and precise object planning. During the object planning also innovative and unconventional approaches should be considered. Such approaches could be for example the building of small local solutions like small septic tanks or the hydraulic rehabilitation of sewers by using seepage and infiltration schemes instead of draining unpolluted storm water.

The economical targets should be also defined within a long-term rehabilitation strategy using a condition forecast. Each single object requires a dynamic cost-benefit analysis in which not only direct, but also indirect effects and social costs should be considered to evaluate sustainability.

If elements of sanitary networks meet the targets shown in table 1, they can in general be called sustainable. Nevertheless the technologies used for the building and the rehabilitation of sanitary networks differ very much in their sustainability aspects. In the following chapters such aspects will be closer analysed.

4   SUSTAINABILITY OF BUILDING AND REHABILITATION TECHNOLOGIES

For the evaluation of the sustainability of single objects have to be defined different criteria than the general targets of sanitary engineering. The 3-pillar-approach has to be projected from network level to single objects. On this level, there are two main aspects that have influence on sustainability. This is the building technology on the one side and the pipe material on the other. Both aspects can be independently analysed according to their sustainability. The overall evaluation of an object is gained by combining the evaluations of the separate aspects technology and pipe material.

Table 2. Criteria for sustainability evaluation of pipe material (selection)

Environmental criteria	Social criteria	Economical criteria
Pollutant emission	Health risks	Costs
Energy consumption	Use of resources	Useful lifetime
Use of resources Use of resources		

Table 3. Ecological sustainability of pipe materials (Stein & Brauer 2004)

Pipe material	Spec. Energy consumption during production process	Emission of CO_2 during production process
Concrete	low	low
Reinforced concrete	medium	medium
Clay	medium	low
Coated ductile iron	high	high
PVC	high	high
PE-HD	high	high

4.1 *Pipe material*

The material focused sustainability analysis should be also done obeying ecological, social and economical criteria. Some criteria (for example the use of resources) can not be sorted clearly in one of the groups, since they touch all aspects more or less. A selection of possible criteria is shown in table 2.

Especially interesting from the ecological point of view are the emissions of pollutants and the energy consumption during the production process of the pipes. Studies for analysing these aspects have been undertaken and show for example the results as stated in table 3 (Stein & Brauer 2004).

Social criteria for evaluating the sustainability of pipe materials rather don't play an important role, since they are usually fixed in regulations and standards. For example, materials which could be possibly risky for public health like lead or asbestos are forbidden. All materials that are used today fulfil these requirements. Of further importance are the used raw materials with respects to their amount available for human use. While the basic materials for concrete and clay pipes are continuously available, crude oil as the basic material for PE and PVC pipes is becoming more and more a rare raw material and should be rather used for other purposes.

For economical analysis of pipe material sustainability can be used the costs per meter and also the useful lifetime of the pipes. Each material has advantages and disadvantages in its specific properties and is therefore more or less adequate for a specific use. The useful lifetime of new build pipes from all materials is supposed to be at least 80 years. The price per meter depends on the prices for raw materials, energy and the necessary effort for production. There is a wide diversity of pipe prices on the market, so further information are not given in this paper.

There are already some studies published which show possible approaches how to analyse the different aspects of the sustainability of pipe materials (see Stein & Brauer 2004, Müller & Baudrit 2009).

4.2 *Building/rehabilitation technology*

There are three groups of technologies: repair, renovation and renewal. Repair and renovation are often done trenchless, depending on the pipe dimension. For new building of pipes and renewal of old pipes there are also trenchless technologies available as alternatives for open trenching technologies. The technologies are not explained in detail in this paper. The shares of the groups of technologies for German pipeline construction sites are shown in table 4 (Berger & Falk 2011).

Table 4. Shares of groups of technologies (Berger & Falk 2011)

| Year | New Building/Renewal | | Renovation | Repair |
	open trenching	trenchless		
2001	48.0 %	5.0 %	17.0 %	30.0 %
2004	40.1 %	8.8 %	26.1 %	25.0 %
2009	35.6 %	8.1 %	20.1 %	36.2 %

Table 5. Criteria for sustainability evaluation of building or rehabilitation technology (selection)

Environmental criteria	Social criteria	Economical criteria
Impact on (ground) water	Impact on traffic	Direct costs
Impact on soil	Impact on trade/business/industry	Indirect costs
Impact on flora and fauna	Impact on public life	Life-cycle costs
Impact on air	Useful lifetime Use of resources	Use of resources
Use of resources		

Table 6. Carbon dioxide emission depending on technology (Hölterhoff 2008)

Example construction site	Trenchless technology	Open trenching technology
Sewer DN 400, depth 3 m, length 200 m	CO_2 emission $\sim$ 10.2 t	CO_2 emission $\sim$ 29.3 t
Sewer DN 600, depth 4,5 m, length 250 m	CO_2 emission $\sim$ 22.2 t	CO_2 emission $\sim$ 59.2 t

Obviously the share of building in open trenching is decreasing over the years. It is analysed in the following if this trend indicates a sustainable development or not.

Again one may define criteria for the evaluation of sustainability from the three aspects environment, society and economy. A selection of possible criteria is shown in table 5.

For the evaluation of technologies with respect to their ecological aspect of sustainability the impacts on water, soil and air as well as flora and fauna should be analysed. Such impacts could be for example the destruction or pollution of natural soil, the chemical pollution of flora and ground water, the lowering of the ground water level or the pollution or air by the construction site and the chosen technology

The open trenching technology needs a lot of excavation and thus causes damages of roots and a disturbing of the natural soil. Also lowering of ground water can be necessary, depending on the depth of the excavation. Generally, a construction site with open trenching technology causes air pollution from the exhaust fumes of construction machines like lorries, excavators and so on. In (Hölterhoff 2008) an approach is shown for a rough calculation of the emission of carbon dioxide, which not includes the higher emissions due to traffic disturbances and rerouting. An extract from this study can be seen in table 6.

The emission of carbon dioxide is more than twice as high by applying open trenching technologies. The decisive factor for this is the necessary excavation and thus the transport of soil.

Trenchless technologies like burst-lining or micro-tunnelling can also damage roots. Due to the application of chemicals like injections for pre-sealing, resins or solvents, the groundwater and the soil could be polluted. However, all technologies have to be approved by governmental institutions – like for example DIBT in Germany – and thus are analysed with respect to their environmental safety.

Figure 2.   Trenchless construction site in Dresden, Germany.

The social impacts of a construction site show especially in the disturbance of traffic, trade and business. Obviously trenchless technologies have advantages, since they just need a starting and a finishing pit, but no trench in between. This means not only shorter construction times, but also less disturbing transports of soil. Furthermore, there are less additional annoyances like noise or dust. Another advantage of trenchless technologies is the availability of parking spaces and traffic ways for residents, deliverer and rescue vehicles.

Figure 2 shows an exemplary construction site in Dresden. There is an egg shaped sewer in a depth of 8 m under the street, indicated by the black dotted line. The renewal is done by micro-tunnelling, so the residents can still use the street and the shops may stay open.

The disturbances caused by using public space for construction sites do have also economical and ecological consequences. Traffic jams and reroutes cause a even higher carbon dioxide production. Depending on assumptions for traffic jam duration, fuel consumption of the affected cars and the duration of construction sites, the study from Hölterhoff (Hölterhoff 2008) calculates for an exemplary construction site a higher carbon dioxide exhaust up to 1000 t. This could be significantly reduced by applying trenchless technologies, so that the construction time would be shorter and the traffic disturbances are less.

To evaluate the economical aspect of sustainability it is necessary to watch not only the direct costs, but also to take the indirect or social costs into account. The direct costs are caused straight by the construction project itself and have to be paid by the customer, for example the municipal utility. In (GSTT Information Nr. 11 1999) the German Society for Trenchless Technologies (GSTT) compares the cost shares of open trenching and trenchless projects. For projects built with open trenching, the shares of pit lining, excavation, backfilling and replacement of street surface add up to about 70% of the whole project costs. This share is only about 22% for trenchless technologies. With increasing depths the share is still raising for open trench technology, while they stay almost constant for trenchless projects. In general it can be stated that trenchless technologies become more and more advantageous with increasing depth.

The social or indirect costs affect third parties. Usually nobody is accounted for, since they are very difficult to monetise. Examples are costs for changed traffic ways (e.g. higher fuel consumption, delay), impacts on residents (e.g. turnover losses, dust and noise emissions) or for long-term damages on other assets (e.g. reduced useful lifetime of pavements or nearby pipes). It can be assumed, that the indirect and social costs usually reach up to the same amount as the direct project costs.

The total costs of a project are the sum of direct and indirect costs. By applying a dynamic cost comparison method the most efficient and economic combination of material and technology can be found. It is important to take all cost shares into account as life-cycle costs. Therefore, the costs have to be calculated over the whole useful lifetime including operational costs and removal expenses if necessary.

5 SUMMARY

The sustainability of building projects for pipe rehabilitation can be determined by analysing the environmental, social and economical aspects. Thus, a multi criteria analysis of the project with respect to the three aspects of sustainability is necessary. For example a formalised weighing and ranking procedure can be used for such purposes, as it is described in (Plenker & Schmidt 2003). This helps to make transparent decisions. Unfortunately, the current practice focuses too often just on ostensibly costs aspects without scrutinising further criteria.

A real evaluation of sustainability can not be done overall but depends on the individual project. Nevertheless, there are some general trends for the material choice as well as for the technology choice. For material choice this means, that only in cases where it is really necessary, e.g. in pressure pipe networks, resource and energy intensive materials should be considered. For technology choice it is obvious, that trenchless technologies are in all aspects advantageous compared with open trenching. This is not valid for coordinated projects, where for example pipes from more than one utility are laid and also the street is rebuilt.

The consequences of non-sustainable acting show usually rather medium- or long-term, hence politics and society face the challenge to insistently put the sustainability aspects into the focus of decision makers in the area of sanitary engineering.

REFERENCES

Berger, C. Falk, J. 2011: Zustand der Kanalisation in Deutschland. Ergebnisse der DWA-Umfrage 2009. *KA Korrespondenz Abwasser*, Abfall 2011 (58) Nr. 1. DWA, Hennef, 2011.
GSTT Information Nr. 11. 1999: Kostenvergleich offener und geschlossener Bauweisen unter Berücksichtigung der direkten und indirekten Kosten beim Leitungsbau und der Leitungssanierung. GSTT, Berlin.
Hölterhoff, J. 2008: *Vergleich Kanalsanierung in offener und geschlossener Bauweise anhand ihrer umweltbezogenen Auswirkungen*. Presentation, 7. Kanalsanierungstage der DWA, Dortmund, 2008.
http://www.fbsrohre.de/fileadmin/user_upload/files/rwa_expertise_oekobilanz.pdf
Müller, B. Baudrit, B. 2009: Nachhaltigkeitsbewertung von Kunststoffrohren bei der Kanalsanierung. *3R international* (48), Heft 10/2009, Vulkan-Verlag Essen. S.557–561.
Plenker, T. Schmidt, T. 2003: Computergestützte Auswahl von Sanierungsverfahren für Abwasserkanäle. *KA Korrespondenz Abwasser*, Abfall 2003 (50) Nr. 2. GFA – Gesellschaft zur Förderung der Abwassertechnik e.V. Hennef, 2003. S. 166–171.
Stein, D. Brauer, A. 2004: *Leitfaden zur Auswahl von Rohrwerkstoffen für kommunale Entwässerungssysteme – Teilexpertise Umweltverträglichkeit (Ökobilanz)*. Study for the Fachvereinigung Betonrohre und Stahlbetonrohre (FBS), Bonn, 2004.
United Nations General Assembly 1987 *Report of the World Commission on Environment and Development: Our Common Future.* Transmitted to the General Assembly as an Annex to document A/42/427 – Development and International Co-operation: Environment.
United Nations General Assembly 2005. *2005 World Summit Outcome*, Resolution A/60/1, adopted by the General Assembly on 15 September 2005.

Underground Infrastructure of Urban Areas 2 – Madryas, Nienartowicz & Szot (eds)
© 2012 Taylor & Francis Group, London, ISBN 978-0-415-68394-4

Hydrodynamic simulation of sewage inflow control to the *Przemyśl* town's left-bank pumping station

D. Słyś, J. Dziopak & J. Hypiak

Department of Infrastructure and Sustainable Development, Rzeszów University of Technology, Poland

ABSTRACT: Intensive development of urban areas observed nowadays represents a challenge for the existing storm drainage and combined sewerage systems. Progressing impermeability of the catchment and construction of drainage systems in new areas results in temporary overloads in the existing drainage systems and facilities cooperating with them. These unfavorable phenomena require undertaking highly capital-intensive modernization projects. In this paper, based on the case study of the town of *Przemyśl*, we present results of hydrodynamic simulation performed for reconstructed combined sewage network cooperating with a system of storage reservoirs and the sewage pumping station. In the framework of the study, an analysis was carried out concerning effect of selected variants of control algorithms for these facilities on the profile of certain hydraulic parameters in the drainage system.

1 INTRODUCTION

Rapidly progressing urbanization processes in towns result in intensification of rainfall wastewaters surface runoff. This in turn is the cause of depletion of volume reserves available in existing sewers, flooding of sanitary network facilities, occurrence of the of the so-called "urban floods" and flood-type phenomena in urban water courses. Such situation requires undertaking of capital-intensive investment activities aimed at limitation of rainfall wastewater flows and controlling them by using appropriate wastewater storage facilities (Słyś & Dziopak 2009).

European standard EN 752 (PN-EN 752) requires that the investors are obliged to perform an analysis of sewerage system operation conditions before implementation of modernization activities. For this purpose, both commercially available and free software tools can be used allowing to simulate operation of a sewerage system in conditions close to actual one within a system approach.

This paper presents a hydrodynamic analysis of the effect of construction of two storage reservoirs on operation of the combined sewerage in *Zasanie* residential quarter of the town of *Przemyśl*. An assessment was performed of possibility to control effectively the wastewater flows within the sewerage system and inflows to the *Przemyśl* town's left-bank sewage pumping station for selected storage reservoirs system operation control algorithms.

2 FORMULATION OF THE PROBLEM

The subject of the research work consisted in analysis of operation of the combined sewage system in *Zasanie* residential quarter within the area of *Przemyśl* town for selected variants of cooperation within a system of storage reservoirs and sewage pumping stations. In its existing state, the sewerage network consist of a system of combined sewage channels, a storm overflow, and a pumping station that transfers sewage to the treatment plant located on the opposite bank of *San* river.

During dry weather periods, the sewage system transfers household and industrial wastewaters from *Zasanie* quarter with total surface area of 632.88 ha to the municipal wastewater treatment plant through a pumping station with maximum capacity of 900 dm³/s.

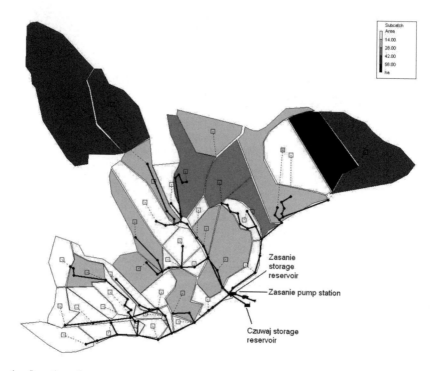

Figure 1.   Location of storage reservoirs and the pumping station in the *Zasanie* quarter of *Przemyśl* town.

As a result of physical development of the catchment area, in the periods of intensive precipitation, many cases of hydraulic overflow in main riverbank collecting sewers and the *"Zasanie"* sewage pumping station are observed, accompanied by a number of discharges through combined sewage storm overflow to *San* river exceeding statutory limits (Słyś & Dziopak 2011).

In order to reduce the frequency of these negative occurrences caused by insufficient throughput capacity of the existing sewage system or possibly eliminate it at all, a conception for extension of the quarter's drainage system was developed (Dziopak & Słyś 2008a, 2008b) consisting in modernization of the sewage pumping station and construction of two storage reservoirs, *"Zasanie"* and *"Czuwaj"*. The facilities will have capacity of about 10,000 m³. Figure 1 presents location of storage reservoirs and the sewage pumping stations against a plan of *Zasanie* and main sewers of the quarter.

3   CATCHMENT HYDRODYNAMIC MODEL

To analyze operation of the *Zasanie* quarter's drainage system, a hydrodynamic model of the catchment and the drainage system was developed in *Storm Water Management Model* software environment (*SWMM* Manual 5.0). For this purpose, technical documentation of the drainage system and related facilities was used, ortophotographic maps, data from geodetic surveys as well as results of measurements of precipitation and wastewater flows in the sewerage system, that were used for calibration of the developed model.

Moreover, a second version of the drainage system model was developed that was complemented with facilities provided in the system extension plan, i.e. construction of two storage reservoirs for accumulation of excess combined sewage in the course of rainfall flows. The adopted data characterizing the simulation model are summarized in Table 1.

Simulation calculations were carried out for a set of data on actual precipitation for the years 2007 and 2008, obtained from measurements registered by three rain gauges located in the catchment area. Input data for the hydrodynamic model characterizing the catchment and the drainage system

Table 1. Data characterizing the hydrodynamic model developed in *SWMM* program

Parameter	Parameter value/unit
Flow units	dm³/s
Infiltration model	Horton
Flow routing	dynamic wave
Report	10 min
Wet weather	10 min
Dry weather step	10 min
Link offsets	depth

Table 2. Input data for hydrodynamic model of the *Zasanie* catchment

Parameter	Unit	Value
Subcatchment impermeability ratio	%	10–60
Catchment surface area	ha	632.88
Reduced catchment surface area	ha	248.82
Hydraulic width of the catchment	m	171–2114
Manning's coefficient for impermeable surfaces	–	0.015
Manning's coefficient for permeable surfaces	–	0.30
Permeable surfaces retention height	mm	1.5
Impermeable surfaces retention height	mm	7.0
Minimum infiltration rate	mm/h	20
Minimum infiltration rate	mm/h	90
Infiltration intensity decay constant	1/h	4
Manning's coefficient for concrete conduits	–	0.02

were determined on the grounds of data obtained from the drainage system manager and from published sources (American Society..., Skotnicki & Sowiński 2009). The data are presented in Table 2.

4 SYSTEM OPERATION VARIANTS

The conception of reconstruction of the existing sewerage system in the "*Zasanie*" quarter assumes that two facilities will be provided for temporary storage of rainfall wastewaters. In view of their location, different roles to be played in the sewage system were assigned to them. The main task of the "*Czuwaj*" storage reservoir is to provide a relief to the main sewage collector sever that runs along *San* river towards the "*Zasanie*" sewage pumping station. At the same time the reservoir, in specific hydraulic conditions, allows to reduce wastewater flow to the pumping station through accumulation of a portion of the combined sewage flow.

The fundamental function of the "*Zasanie*" storage reservoir is to provide hydraulic relief to the left-bank "*Zasanie*" sewage pumping station and storm overflow located immediately upstream in such a way that statutory requirements concerning frequency of combined sewage discharge to *San* river will be met.

In the conception development stage, three basic variants were adopted for controlling operation of storage reservoirs, for which simulation tests were then performed.

Variant 1 – "*Zasanie*" and "*Czuwaj*" storage reservoirs operate independently based on separate filling and emptying process control algorithms. The "*Zasanie*" storage reservoir is emptied depending on the pumping station's capacity, while the "*Czuwaj*" storage reservoir is emptied by pumping starting from the moment of completion of its filling.

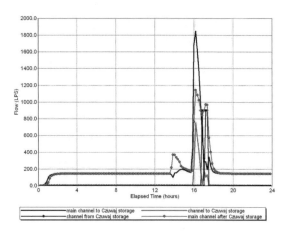

main channel to Czuwaj storage	channel to Czuwaj storage
channel from Czuwaj storage	main channel after Czuwaj storage

Figure 2.   Flow rate in inflow and outflow channels of the "*Czuwaj*" storage reservoir for overflow edge height of 0.6 m and the rainfall of June 23, 2008.

Variant 2 – "*Zasanie*" and "*Czuwaj*" storage reservoirs cooperate with the "*Zasanie*" sewage pumping station. The "*Czuwaj*" storage reservoir is subject to emptying in the first place. The "*Zasanie*" storage reservoir is emptied after emptying the "*Czuwaj*" storage reservoir.

Variant 3 – the "*Zasanie*" and "*Czuwaj*" storage reservoirs cooperate with the "*Zasanie*" sewage pumping station. The "*Zasanie*" storage reservoir is subject to emptying in the first place. The "*Czuwaj*" storage reservoir is emptied after emptying the "*Zasanie*" storage reservoir.

5   SIMULATION RESEARCH

5.1   *Variant 1*

The reason for which the "*Zasanie*" storage reservoir was introduced to the sewage system modernization plans was providing a relief to the "*Zasanie*" sewage pumping station. On the other hand, the use of the "*Czuwaj*" storage reservoir was aimed at provision of a relief to the quarter's main sewage collection sewer located upstream the storm overflow and the pumping station. In *Variant I*, the following operation scheme was adopted for storage reservoirs:

- the "*Zasanie*" reservoir is filled by pumping starting from the moment when the flow rate of sewage transported to the pumping station exceeds 900 dm^3/s;
- the "*Zasanie*" reservoir is emptied gravitationally, when sewage inflow to the pumping station is lower than its maximum capacity amounting to 900 dm^3/s;
- the "*Czuwaj*" storage reservoir is being filled gravitationally, when the filling of the main inflow channel to the reservoir exceeds an assumed overflow edge level equaling 0.60 m or 1.1 m, respectively;
- the "*Czuwaj*" storage reservoir is emptied by pumping, when filling level of the main inflow channel to the reservoir drops below an assumed overflow edge level equaling 0.60 m and 1.1 m, respectively.

The adopted heights of the distribution overflow edge located upstream the "*Czuwaj*" reservoir result form the wastewater mass balance. Overflow edge location height equaling 0.6 m and more provides protection against wastewater discharge to the "*Czuwaj*" storage reservoir in the sewage rainfall-free flows.

In view of the adopted hydraulic system of "*Zasanie*" pumping station storage reservoir it was assumed that the pumping system filling the "*Zasanie*" reservoir does not operate in the course of its emptying.

Figures 2 and 3 present simulation results for sewage flow rates in inflow and outflow channels of the "*Czuwaj*" storage reservoir for the overflow edge located at height of 0.6 m and 1.1 m, respectively.

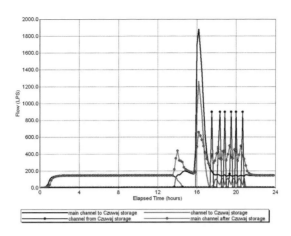

Figure 3.   Flow rate in inflow and outflow channels of the "*Czuwaj*" storage reservoir for overflow edge height of 1.1 m and the rainfall of June 23, 2008.

Further, Figures 4 and 5 shows the effect of different distribution overflow edge location heights of the "*Czuwaj*" reservoir on the wastewater flow rate in inflow and outflow channels of the "*Zasanie*" storage reservoir.

On subsequent Figures 6 and 7, levels of filling in wastewater storage reservoirs and in the pumping station are presented for the overflow edge located at the level of 0.6 m and 1.1 m above the main sewer's bottom.

Results of hydrodynamic calculations for *Variant I* that were carried out for most intensive and long-lasting rainfalls indicate that storage reservoirs effectively limit wastewater flow rates and counteract hydraulic overloads in both collection severs and pumping station.

Depending on height of overflow edge of the "*Czuwaj*" reservoir, different channel volume reserves were obtained depending on rainfall profile and intensity, while the largest values were always obtained for lower position of the distribution overflow edge, according to the general rules of wastewater balance on overflows.

Test results indicate also that the "*Zasanie*" storage reservoir effectively relieves the sewage pumping station. For the overflow edge height of 0.6 m in simulation tests performed for rainfalls from the years 2007–2008, no overflows were observed either in the pumping station or in the storage reservoirs. On the other hand, for the overflow edge situated at height of 1.1 m some short-term overfillings of the pumping station's reservoir occurred resulting in swelling in its inflow channel. Figure 8 shows result of simulation for the filling level in the "*Zasanie*" sewage pumping station inflow channel with diameter of 1.0 m for selected rainfalls.

5.2   *Variant 2*

Variant 2 of the storage reservoirs operation control algorithm assumed that emptying processes for both reservoirs will be correlated with operation of the sewage pumping station. It was also assumed that the "*Czuwaj*" reservoir will be emptied first. For *Variant II*, the following operation conditions were adopted for the reservoirs:

- the "*Zasanie*" storage reservoir is filled by pumping starting from the moment when the sewage flow rate to the "*Zasanie*" pumping station exceeds its maximum capacity of 900 dm^3/s;
- the "*Zasanie*" storage reservoir is emptied gravitationally, when the inflow of wastewaters to the pumping station is slower than its maximum capacity and when the "*Czuwaj*" storage reservoir is emptied;
- the "*Czuwaj*" storage reservoir being filled gravitationally, when the level of filling the main inflow channel to the reservoir exceeds the overflow edge level equaling 0.60 m;
- the "*Czuwaj*" storage reservoir is emptied by pumping, when the "*Zasanie*" sewage pumping station operates at a capacity less than the maximum and when the filling level of the main inflow channel to the reservoir will be lower than the overflow edge level.

237

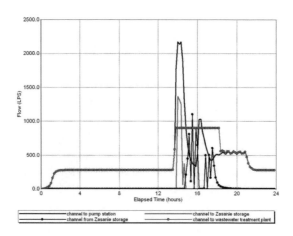

Figure 4.   Flow rate in inflow and outflow channels of the "*Zasanie*" reservoir for overflow edge height of 0.6 m and the rainfall of June 23, 2008.

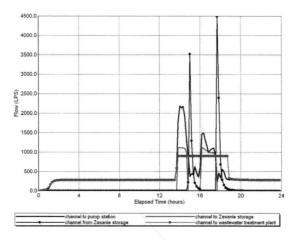

Figure 5.   Flow rate in inflow and outflow channels of the of the "*Zasanie*" reservoir for overflow edge height of 1.1 m and the rainfall of June 23, 2008.

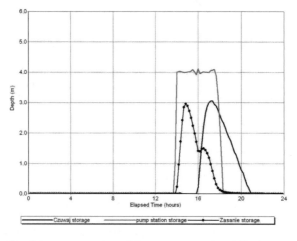

Figure 6.   Filling depths of storage reservoirs and the sewage pumping station's reservoir for overflow edge height of 0.6 m and the rainfall of June 23, 2008.

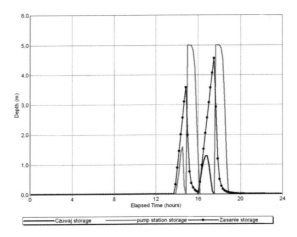

Figure 7.   Filling depths of storage reservoirs and the sewage pumping station's reservoir for overflow edge height of 1.1 m and the rainfall of June 23, 2008.

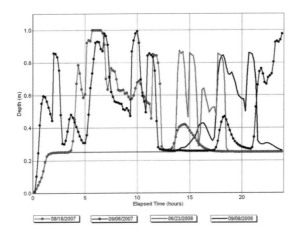

Figure 8.   The filling level in the "*Zasanie*" sewage pumping station inflow channel with diameter of 1.0 m for the overflow edge position height of 1.1 m for selected rainfalls.

It was assumed in this variant that the "*Czuwaj*" storage reservoir will be filled after exceeding of the filling level in the main inflow channel equaling 0.6 m, for which the whole rainfall-free flow is transferred directly to the pumping station.

Analysis of results obtained in the course of numerous simulations allows to conclude that in *Variant I* the process of sewage accumulation in "*Zasanie*" and "*Czuwaj*" storage reservoirs proceeds almost identically as in *Variant I.*

Differences in operation of the system can be observed in the case of introducing a control algorithm for storage reservoirs emptying different than this of *Variant I* and conditioned by the pumping station's throughput capacity. In *Variant II*, one can observe reduction of the sewage pumping station operation time at maximum capacity and increase of the filling level of both storage reservoirs an the pumping station's reservoir. Retention volume is used in this case to a larger extent than in the case of the reservoirs operating independently. Thanks to cooperation of storage reservoirs and this of the pumping station, favorable conditions for sewage flows in the main collector of the system were achieved. Figure 9 shows simulation results for storage reservoirs and the sewage pumping station filling up levels for a selected rainfall-wastewater runoff occurrence.

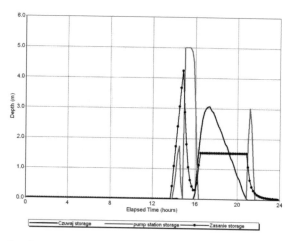

Figure 9. Filling depths of storage reservoirs and the "*Zasanie*" sewage pumping station reservoir for and the "*Zasanie*" pumping station's reservoir for Variant II and the rainfall of June 23, 2008.

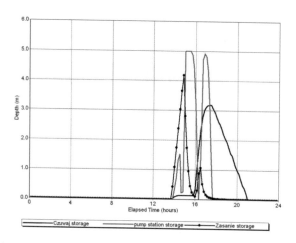

Figure 10. Filling depths of storage reservoirs and the pumping station's reservoir for Variant III and the rainfall of June 23, 2008.

5.3 *Variant 3*

In *Variant III* it was assumed that both storage reservoirs will cooperate with each other and with the pumping station, while the "*Zasanie*" storage reservoir will be emptied in the first place.

In *Variant III*, the following operation conditions for the system of storage reservoirs were adopted:

- the "*Zasanie*" storage reservoir is filled by pumping when the rate of sewage inflow to the pumping station exceeds its maximum capacity amounting to 900 dm³/s;
- the "*Zasanie*" storage reservoir is emptied gravitationally, when sewage inflow to the pumping station will be less than its maximum capacity;
- the "*Czuwaj*" storage reservoir is being filled gravitationally, when filling level of the main inflow channel to the reservoir will exceed the overflow edge level equaling 0.60 m;
- the "*Czuwaj*" storage reservoir is emptied by pumping, when the "*Zasanie*" sewage pumping station operates at capacity below the maximum, while filling level of the main inflow channel to the reservoir will be less then 0.60 m, and the "*Zasanie*" storage reservoir will be emptied.

240

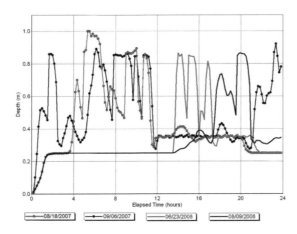

Figure 11. Filling level of the sewage pumping station inflow channel with diameter of 1.0 m for Variant III and selected rainfalls.

In *Variant III*, sewerage system operating conditions were similar to those of *Variant II*. Main collectors have been effectively relieved and did not need to operate under pressure. Only short-term overloads occurred in the "*Zasanie*" sewage pumping station inflow channel. Filling depths of storage reservoirs did not reach their maximum filling levels.

Analysis of operation of the "*Zasanie*" sewage pumping station allows to conclude that the period of its operation at maximum capacity was close to this observed in *Variant I*, however the pumping station's reservoir was for a period of time filled in full, contrary to *Variant I*. Such situation was caused by sewage outflow from the "*Zasanie*" reservoir that was subject to full emptying in the first place. Figure 10 presents wastewater filling levels in storage and the pumping station's reservoirs, while Figure 11 shows the filling level for the sewage pumping station inflow channel for selected extreme rainfalls.

6 CONCLUSIONS

It follows from analysis of the obtained results of numerical hydrodynamic tests carried out for selected variants of control algorithm for operation of the storage reservoirs system and the sewage pumping station located within the sewage system of *Przemyśl* town that effective control of wastewater flows require development of appropriate algorithms controlling operation of these facilities in order to synchronize processes of filling and emptying them.

Comparison of simulation calculations results for the analyzed variants allows to conclude that introduction of cooperation between storage reservoirs and the sewage pumping station had no significant effect on improvement of hydraulic conditions for sewage flow in main collecting sewers, but remained very important for hydraulic relieving and stable operation of the "*Zasanie*" sewage pumping station.

In the case of *Variant I*, where storage reservoirs operated independently, a more advantageous solution would consist in location of the distribution overflow edge at height 0.6 m in view of larger volume reserves that were created in the sewers and effective utilization of storage volume of the reservoirs without overfilling them. Moreover, it was observed that no cases occurred of pressure operation in the sewage pumping station inflow sewer, contrary to situation observed when the overflow edge was located at height of 1.1 m.

Taking into account the results of hydrodynamic simulations obtained for the analyzed variants of storage reservoirs control algorithms one can conclude that the most favorable solution for modernization of the existing sewage system is *Variant II*. For a number of rainfall-sewage flow occurrences, high effectiveness of this option was observed in hydraulic relieving of both network and the pumping station, as well as advantageous effect of synchronous operation of storage reservoirs in the process of emptying them on reduction of wastewater volume pressure-transported in the sewerage system.

Bearing in mind high costs related to investments in sewage networks and significant changes of hydraulic conditions occurring after modernization of such systems, development of a hydrodynamic model of the town should be a necessary and indispensable component of each investment process aimed at extension of urban sanitary infrastructure.

REFERENCES

American Society of Civil Engineers. 1996. Task Committee on Hydrology Handbook. ASCE Publications.
Dziopak, J. & Słyś, D. 2008b. *Significance of the planned storage reservoirs in the combined sewage system of the town of Przemyśl in Zasanie quarter and their effect on San river water quality status.* Water Supply and Sewage Company in Przemyśl.
Dziopak, J. & Słyś, D. 2008a. *Opinion on the study entitled "A technological conception of Zasanie sewage pumping station in Przemyśl" and development of an optimum design variant for the investment.* Water Supply and Sewage Company in Przemyśl.
EPA SWMM v.5.018. Manual v.5.0. US Environmental Protection Agency. http://www.epa.gov/ednnrmrl/models/swmm/.
PN EN 752. *Drain and sewer systems outside buildings.* The Polish Committee for Standardization.
Skotnicki, M. & Sowiński, M., 2009. *Verification of a method used for hydraulic width of subcatchments based on a selected examples of urban catchment.* Prace Naukowe Politechniki Warszawskiej. Inżynieria Środowiska 57: 27–43.
Słyś, D. & Dziopak, J. 2011. *Development of Mathematical Model for Sewage Pumping-Station in the Modernized Combined Sewage System for the Town of Przemyśl.* Polish Journal of Environmental Studies 20(3): 109–119.
Słyś, D. & Dziopak, J., 2009. *Assumption for optimization model of sewage system cooperating with storage reservoirs.* Underground Infrastructure of Urban Areas: 249–256. CRC Press.

Underground Infrastructure of Urban Areas 2 – Madryas, Nienartowicz & Szot (eds)
© 2012 Taylor & Francis Group, London, ISBN 978-0-415-68394-4

Hydrodynamic modeling of detention canal

D. Słyś, J. Dziopak & J. Stec
Department of Infrastructure and Sustainable Development, Rzeszów University
of Technology, Poland

ABSTRACT: As a result of intensive urban development the existing sewage systems have to meet the higher demands in the field of technical and environmental terms. One of the most important issues that affect the financial efficiency of the construction and operation of sewage systems and facilities, interacting with them, is the regulation of the flows of rain water, that is carried out mainly by reservoirs of various designs. The paper presents the hydraulic model of innovative solution for detention canal, with the corresponding mathematical model and an example of hydrodynamic modeling of such objects using the software Storm Water Management Model. The analysis shows that detention canal in certain investment conditions may be competitive solution for the sewage retention tanks. The presented solution of detention canal is a subject to patent application.

1 INTRODUCTION

The contemporary urban development is accompanied by the increase of the degree of surface sealing, mainly due to the construction of new buildings and roads. The intensification of surface runoff from the catchment area often leads to adverse effects, such as pressure wastewater flow in sewage networks, or the lack of possibility of wastewater discharge from urban area. Sometimes it brings to the necessity of building the additional network or extension of the existing one. The problems of hydraulic overloading of sewage systems, particularly in urban centers and industrial areas, only to a limited extent, can be solved with the use of equipment and facilities for draining rainwater into the ground. The main obstacles are the following: the lack of suitable land surface for the construction of such facilities, the high cost of purchasing the land and the high pollution of storm water runoff, which requires expensive treatment technologies.

Therefore, the aim should be to limit the amount of rainwater entering the sewerage system, especially in the areas with dispersed and single-family buildings development, where rainwater are not contaminated by large loads of pollutants and can be directly discharged into the ground or used for domestic purposes (Słyś 2009, Słyś & Bewszko 2010). In the case of storm water entering the sewerage system the aim of priority is to regulate their flow intensity, particularly by the use of storage reservoirs of various designs (Dziopak & Słyś 2007, Słyś 2010), interacting with regulators of flows (Kotowski & Wojtowicz 2009, Wojtowicz & Kotowski 2010) and pump systems.

In many cities the major problem is the lack of sufficient building territories for large storage reservoirs. Then one of the possible technical solutions are tubal reservoirs built within the existing collectors or built next to them in the form of by-pass.

The paper presents the hydrodynamic modeling of the original solution of detention canal, which is the subject of patent application (Słyś & Dziopak 2010).

2 PHYSICAL MODEL OF THE CANAL

The detention sewage canal is a part of storm-water or combined network of a definite length and diameter with consideration of their detention capacity. Appropriate determination of the required canal capacity is a priority in terms of determining the investment costs and the choice of wastewater storage option.

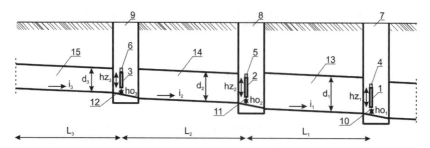

Figure 1. Detention canal model. 1 – the lowest piling partition, 2 – central piling partition, 3 – the highest piling partition, 4, 5, 6 – emergency overflows of piling partitions, 7, 8, 9 – piling partitions' chambers, 10, 11, 12 – through-flow holes of piling partitions, 13 – lowest storage chamber, 14 – middle storage chamber, 15 – highest storage chamber.

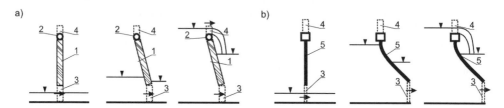

Figure 2. Operation scheme of piling partitions during rainless flows, flows subjected to piling and emergency flows for the types: a) articulated rigid barrier, b) flexible barrier. 1 – rigid piling partition, 2 – partition joint, 3 – through-flow partition hole, 4 – emergency overflow, 5 – flexible piling partition.

Canal capacity calculations are based on mass balance equations for wastewater flowing into the various chambers of the canal and flowing out of them and can be carried out according to the stationary methods or using the models of kinematic and dynamic waves.

Detention canal used in on-line solutions and in by-pass sewage collector consists of a separate storage chambers located in series relative to each other. Between the storage chambers there are the revision chambers located, in which the piling partition are suspended. In the lower part of the partitions the through-flow holes are located, which allow the flowing of wastewater during the periods without precipitation and the flowing of the reduced amount of rain water. On the other hand, in the upper parts of the piling partitions the emergency overflows are located, operating at the time when in storage chambers the emergence level of filling is reached. The scheme of detention canal of the elaborated solution is shown in figure 1.

The developed solution of detention canal assumes that, depending on the canal construction, the partitions piling the wastewater in storage chambers, can be made as articulated rigid barrier and/or as a flexible barrier. Mode of action of these barriers during the successive phases of wastewater accumulation in the detention canal is shown in Figure 2.

3 MODEL OF CANAL OPERATION

The phenomenon of gravitational fluid flow in the pipes has been described by the system of quasi-linear equations of hyperbolic type (1) in 1871 by Jean Claude Adhémar Barré de Saint-Venant.

$$\frac{\partial y}{\partial t} + v\frac{\partial y}{\partial x} + \frac{A}{B}\frac{\partial y}{\partial x} = \frac{q}{B}$$
$$\frac{1}{g}\frac{\partial y}{\partial t} + \frac{v}{g}\frac{\partial v}{\partial x} + \frac{\partial y}{\partial x} + S_f - I - \frac{q}{gA}\left(v_q - v\right) = 0$$

(1)

244

where: y – the state (depth), v – velocity, A – flow cross-sectional area, B – width of the mirror, q – distributed side inflow, g – acceleration due to gravity, I – inclination of the bottom of the bed, S_f – hydraulic drop, v_q – component velocity of side inflow down the channel

A mathematical model of De Saint Venant is used for accurate description of wastewater flow in sewage system, as it allows a detailed analysis of the variability of flow and wastewater level in the terms of time of wastewater transport and the length of the canal, as well as the possibility to identify the potential piling of reverse flows. Some parameters of the equation are of non-linear character, so the system of equation has no analytical solution and is solved in practice by numerical methods, using different software instruments, e.g. storm water management model, provided by the Environment Protection Agency. This program was also used to set the results of the detention canal investigation with consideration of actual data on amount of rainfall.

Hydraulic processes in the canal serving for the storage of wastewater during rainfall periods is complex and depends on a number of design parameters, in particular on the canal capacity, the size of piling partitions, the flow holes and drop of the canal. The general mass balance equation of wastewater in detention canal can be described by the equation (2).

$$
\begin{cases}
\dfrac{dh_1}{dt} = Q_2 \cdot F_1^{-1}(h_1, h_2) + Qa_1 \cdot F_1^{-1}(h_1, h_2) - Q_1 \cdot F_1^{-1}(h_1, h_4) \\[2ex]
\dfrac{dh_2}{dt} = Q_3 \cdot F_2^{-1}(h_2, h_3) + Qa_3 \cdot F_2^{-1}(h_2, h_3) - Qa_2 \cdot F_2^{-1}(h_1, h_2) - Q_2 \cdot F_2^{-1}(h_1, h_2) \\[2ex]
\dfrac{dh_3}{dt} = Qd \cdot F_3^{-1}(h_2, h_3) + Qa_3 \cdot F_3^{-1}(h_2, h_3) - Q_3 \cdot F_3^{-1}(h_2, h_3)
\end{cases}
\tag{2}
$$

where: dh_1, dh_2, dh_3 – changes of wastewater level in detention canal sections at the time dt, Qd – the intensity of wastewater inflow into the uppermost section of the canal, Q_4 – the intensity of wastewater inflow from the lowest section of the canal, h_1, h_2, h_3 – wastewater level in detention canal sections at the time dt, Q_1, Q_2, Q_3 – wastewater inflow intensity through the lower holes of piling partitions, Qa_1, Qa_2, Qa_3 – wastewater inflow intensity through upper holes of piling partition, F_1, F_2, F_3 – horizontal surfaces of the sections depending on the filling of detention canal sections by wastewater.

4 CASE STUDY

4.1 *Catchment description*

Basing on the description of hydraulic model of detention canal the hydrodynamic model of its functioning within sewerage system was formulated. In this regard, the software Storm Water Management Model was used, in which actual catchment area of Przemyśl city located in south-eastern part of Poland was implemented. The model area included Zasanie district with an area of 632 hectares and a population of approximately 35,000, which is located on the left bank of the San river – the tributary of the largest Polish river Vistula.

The City of Przemyśl has a gravity combined sewage system. Wastewaters from the city area are transported by the main collectors, which are located on both sides of the river San, to the municipal Wastewater Treatment Plant, which is located at the rout of the rivers San and Wiar.

Wastewater from the catchment area of the Zasanie left bank are carried across the river by pressure pipe using a pumping station. Maximum efficiency of pumps system transporting wastewater during rainy weather is 900 dm³/s. After connecting with the right bank's collector in decompression well, the wastewater flows gravitationally to the equalizing reservoir, located before WWTP, with usable capacity of about 9,000 m³ where the excess of wastewater is accumulated. The treated wastewater is discharged to the San river.

The main problem requiring the quick solution is the way sewage system operation during rainfall in the district Zasanie. Due to the extension of the city and the increase of the areas with impermeable surface, there has been an intense increase in the amount of storm wastewater discharged to the main collector of the district, which is hydraulically overloaded very frequently. In connection with this fact there is a need of the construction of storage facilities for retention of the excess of storm water.

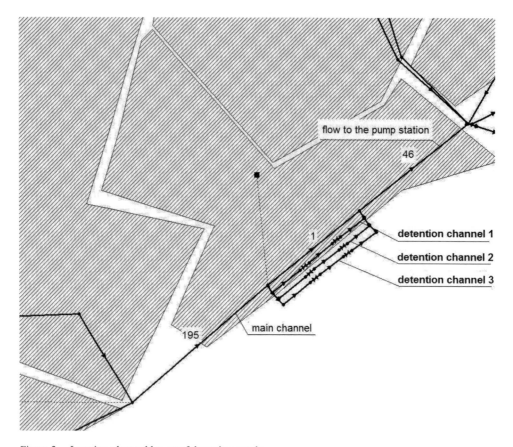

flow to the pump station

46

detention channel 1

detention channel 2

detention channel 3

1

195

main channel

Figure 3. Location plan and layout of detention canals.

The paper presents the results of hydrodynamic analysis of detention canal operation, which is located within the by-pass of main collector in Zasanie district. The study based on the measurements of precipitation in the period of 2007–2008 obtained from three pluviometers located in the studied catchments.

4.2 *Hydrodynamic analysis*

The observations of hydraulic conditions in sewerage system and the obtained balance of wastewater in Zasanie catchment area allowed to determine the required storage capacity of the facility unloading the main sewer collector of the district. Calculated retention capacity should be about 3000 m³. For the detention canal construction the use of pipes with a diameter of 2 m located in three parallel strings, each with a length of 300 m is expected. The total volume of the reservoir constructed in such way is 3203 m³. The localization of detention canal chambers is presented in figure 3.

Hydrodynamic studies or several rainfall events was carried out based on the developed and calibrated model of Zasanie catchment. In order to demonstrate the advantages of the innovative solution of the detention canal the comparative analysis of two scenarios was made.

Option 1 is based on the use of the tubular reservoir in parallel routs of pipes without piling.

Option 2 is based on the use detention canal in the system of parallel pipe routs with piling partitions according to patent application (Słyś & Dziopak 2010).

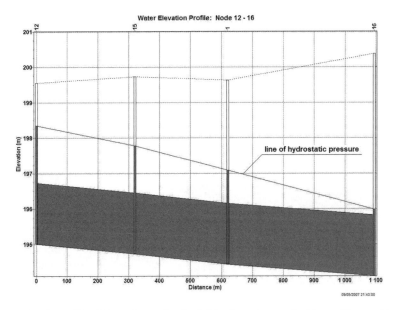

Figure 4. Hydraulic profile of the main canal unloaded by detention canal without piling partitions for the rainfall on 09/05/2007.

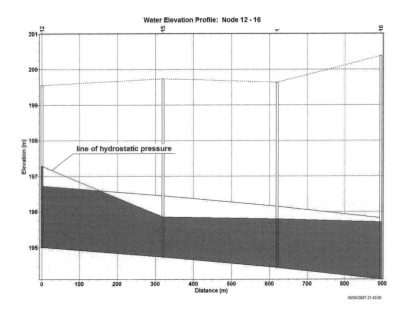

Figure 5. Hydraulic profile of the main canal unloaded by the detention canal with piling partitions for precipitation on 09/05/2007.

Figure 4 shows the maximum filling of the main canal by wastewater for the tubular reservoir without partitions, and in Figure 5 – the filling of the main canal for the tubular reservoir with piling partitions for one of the analyzed extreme rainfall.

Analysis of the results of simulations carried out for the highest storm water flows from the catchment shows, that the use of detention canal with the partitions allows a significant reduction

247

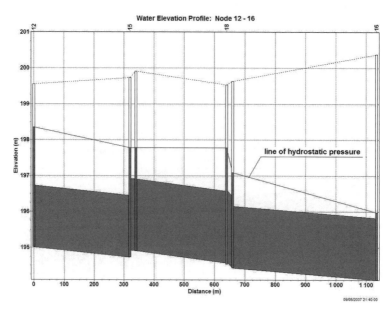

Figure 6. Hydraulic profile of the main canal and detention canal without piling partitions for the rainfall on 09/05/2007.

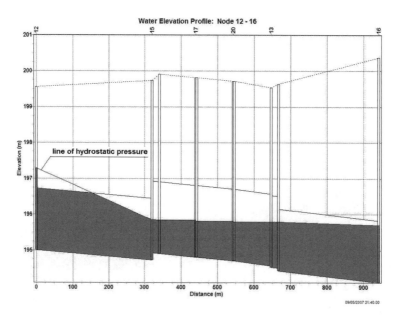

Figure 7. Hydraulic profile of the main canal and of detention canal with piling partitions for the rainfall on 05.09.2007.

of wastewater flow and the filling level in main collector. At the same time there has been favorable trend of the filling of detention canal under the use of piling partitions.

The results of calculations of maximum filling levels in detention reservoir without the piling partitions and in the reservoir equipped with piling partitions are respectively shown in Figure 6 and 7.

Link Depth

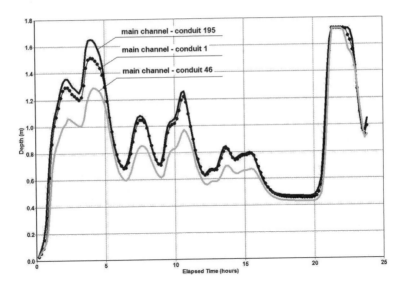

Figure 8. Filling of the existing non-unloaded main canal.

Link Depth

Link 195 —●— Link 1 ——— Link 46

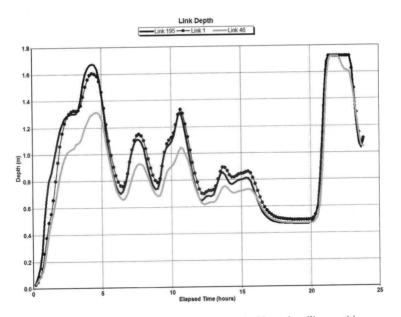

Figure 9. Filling of the main canal unloaded by detention canal without the piling partitions.

The obtained results of simulation in the range of main collector filling and wastewater flows in the existing network show a beneficial effect of detention canal with piling partitions on pressure flow limitation in the investigated collector.

Figure 8 shows in-time changes of filling of the existing non-unloaded main collector for selected rain. However, Figures 9 and 10 show the changes of filling for the same rainfall respectively for I and II options of detention canal.

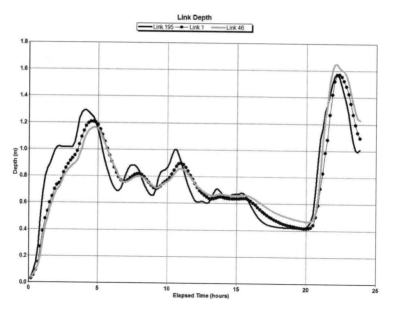

Figure 10. Filling of the main canal unloaded by tubal reservoir with piling partitions.

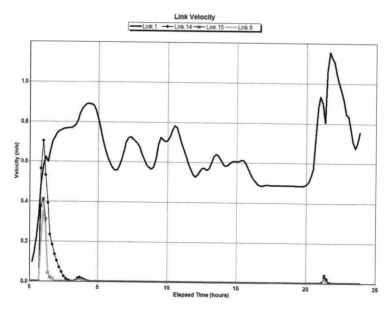

Figure 11. Wastewater flow velocity in the main canal and retention canals of tubal reservoir without piling partitions (Option 1).

Analyzing the results of the researches for more number of rains it can be concluded that the detention canal, made in accordance with option I, at the initial stage of storm flow has an unloading effect, related to the process of its filling and wastewater outflow from the main collector. However, in the later phase such solution of detention canal for certain hydraulic conditions can extend the duration of wastewater pressure flows in main canal.

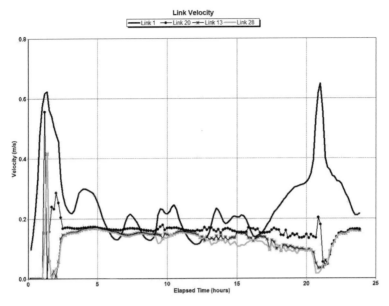

Figure 12. Wastewater flow velocity in the main canal and retention canals of tubal reservoirs with piling partitions (Option 2).

The flow of wastewater through detention canal is characterized by low velocities that can result in the sedimentation of pollutants transported by wastewaters. This phenomenon necessitates the need of much more intense canal flushing. Figure 11 shows the velocity of wastewater flow.

Option II solution of the detention canal is far more effective in equalizing of wastewater flow and reducing of the pressure effect of sewage network, making it also possible to obtain much higher wastewater flow velocity and to reduce the sedimentation of pollutants in the canal. Figure 12 presents the results of simulation of wastewater flow velocities for Option 2 in the retention canals and the main canal.

5 CONCLUSIONS

The paper presents the results of simulation researches using the hydrodynamic model of urban catchment and innovative solutions of detention canal. The researches were performed for the real data on amount of rainfall. The purpose of the analysis was to present the primary advantages of the developed solution of the object for wastewater accumulation within sewage system in relation to currently used classical solutions of tubal reservoirs.

Hydraulic analysis performed for a wide range of input data made it possible to formulate a number of general conclusions.

1. The solution of detention canal allows on much more effective reduction of wastewater flow rates and filling and their stabilization over long periods of time, than when using single-chamber tubal reservoirs of the same accumulation capacity.
2. The decisive importance for the process of wastewater retention has the usage of piling partitions within the hydraulic scheme of the reservoir
3. On the basis of the obtained results it was showed that the detention canal solution limits the pressure flow in the main sewage collector for longer periods of time than traditional tubular reservoir, which is characterized by off-loading action only during the process of its filling.
4. Under retention of wastewater in the detention canal more favorable conditions for its self-purification appears. In traditional canal due to suppressing of wastewater outflow in sewage system the flow velocity reduction below the speed of sand grains floating was observed, which may affect their sedimentation and bring to the necessity of additional flushing equipment usage.

The proposed technical solution of detention canal is a real alternative to traditional construction of storage reservoirs. This concerns in particular the areas where there is the lack of sufficient space for building of concrete chamber reservoirs, and the height profile of the sewage network allows the use of gravitational objects.

REFERENCES

Dziopak J. & Słyś D. 2007. *Modelowanie zbiorników klasycznych i grawitacyjno-pompowych w kanalizacji.* Politechnika Rzeszowska, Rzeszów 2007.

Kotowski A. & Wojtowicz P. 2009. Analysis of Hydraulic Parameters of Conical Vortex Regulators. *Polish Journal of Environmental Studies* 19(4): 749–756.

Słyś D. 2010. Application of numerical simulation in design of innovative Kalipso-type sewage tank. *Environment Protection Engineering* 3: 113–126.

Słyś D. 2009. Potential of rainwater utilization in residential housing in Poland. *Water and Environment Journal* 23(4): 318–325.

Słyś D. & Bewszko T. 2010. LCC Analysis of rainwater utilization system in multi-family residential buildings. *Archives of Environmental Protection* 36(4): 107–118.

Słyś D. & Dziopak J. 2010. *Retencyjny kanał ściekowy.* Patent Application P-391198. The Patent Office of The Republic of Poland. Warsaw. Poland.

Słyś D. & Dziopak J. 2011. Development of Mathematical Model for Sewage Pumping-Station in the Modernized Combined Sewage System for the Town of Przemyśl. *Polish Journal of Environmental Studies* 20(3): 109–119.

Wójtowicz P. & Kotowski A. 2009. Influence of design parameters on throttling efficiency of cylindrical and conical vortex valves. *Journal Of Hydraulic Research* 47(5): 559–565.

Underground Infrastructure of Urban Areas 2 – Madryas, Nienartowicz & Szot (eds)
© 2012 Taylor & Francis Group, London, ISBN 978-0-415-68394-4

Construction of pressure pipelines using the modern CC-GRP pipes

Z. Suligowski, M. Orłowska-Szostak & M. Szostak
Gdansk University of Technology, Faculty of Civil and Environmental Engineering

M. Cacaveri
Hobas Tubi SRL, Italy

D. Kosiorowski
Hobas System Poland, Poland

ABSTRACT: This paper covers current questions of developing the technology of trenchless laying pipelines with the use of microtunnelling. The paper presents successful trenchless installations of pressure pipelines with application of the CC-GRP pipes. The successful performance of the pipeline projects was possible thanks to application of the pipes since these possess the characteristics necessary for jacking as well as for pressure pipes. The paper describes performance of the projects of laying a treated water outlet pipeline under the lagoon of Venice in Italy. In the Italian project they used the method of microtunnelling to make a 351-m-long pipeline under the sandbar of Lido using pipes PN 06 with an external diameter of 1720 mm. A very large allowed jacking force of the pipes and the use of appropriate technology made it possible to perform the whole work from one jacking pit without intermediate jacking pits and intermediate jacking station.

1 INTRODUCTION

Laying pipelines using the method of microtunnelling was initiated in the USA and Japan. The first pipeline to be laid using this method was in Japan by the company of Komatsu in 1975. In Poland the technology of microtunnelling was used for the first time in 1997 as a part of the construction of new sewerage system in the town of Toruń. The sewage network was 973 m in length and 1600 mm in diameter.

The microtunnelling technology has been presented in many publications of technical literature. The description below was mainly prepared based on (Kuliczkowski et al. 2010, Madryas et al. 2002, PJA 1996, Zwierzchowska 2006).

The method is distinguishes itself from among the other no-excavation methods by the considerable level of computerisation and automation.

First, two pits are prepared, the starting pit and the receiving pit. Next, the microtunnelling equipment is installed on site. The equipment consists of (see Fig. 1): control container, jacking station, lubrication system, drilling fluid system with pumps, feed device and slurry head. Finally, cable and pumping systems for the removal of soil and lubrication fluid are performed.

A jacking station consists of hydraulic jacks and a jacking ring. The microtunnelling device is placed on special guide bars called the mount, which are present in the jacking shaft. The device is jacked into the ground by means of the jacking station. In the back part of the jacking shaft there is a thrust plate against which the hydraulic jacks are pressed when they are extended. The number and size of the hydraulic jacks is so selected as to ensure appropriate jacking forces. The task of the thrust plate is to distribute the force of the reaction of the thrust exerted by the hydraulic jacks and to transfer it to the ground behind it. The head of the microtunnelling machine is directly followed by the jacking pipes pushed into the ground, making up the pipeline constructed.

Figure 1. Pipeline and required facilities during construction (Kuliczkowski et al. 2010) 1. Jacking shaft; 2. Reception shaft; 3. Drilling head; 4. Installed pipes; 5. Frame with hydraulic jacks; 6. Shaft pump; 7. Drilling fluid conduits; 8. Control container; 9. Feed device; 10. soil separation plant; 11. Muck 12. Jacking pipe storage; 13. Crane.

During pipe jacking, sequential pipes are added until the drilling head reaches the reception shaft. During the pipe jacking the head's cutting shield breaks down ground at the microtunnel's face.

Depending on existing ground conditions, various types of cutting shield are applied: shields with cutting rollers in case of rocks or rocky soils, shields for cohesive soils which are equipped with cutting picks and shields with large inlet openings for loose soils. The ground, after having been broken down, is directed to the crushing chamber where it is comminuted by a cone crusher and larger stones are broken up. The muck is mixed with bentonite or polymer-bentonite slurry and transported to the surface through a system of conduits where it is separated from the slurry in the soil separation plant, including gravity separators, screens and hydrocyclones. Moreover, the drilling fluid's (slurry's) additional task is to prevent an outflow of groundwater from the surrounding ground which could cause the creation of caverns and ground failures at the microtunnel face which, in turn, could impact the surface. The drilling fluid circulates in a closed cycle between the head and the surface, being forced by two pumps: supply pump and return pump.

The right direction and alignment of the pipeline under construction are ensured by the drive control system which possesses a camera which transfers an image to be displayed to the systems' operator. The system uses a computer-based control of a drilling course, which is composed of: control container with operator's console, laser theodolite, electronic laser beam receiver, hydraulic jacks for steering the cutting shield and computer. An alternative to the laser theodolite is a gyro-compass-based solution composed of: a gyro-compass and a water level.

For the purpose of reducing friction between the outside surface of the pipes and the surrounding ground during jacking, a lubrication system is applied using a solution of bentonite and lubrication polymers. The slurry used to lubricate the outside surface of the pipeline is pumped through a conduit which is led inside of the pipeline constructed. Then it is supplied from the feed conduits to dividers and next to nozzles placed in the jacking pipes. The system requires the usage of intermediate jacking stations. The intermediate jacking stations are composed of a set of hydraulic jacks, usually placed in a steel casing which is adjusted to the outside diameter of the pipeline. The jacks push the jacking ring inside the station. In case the intermediate jacking stations are used the whole length of the constructed pipeline is divided into segments lying between the stations. The individual segments are jacked one after another using their corresponding intermediate jacking stations. The last segment is jacked by the main jacking station. It is equal to several smaller jacking processes. The use of the intermediate stations reduces the potential number of pipe damage occurrences as the maximum forces exerted on the segments depend on the number of pipes and coefficient of friction within a given segment. The intermediate stations are controlled from the

Table 1. Examples of microtunnelling applied to constructing pressure sewerage pipelines (PJA 1996).

Construction site	Year of construction	Length [m]	Outside diameter [m]	Working pressure PN [bar]
Orlando, USA	1998	1400	975	7
Sete, France	1999	230	1099	4
Tayside, England	1999	200	1447	3
New York, USA	1999	1000	1292	9
Honolulu, USA	2000	700	1453	4
Magdeburg, Germany	2001	300	1099	4

operator's console independently. After the machine has reached the reception shaft the head is taken out, the equipment in the starting shaft is disassembled and the shafts itself filled.

2 JACKING PRESSURE-PIPELINES – HISTORICAL OUTLINE, CHARACTERISTICS

The technology of microtunnelling may also be used for laying the pipelines of pressure sewerage as well as water supply pressure mains. Thanks to the special properties of the CC-GRP material the pipes can be jacked and later perform the role of the pressure pipeline. The first project with pressure CC-GRP jacking pipes was performed in Orlando (USA) in 1998. They used the Ø 1400-mm CC-GRP Hobas-type pipes there. 8 segments of the pipeline were laid at a depth of 5.5 m, groundwater table about 1.5 m above the pipeline and in sand difficult to condense. The longest segment was 310 metres long, no intermediate jacking stations were applied. The basic characteristics of the first pressure pipelines made of the pipes manufactured by Hobas are presented in Table 1.

Laying pipelines with the help of the microtunnelling method is often performed in difficult ground conditions at considerable depths, e.g. discharge outlets into sea. While jacking, the pipes undergo large loadings. Therefore, fastness to longitudinal loadings, which take place during the jacking process, is particularly important in selecting the appropriate pipes. Calculations should be performed in accordance with the guideline ATV-A 161 for both the pipes to be jacked along a straight line and a bend. In the process of selecting pipes for jacking, there are also many other factors taken into account, such as: mechanical strength, ways of joining and quality of outside surface. After installation, the pipes have to meet the severe requirements of daily operation, i.e. they must be tight, corrosion-proof, hydraulically smooth and resistant to high pressure

3 JACKING PRESSURE PIPES OF TYPE CC-GRP MANUFACTURED BY HOBAS, PIPE REINFORCEMENT

HOBAS jacking pipes are made of unsaturated polyester resins, cut glass fibre, quartz sand and a mixture of calcium carbonate and resin. In the process of automated centrifugal casting all the components are delivered into a rotating mold by means of a feeder and afterwards the rotation speed is increased. The structure of the pipes created from the outside layer to the inside layer at a pressure from 30 to 70 bar ensures for the appropriate density and hardening of the material. The process of centrifugal casting ensures round cross-section of the pipe and equal thickness of pipe wall over its perimeter. The layered structure enables optimum transfer of stresses. The layers reinforced with glass fibre allow for the transfer of perimeter stresses coming into being as a result of loading exerted by the surrounding ground, traffic impacts, inside pressure of the transported liquids. The middle layer, with a considerable quantity of mineral fillers, ensures an appropriate compressive strength which is especially important in the process of pipe jacking and pushing. Thanks to the spatial chemical bonds of resin the pipe, being made of thermoseting material, behaves in a stably way even in increased temperatures.

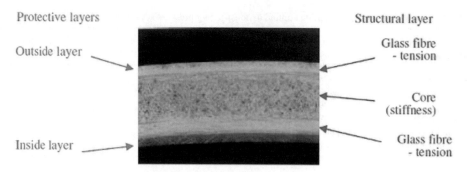

Figure 2.   Layered structure of the HOBAS jacking pipes (http://kolektorczajka.pl).

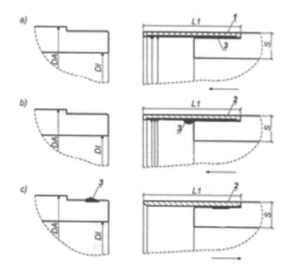

Figure 3.   Couplings for the GRP pipes a) Stainless coupling b) and c) GRP coupling with GRP 2 – Ring from GRP, 3 – Gasket (Madryas et al. 2002).

The pipe couplings used in microtunnelling are characterised by such a construction that their outside diameter is equal to the outside diameter of the pipes which they are to be connected. The couplings applied to the low-pressure pipes manufactured by Hobas, in the scope of up to PN03 with diameters from DN 150 to DN 1500, are made of stainless steel or polyester resin. The coupling of stainless steel consists of a ring of stainless steel and gasket made of an ethylene-propylene-diene terpolymer. The FWC couplings used to join the high-pressure pipes for jacking are manufactured of a composite similar to the composite of the pipes by winding method.

The FWC couplings are provided with an elastomeric seal of an ethylene-propylene-diene ter-polymer. The seal is an integral part of the coupling and can easily be adjusted. Thanks to that the pipe joints are tight, keeping up to the parameters of the pipes themselves. Tests of tightness and pressure tests of the pipe couplings performed in compliance with the requirements of the standards EN 1119, ISO 8639, ASTM D 4161 (loading conditions as per the standards EN and ISO are more rigorous). When the tests were performed in conditions of angular deviation and transverse centre line shift caused by shear loads, no leaks were observed. The pressure testing included tests of influence of long lasting and periodically changing pressure on the pipes' parameters. A pipe with the PN10 class of nominal pressure was subjected to changing pressure where the inside pressure changed in a range from 7.5 bar to 12.5 bar with a frequency of 19 cycles per minute. The bursting tests performed on test pieces of the pipes did not show any changes to their strength after 1 million cycles.

HOBAS jacking pipes are characterised by a high longitudinal compression strength which allows the pipes to be jacked although their wall's thickness is smaller in comparison to pipes made of other materials. This allows reaching smaller outside diameters and lower weights than those of the pipes with the same diameter manufactured by other producers, thus using smaller machinery which reduces costs of purchasing or hiring equipment as well as power consumption. A sequential advantage is the reduction in volume of muck to be removed when compared with pipes made of other materials.

Application of pipes with smaller diameters is also possible thanks to the increased through-put of media in the case of Hobas pipes, being a result of their lower coefficient of wall surface roughness, which has experimentally been proved for the pipelines being in service for the last several years. Measurements showed that the hydraulic roughness of wall surface amounted to k ~ 0.01 mm. Thanks to the very smooth inside wall surface of the CC-GRP pipes, they are characterised by excellent properties such as: large throughput and slight flow resistance relative to those of other products with similar dimensions. The inside layer, rich in resin, protects the pipe wall's inside against even very strongly corrosive media. The pipe wall inside is also not susceptible to tuberculation and incrustation. In Hobas pipeline systems there have been noted neither failures nor any throughput reduction resulting from sediment gatherings inside the pipes.

The CC-GRP pipes also possess a very smooth outside surface that ensures low friction between the pipes and surrounding ground during the jacking process which, in turn, allows for the application of large distances between the jacking and reception shafts and avoidance of the usage intermediate jacking stations. The outside surface of the pipes is permeable to a very slight degree so that the pipes absorb some small quantity of water and as a result they do not stick to wet soils. Thus, the pipes offer a relatively low resistance, even at the beginning of jacking after a longer time of standstill. In some cases, depending on ground conditions and groundwater level, it is necessary to reduce the friction between the pipes and surrounding ground so as to increase the distance between the jacking shaft and the reception shaft. In the case of accessible pipes, bentonite-and-water slurry lubrication is applied. The jacking pipes adapted to supplying the bentonite-and-water slurry are fitted with holes with nozzles and non-return valves to which there are connected conduits supplying the lubricating slurry. After completing the jacking process, the holes are plugged with appropriate stoppers and specially laminated so as not to worsen hydraulic parameters of the pipe. When the indispensable jacking force exceeds the value permissible for a given pipe, which makes it impossible to push the pipeline over the whole distance from one station, then an intermediate jacking station has to be applied. Use of an intermediate jacking station increase safety of the jacking process as well as gives a substantial advantage in the form of cheaper pipes (a smaller pipe wall thickness is required). The intermediate jacking stations, in the case of pressure pipes, are applied for accessible pipelines with a diameter of $DN \geq 1000$ mm, this is because of the possibility of later dismantling the hydraulic jacks and thrust rings. Such connection is laminated from inside with the use of glass fibre mats and polyester resin with the object of obtaining a fully tight joint of the pipes.

Depending on pressure class and diameter of pipes, there are various couplings for connecting the Hobas system with other materials, laminated joints or flanged joints. For the purposes of water supply, constructing forcing collectors and other pressure applications, it is possible in definite cases to apply, beyond the HOBAS GRP pressure fittings, pipe fittings and accessories made of cast iron or spheroidal cast iron that are connected to the CC-GRP pipe systems by means of flanged joints. It is also possible to use fittings of steel. The assortment of fittings includes elbows of arbitrary angle, symmetrical and asymmetrical reducers as well as T-pipes and pipe laterals with the same, or reduced diameter.

The Hobas jacking pipes comply with the requirements set out in the standards for the GRP material – no matter what manufacturing process are applied. These are the following standards: ISO 10639, ISO 10467, PN-EN 1796 and PN-EN 14364. In order to present product properties in more detail, a special specification of pipe quality was introduced for the GRP pipes cast centrifugally. The requirements of the specification define over-standard parameters in order to ensure high quality of product. HOBAS possesses the octagonal quality marking issued by TÜV OKTAGON, and many other approving certificates.

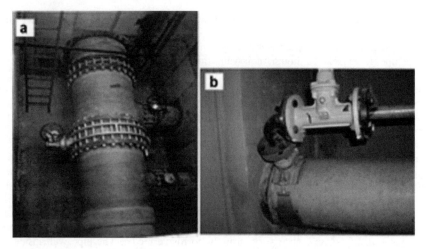

Figure 4. a) Valve chamber with dewatering T-pipes, b) Hawle pipe saddle on Hobas pipe (http://kolektorczajka.pl).

4 SAMPLE PROJECT: THE VENETIAN LAGOON IN ITALY – JACKING OF HOBAS PRESSURE PIPES UNDER THE LIDO SANDBAR IN VENICE

The agriculture and industry development in the Italian region of Veneto causes a more and more substantial pollution of the Venetian Lagoon. In 2000, in order to save the largest in Italy and one of the most famous in the world lagoons the authorities of the region developed a comprehensive plan for preventing further pollution and conditioning the waters running off the area of the lagoon's drainage area. The project of Fusina (originally: P.I.F. – Progetto Integrato Fusina) included the rebuilding of the water treatment plant in the town of Fusina where the sewage and rain water from the area inhabited by a population of circa 350 thousand people, as well as industrial effluents and polluted groundwaters from the port of Marghera are purified. Part of the purified waste water is discharged into sea by means of a 20-km long pipeline with a diameter of DN 1400. The pipeline conveys water over a distance of 10 km from the establishment in Fusina to the Lido sandbar in Venice and the sequential 10 km to the sea where tankers stay. In order to lay the pipeline under the Lido sandbar, they applied microtunnelling technology while constructing a 351-m long segment of the pipeline using pipes with a diameter of DN 1400 and pressure class of PN 6.

The sea outlets for purified waste water are currently a standard and common solution. The use of jacking pipes with the object of laying the pipeline under the Lido sandbar allowed for reduction of environmental interference to a minimum. The company of Mantovani company decided, after a long search and penetrating analyses, to use the HOBAS CC-GRP jacking pipes as they possess all the features indispensable for both pressure and jacking pipes, i.e.: considerable mechanical strength and optimal hydraulic parameters. Usually, in order to meet all the requirements it is necessary to apply two different pipe systems. The pipeline was laid at a depth of circa 5.5 m below sea level. The jacking shaft and reception shaft were situated at sea, so in order to construct them two artificial peninsulas were built around them. The total length of the pipeline amounted to 351 m. The pressure pipes selected for the jacking technology used there were 3 metres long and of an outside diameter of 1720 mm and with a wall thickness of 85 mm. Their maximum jacking force amounted to 6,926 kN. The applied system was provided with high-class pipe couplings that are characterised by an extraordinary tightness and can withstand a working pressure from 6 to 10 bar. They met the design requirements of the project and ensured successful construction of the pipeline under the Lido sandbar.

The pipeline was laid in sandy ground, which is inconvenient with respect to the necessary jacking force. Yet, the very smooth outside wall surface of the pipes allowed application of relatively small forces and thus the pipeline segment for which the jacking technology was planned was laid without intermediate jacking stations. The fact that at the 221st metre of it (it corresponded to the 74th pipe)

Figure 5.   View of a construction site (to the left); Fully equipped intermediate station (to the right).

Figure 6.   View of the jacking shaft with the brake mounted on the jacking pipe.

the indispensable jacking force amounted to merely 900 kN confirms that the friction resistance of the pipes was really very low. After 14 hours of standstill, when jacking was started again at the beginning the 75th pipe required a pushing force of 2040 kN, whereas at the end of the pipe only a pushing force of 1040 kN was large enough. Thanks to that only the jacking and reception shafts were used and it was possible to resign from usage the three intermediate stations that had been constructed between the following pipes 17–18, 40–41 and 75–76. As a result, a considerable time and cost savings were attained.

Because of the buoyancy there is a possibility that a pipeline will be pushed out of the ground. Therefore, a brake was applied in the jacking shaft that blocked the possible retreat of the pipeline when a sequential pipe is prepared, as well as cabling and drilling fluid conduits are reinstalled.

A coupling was placed at the pipe's forehead which made it possible to exert a maximum jacking force in case of such a necessity. If the couplings were mounted at the other end of the pipe it would be necessary to use a spacer ring with a diameter smaller than the pipe's diameter so the maximum possible jacking force would have to be reduced. In order to diminish the friction between the pipes and ground, the bentonite-and-water slurry was applied as a lubrication measure. The pipes were fitted with openings with nozzles and non-return valves. The lubricating nozzles were present in every fourth pipe (three over pipe perimeter). When jacking had been completed the openings were plugged with stoppers and laminated.

The pipeline's construction was executed with the use of the MTBM (Micro Tunnel Boring Machine), machinery with slurry system. The low hydraulic roughness of the CC-GRP pipes and thanks to that their smaller outside diameter in comparison to pipes with the same inside diameter

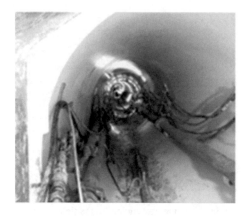

Figure 7.   Pipeline inside view showing technological lines and conduits (left), Pipe with lubricating nozzles (right).

but made of other materials, allowed the use of smaller jacking machinery. That resulted in a smaller labour force, as well as reduced volume of excavated soil (muck) to be removed – which allowed for the minimising costs and time of the project which was executed in less than one month. After completion, the pipeline underwent pressure testing. During 24 hours the pipeline was subjected to a pressure of 5 bar. The testing was successful. No operation irregularities were noted.

5   SUMMARY

Apart from the fundamental advantages following from the structure of the pipelines used in the technology of microtunnelling the most important advantages of microtunnelling with the use of the CC-GRP pressure pipes are as follows:

– Parameters of offered pipes; the pipelines excellently meet double requirements – for pressure pipes and for jacking pipes (wide range of stiffness).
– Relatively low weight and practical couplings; depending on their diameter and pressure class they may be joined by means of various connecting pieces, flanged joints or by laminating. It is also possible to obtain an angular deflection at the pipe couplings which allows for jacking pipelines both along a straight line or a bend
– Very smooth both inside and outside pipe wall surface ($k \leq 0.01$ mm).
– Small absorption of water by the pipe wall's outside surface; at the beginning of jacking pipes, even after a longer standstill, the pipes show a relatively small friction between pipe and the soil during jacking.
– High chemical resistance. High corrosion and abrasion resistance. Installation independent from atmospheric conditions. Operation period up to 100 years.
– Series of other, very improving construction work progress, technical solutions described in the paper (e.g. the brake in the jacking shaft described in the paper that protects against jacking pipe retreat, easiness of cutting and adjusting pipes at the building site, wide assortment of pipe connectors).

The excellent pressure parameters of the CC-GRP jacking pressure pipes manufactured by means of the centrifugal method have been confirmed, in practise, in many large projects of constructing pipelines using the method of microtunneling.

REFERENCES

HOBAS. 2011. *The CC-GRP pipes – properties, tests*, advantages. HOBAS System Polska.
HOBAS. 2010a. *Jacking pipes and microtunnelling*. HOBAS System Polska.

HOBAS. 2010b. *Hobas for the Venetian Lagoon in Italy*. HOBAS PipeLine.

http://kolektorczajka.pl/ – the web site for the collector of the Water Treatment Plant Czajka in Warsaw.

Kuliczkowski A., Kuliczkowska E., Zwierzchowska A., Zwierzchowski D., Dańczuk P., Kubicka U., Kuliczkowski P., Lisowska J. 2010. *Trenchless technologies in environmental engineering*. Wydawnictwo Seidel-Przywecki.

Madryas C., Kolonko A., Szot A., Wysocki L. 2006. *Microtunnelling*. Wrocław. Dolnośląskie Wydawnictwo Edukacyjne.

Madyras C., Kolonko A., Wysocki L. 2002. *Construction of sewers*. Wroclaw. Publishing House of the Wroclaw University of Technology.

PJA.1996. *Guide to best practice for the installation of pipe jacks and microtunnels*. Pipe Jacking Association.

Zwierzchowska A. 2006. Mikrotunelowanie i przeciski hydrauliczne. Warsaw. *Nowoczesne Budownictwo Inżynieryjne:* 32–37.

Zwierzchowska A. 2006. *Trenchless technologies for building gas, water and sewage networks*. Kielce. Publishing House of the Kielce University of Technology.

Underground Infrastructure of Urban Areas 2 – Madryas, Nienartowicz & Szot (eds)
© 2012 Taylor & Francis Group, London, ISBN 978-0-415-68394-4

Possibility of increasing precipitation water retention in agglomerations based on examples from Warsaw experience

Z. Suligowski & M. Orłowska-Szostak
Gdansk University of Technology, Faculty of Civil and Environmental Engineering, Poland

O. Łużyński
Hobas System Poland, Poland

ABSTRACT: The paper covers issues related to the rainwater management in urban areas. The growth of the intensity of rainwater runoff requires solutions that increase retention abilities of the urban drainage area. Application of covered water storage reservoirs are becoming a necessity. Pipe reservoirs are occupying a particular place among them. Solutions deployed in the Warsaw agglomeration have been discussed. It has been shown, that for areas with a great usage intensity a preferred approach can be pipe reservoirs built with an application of microtunelling technology.

1 INTRODUCTION

The questions of precipitation water retaining in urban areas are rated among problems of strategic importance to colonization. The fundamental problem is the increase of area of sealed land surface of catchment areas within urbanised localities. At the same time, it is observed an increase in the intensity of storm precipitations essentially impacting the technical conditions. Although the magnitude of yearly precipitation does not change often, yet it gets more concentrated in shorter time periods. It results in more intensive run-offs and makes the so far design assumptions not up to date any more. The phenomenon is an objective fact and practically cannot be avoided.

There are also secondary consequences of those changes in the precipitation water run-off conditions while the traditional canalising and sewerage solutions are applied. These are as follows:

– Increasing disturbances in water circulation when compared with what is observed in the nature (in natural environments only ca. 10% of precipitation water runs off directly to an water receiver whereas it is from 80 to 90% in urbanised catchment areas,
– Violent increase in water levels in receives, and in consequence propagation of flood phenomena,
– Statistical increase in hydraulic loading of existing elements of systems devoted to precipitation water management.

In case of general sewerage systems there is an additional hazard being a consequence of too frequently switching of the storm overflows. Under extreme conditions, they have to operate non-stop, draining sewers without purification.

In addition to the environmental consequences, continuation of the traditional policy without worsening the settling conditions is impossible as:

– adaptation of these solutions to such condition changes is costly and often practically unfeasible,
– negative phenomena will recur more and more frequently,

Finally it may appear that exclusion of some areas from use will be necessary.

The only rational solution is to increase retention capacity of the catchment area of the urbanised area under consideration. Here, the following is feasible:

– Elongating the precipitation water run-off time through catching water in tanks,
– Facilitating water infiltration into ground,
– Parallel use of both of the methods.

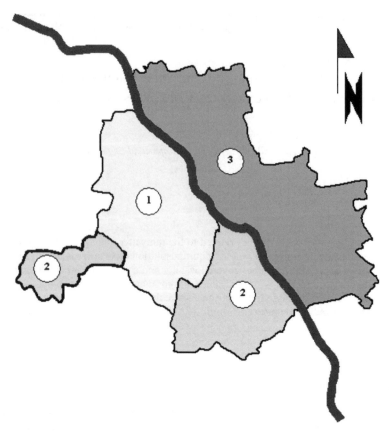

Figure 1. Arrangement of sewerage catchment area in the Warsaw locality: 1 – combined, 2 – separate, 3 – mixed.

Stopping run-off and directing water to other use is perhaps attractive, yet technically ineffective.

In areas intensively used, the fundamental problem is lack of place, in turn the climate conditions in Poland do not allow seriously considering the evaporation reservoirs. The problem with fog and rime limits the purposefulness of using open reservoirs in the neighbourhood of mayor roads.

2 SPECIFIC PROBLEMS OF WARSAW

The sewerage in Warsaw operates in a characteristic spatial configuration shaped by historical conditions (Fig. 1). The sewage system on the right-hand (eastern) bank of the Vistula operates in two different ways – in older areas there is combined sewage system and in newer ones separate sanitary and storm sewerage systems. As for the left-hand (western) bank of the Vistula, the most urbanised area which is the city centre possesses a combined sewerage system. New housing estates located in the south of the city are served by separate systems. In the west of the city, the housing estate Ursus is connected to the sewerage of the town of Pruszków. Existence of a combined sewage system serving a mayor part of an agglomeration produces an additional hazard to the reservoir water by the waters from storm overflows.

The superior thing is that the whole agglomeration is situated within the catchment area of the Vistula. Canalising waste waters in a city essentially influences flow conditions in local courses and finally in the Vistula. The right-hand, considerably lower, bank is more susceptible to changes of the Vistula's water level. It is true that there are protective embankments there but when the river water reaches its higher levels the phenomenon of backwater occurs (Fig. 2). Then, through outlets

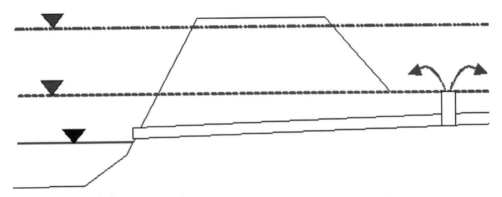

Figure 2.   Phenomenon of backwater resulting from high a water level in the Vistula.

and gullies water floods lower located grounds. An appropriate separation of the existing sewage system situated in the higher located grounds and in the zone of direct hazard might be a good practical solution here. The run-off from the grounds lowly situated when the river water level is high should be stopped or pumped (polder system).

In areas where combined sewerage system is present there are often problems in lower storeys. The state of entire emptying and filling up connected with flooding are the same operation conditions of each sewerage carrying waste water. Hence, it is possible to say about a level as "safe" only with reference to the level of a street (Fig. 3). The problem may mainly be solved by appropriate application of anti-backwater valves.

A sequential problem is the streams existing in the city which are indirect receivers of precipitation water. At present, they do not possess a sufficient reserve of carrying capacity. The first floods are likely to take place at a considerable distance from the main receiver that is to say the Vistula river, the flows are almost critical. The problem also concerns areas of relatively lower intensity of use. An additional consequence is frequently recurring possibility of hazard to areas of special value.

Finally, within the area of Warsaw agglomeration the existing conditions do not allow further sewering precipitation water. It is necessary to implement diverse solutions that are adapted to local possibilities. In the situation that the problem of waste water treatment plants will be regulated shortly, precipitations waters become to be a problem of strategic importance.

3   RETENTION COMBINED WITH INFILTRATION

Under the Warsaw conditions, retention combined with infiltration was first and foremost applied in the southern area which is characterised by a lower urbanisation degree and dispersed building. Interesting experience has been obtained during the activities undertaken within the catchment area of Potok Służewiecki (catchment area of the Wilanówka river). The result is that a programme has been undertaken for the housing estates because the traditional anti-flooding solutions appeared to be too costly. In the alternative solutions there were applied both natural materials (gravel, sand) and water infiltration boxes.

Under conditions where the whole precipitation water is directed to the ground (lack of storm water drains and any reservoirs to catch precipitation water) an infiltration system composed of perforated pipes with gravel filter around, which also constitutes a retention and infiltration tank was used to drain water from internal streets. The solution is needed because of the insufficient permeability of the soils. The street gullies are connected to inspection chambers on water infiltration drains.

Within another hosing estate, a precipitation water infiltration system was installed in the ground to unload the melioration canal running along the estate's boundary. Part of the system draining water from the streets and roofs on the street side of buildings has been operating since 1997

265

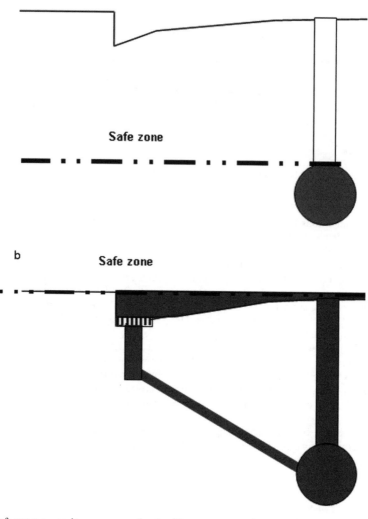

a

Safe zone

b   Safe zone

Figure 3.   Safe zone: a – sanitary sewerage, b – (each) sewerage carrying precipitation waters.

without a problem whereas the rainfall water draining to the inside with a surface area of 0.78 ha was executed improperly, worked faultily causing flooding cellar and had to be improved.

In the case of using infiltration installations, the problem is how to adapt the devices to existing conditions. In particular, infiltration wells do not prove and the use of infiltration boxes should depend on actual external loads. The undersoil should have properties allowing receiving the run-off water. The materials used (filter, geofabric) must be adapted to the locally existing conditions and correspond with the requirements characteristic of their manufacturer. An alternative solution that could applied instead of infiltration boxes is the use of chambers characterised by a number of advantageous solutions.

4   RETENTION – PIPE TANKS

Pipe tanks composed of the large-diameter GRP-type pipes (pressure class PN1, stiffness class SN 10000, of DN 1800, 2000, 2400 or 3000 mm, inspection shafts on DN 1200 mm tank) are used in the drainage systems of the communication routs in Warsaw. The general rule of constructing

Figure 4.   General principle of constructing a pipe retention tank, e.g. of materials from HOBAS.

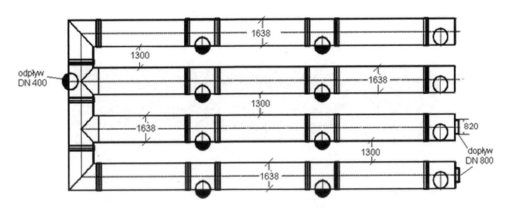

Figure 5.   Schema of pipe retention tank, e.g. of materials from HOBAS.

pipe tanks is presented in Figure 4 und Figure 5. The tanks have the task, apart from retention, to partially clear the precipitation water by preliminary settling of solid matter. The dimensions and position of such an inspection shafts (Fig. 6) allow, as distinct from a number of the tanks constructed based on small infiltration elements tightly shielded by a geomembrane, carrying on normal operation. The tanks based on chambers or boxes with modified structure are an exception.

In the years 2009–2011 in Warsaw in an open trench there were installed pipe tanks with a total length of 3.3 km. In the framework of the project for rebuilding the Jerozolimska Avenue there were constructed 2 tanks (DN 2400, 90 m in length and DN 2000, 22 m in length) with a total volume of 440 m³. The pipes were laid at a depth of 4.8 m under difficult soil and groundwater conditions (water table ca. 0.3 m above mounting level) in a soil difficult to compact.

Sequential tanks were constructed in Warsaw in the framework of building the dewatering system of the expressways S2 and S79. The dewatering system of S2 includes 11 tanks composed of DN3000 pipes with a total length of 2.5 km and unitary volume of varying from 650 to 3500 m³, in total they are capable of retaining over 17,000 m³ of water (Fig. 7). The period of performance of the S2 project extends over the years 2010–2011. The S79 project was carried out in the years 2009–2010, and in its framework they constructed 5 tanks of the DN1800 pipes. They were constructed as an assembly of 3–8 of parallel pipes with a length of 9–50 m and volume of 82–3500 m³.

267

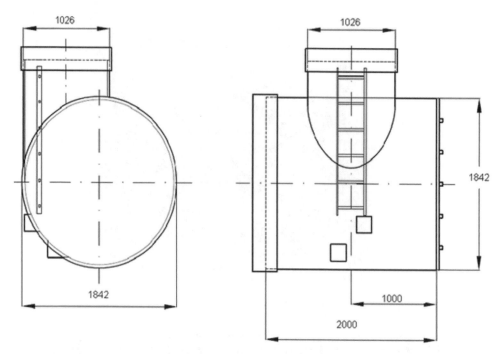

Figure 6. Principle of constructing inspection shaft on tank, e.g. of materials from HOBAS.

Figure 7. Example of a pipe tank constructed in the framework of the S2 project, DN 3000 mm pipes, (3–8) parallel pipes, e.g. of materials from HOBAS.

The particular feature of the pipe retention tanks is their easy installation. At the same time, it is less likely that technological departures take place and in particular use of improper materials as distinct from what is often encountered in case of the classical boxes. They were built in the city in the technology of open trench but not in areas of intensive use.

5   POSSIBILITIES OF CONSTRUCTION IN CITY CENTRES

The common problem of all urbanised areas is lack of free space which would allow conducting earth works. It is even more complicated in the case of city centres where – it is difficult to evaluate in practice but – exceeding a trench depth of 3 m would be a great success. The trenchless

Figure 8.   Microtunnelling in Warsaw (Modlińska Street, Ø3000 mm), materials from HOBAS.

technologies, including the classical tunnelling that allows laying very-large-diameter ducts, have been spread practically in the second half of the 20th century. As a result they began to use them to build underground tanks (e.g. in Tokyo) to periodically retain and store precipitation or overflow water.

The specificity of Warsaw consists in the extraordinary compact building development with predominating the relatively old combined sewerage system (Figure 1). Even the high value of the run-off coefficient used in designs by Lindley ($\psi \approx 0.7$ was the standard used for the city centre areas) it becomes obsolete with time. At present the average of the run-off coefficient amounts to $\psi \approx 0.8$, and under the oncoming conditions which will additionally limit infiltration and make run-off even faster it should be $\psi \approx 0.9$. In the Warsaw reality it is of particular importance that over at least 20 last years the city centre building has become more and more compact what caused that the volume of both sanitary and precipitation waters has increased considerably. The symptoms of it have been observed for a longer time.

It is not very likely that it will be possible to make the work in the technology of open trenches. Interesting experience was gained in the projects connected with the process of modernising the sewerage system of Warsaw (Fig. 8, Fig. 9). The record-breaking results reached in microtunnelling and laying CC-GRP Ø3000 (3000/2800 mm) pipes, and in particular:

– execution of a straight pipeline of 930 m in length,
– performance of 2 bends: 249.9 m and 230.67 m in length, with a bend radius of 900 m and 450 m,

269

Figure 9. Chamber on the right-hand bank of the Vistula, October 2010.

speak much of the capabilities of the method based on single jacking shaft. In the closing stage of the construction includes a Ø3600 mm pipeline.

Microtunnelling in areas of compact building development may be considered in two aspects:

- constructing new collectors to take over a part of precipitation water to (limiting its inflow into the combined sewerage system, diminishing the effect of sealing catchment area),
- building underground retention tanks to unload the water receivers which are practically reduced to the Vistula river.

A large-diameter pipe allows for a retention volume of 615 m³ and over 850 m³ per 1 hm of pipe. In result it is possible from one jacking shaft to push into ground at a depth from several to ten and several metres a retention tank with a volume of several thousands of cubic metres, while the execution of the operation does not impair the activities of the users of the area.

Such tanks may also solve the problems the areas protected by embankments, which is especially important on the right-hand (eastern) bank of the Vistula. A solution for such problems could be:

- Designation of a safe area and an area endangered by backwater,
- Application of anti-backwater devices on the boundary line between both the areas,
- Application of tanks taking run-off from the endangered area for a period of time,
- If necessary, construction of new outlets for the safe areas and installation of the anti-backwater valves in the outlets within the endangered areas.

6 SUMMARY

The strategic task in the area of Warsaw and other large agglomerations is to limit the magnitude of precipitation water run-off. Apart from selected locations it may be obtained by constructing retention tanks. In turn, the protection of flood plains in cities requires for the existing water drainage systems and their outlets to be modernised.

The projects performed in the framework of the Warsaw programme showed the essential advantages of the CC-GRP pipes in the function of retention tanks. In road investments performed outside the areas of high intensity of use, the pipe tanks are installed with the use of the technology of open trench as standard. The trenchless technologies are particularly interesting in highly urbanised areas. According to the so far experience, the technology of microtunnelling may be used both to build high volume retention tanks and to modernise the existing sewerage systems.

Underground Infrastructure of Urban Areas 2 – Madryas, Nienartowicz & Szot (eds)
© 2012 Taylor & Francis Group, London, ISBN 978-0-415-68394-4

Analysis of cylindrical microtunnelling working shaft, subjected to horizontal load

W.St. Szajna & P. Malinowski
University of Zielona Góra, Poland

ABSTRACT: Working shafts are subjected to horizontal loads caused by jack's reaction while microtunnelling. The loads result in deformations of the shaft-soil system and generate internal forces in the shaft lining. The paper presents a numerical analysis for designing a cylindrical lining of a working shaft for microtunnelling, loaded horizontally. The objective was to determine internal forces and displacements of the shaft for several structural models and contact conditions. Three shaft models were considered: a spatial model, a shell model and a beam model. A spatial and Winkler's models were used for the subsoil. The linear-elastic model of materials was assumed. The task was formulated in terms of the finite element method. The 3-D task was solved with ABAQUS system, while the beam Winkler's model was solved with the author's program. To determine soil parameters, the dilatometer tests were used.

1 INTRODUCTION

Trenchless technologies develop thanks to the intensive scientific research in this field. (Chapman et al. 2007) present the review of current research on trenchless technologies. The authors of the paper select four categories of the research, thus: 1) planning and monitoring, 2) machines and soil interactions, 3) problems referring to pipes and connections and 4) accompanying works. Among the accompanying works, they select two groups of problems: 4.1) technical problems connected with the strength, stiffness and stability of shafts and technological excavations as well as problems with connections: pipeline-to-pipeline and pipeline-to-shaft, and 4.2) environmental problems.

Characteristics of trenchless methods as well as problems connected with the design and excavation of launching and reception shafts are discussed in (Madryas et al. 2006). The book does not devote much attention to cylindrical shafts.

The hereto paper presents problems connected with designing and utilisation of shafts made of precast caissons used as working shafts for performing pipe jacking and microtunnelling. Such shafts are easy to perform in typical, simple soil conditions and may be used afterwards as sewage wells or inspection chambers.

Shafts made of precast caissons are subjected to various kinds of loadings in particular phases of installation and exploitation: while sinking, while the performance of a microtunnel and in target operation conditions and loads. The objective of the paper is a numerical analysis of a cylindrical shaft subjected to horizontal loads, caused by the reaction of the jack while microtunnelling. In order to design a working shaft, a calculation model needs to be prepared for this phase of loads. The model should allow the determination of internal forces and displacements in the system. The paper does not include the analysis of problems connected with the ultimate load bearing capacity of the shaft-soil system.

The determination of internal forces and displacements in the shaft involves solving the task of the soil-structure interaction. The calculation model includes three essential elements: the model of the shaft structure, the subsoil model and the model of the shaft-soil contact conditions. The requirements placed on the respective elements of the model have been discussed in (Potts & Zdravković 2001). It is particularly difficult to model the real behaviour of the subsoil. A strong nonlinearity within the range of small strains, the domination of plastic strains, the influence of the conditions of the filtration of porous water on the stiffness and strength of the medium, as well as the strength anisotropy are the main factors making the analysis complicated. Additionally, geotechnical point tests, in which only about 1/1,000,000 of soil volume subjected to structural loads is tested, cause that the spatial system of layers is usually presumptive. The fact that the parameters of the sophisticated constitutive models of soil need to be determined in specialised laboratory tests makes the application of these models inconvenient. For example, deriving samples of sand to test their mechanical parameters in a laboratory is extremely expensive. Because of this, simple models are used in designing. However, the proper procedure of the determination of model parameters should be given special attention.

The next important elements of the soil-structure interaction model are the contact conditions. Full contact between the shaft and the soil results in a nonrealistic transition of tension stresses onto the soil. In case of linear-elastic models of soil, it may lead to an incorrect determination of internal forces and displacements.

Since the stiffness of the precast caissons of the shaft (concrete stiffness exceeds the stiffness of the surrounding soil by a couple of orders), the constitutive model of the shaft material is not as essential as the model of subsoil, and thus the assumption of a linear-elastic model for designing is usually justified. The structure of the reinforced concrete shaft is treated as a cylindrical shell. In certain cases it may be modelled as a beam structure.

The hereto paper assumes linear-elastic models of subsoil and reinforced concrete. The contact conditions are modelled as unilateral. Friction between the shell and the soil has been taken into account. The task has been formulated in terms of the finite element method. The subsoil parameters have been determined on the bases of dilatometer tests (DMT).

2 SOIL INVESTIGATION AND SOIL PARAMETERS

Dilatometer tests consist in inserting the dilatometer blade into the subsoil to the required depth and measuring the pressures deforming horizontally a steel membrane. The blade is pushed into the subsoil with the help of rods through which a pressure conduit is carried to supply gas to the blade. At a given depth, the operator increases the gas pressure making the membrane deform towards the surrounding soil. Two pressure values are measured and recorded as p_0 and p_1. Value p_0 refers to the gas pressure compensating the geostatic pressure on the membrane. Value p_1 is recorded when the internal pressure causes the deformation of the membrane by 1.1 mm in relation to the referential surface. To perform the subsequent measurements, the blade is pushed into the next testing level 20 cm deeper. The DMT is a two-parameter test with the displacements control. On this basis, dilatometer indexes: I_d, K_d as well as E_d are determined, which enable the determination of the type of soil, the state of soil, its strength and stiffness (Marchetti et al. 2001).

The following aspects speak in favour of the application of the dilatometer: the direction of the subsoil loading during the test is consistent with the direction of the loading from the structure at the stage of the action of jacking forces, which is essential in terms of anisotropy of the subsoil mechanical features.

The subsequent factor is the value of the deformation of the subsoil during the test. A strong nonlinearity within the range of small deformation requires, the parameters of linear elasticity to be determined for the range of strains referring to the operational deformations (Mayne 2001).

Figure 1 presents the results of subsoil probing with the help of DMT, derived from the bottom of the excavations, at the depth range $z = 2$–10 m. Index I_d is the indicator of the type of soil. For sandy soils, $I_d > 1.8$. The chart implies that silty sand sediments ($I_d < 3.3$) occur exclusively at depth $z \approx 3$ m. Index K_d refers to horizontal stresses in soil, it is also the indicator of the preconsolidation ratio, and more generally, the indicator of the state of the soil. A low value of this parameter at depth 7.5 m reveals a low sand density at this level. The modulus of the dilatometer E_d refers to the value of soil stiffness. Thus, it is used in formulas correlating the results of probing with oedometer modulus, Young's modulus or Kirchhoff's modulus. The last chart presents the values of the internal friction angle. Since there is a tight correlation between the state of sand and its strength, the value of the parameter ϕ has been determined on the basis of K_d.

The presented results of probing have been applied in the elaboration of the subsoil model of the analysed working shaft. On the basis of the dilatometer modulus, the values of Young's modulus E

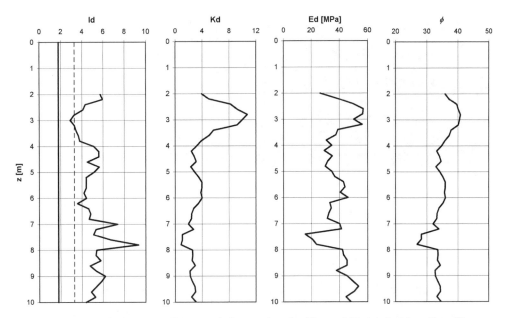

Figure 1. Results of DMT tests; dilatometer indexes and angle of internal friction ϕ of the soil profile.

Table 1. Parameters of elastic soil model.

Layer	z m	E MPa	v –	k MPa
1	0.0–2.0	70.0	0.2	30.0
2	2.0–3.3	66.0	0.2	30.0
3	3.3–7.3	50.2	0.2	30.0
4	7.3–7.9	27.6	0.2	30.0
5	7.9–10	63.7	0.2	–

have been obtained from the relation:

$$E = \frac{E_d}{1 - v^2} \tag{1}$$

where v = Poisson ratio; and E_d = dilatometer modulus.

Winkler's subsoil modulus has been calculated from the formula elaborated for foundation piles (Vesić 1977):

$$k = 0.65 \sqrt[12]{\frac{E \cdot D_s^4}{E_s \cdot I_s}} \frac{E}{1 - v^2} \tag{2}$$

where D_s = external diameter of the shaft; I_s = shaft moment of inertia; and E_s = shaft Young's modulus.

Finally, the subsoil model consists of five layers parameters of which are presented in table 1. Layer 1 consists of compacted soil of the backfill excavation.

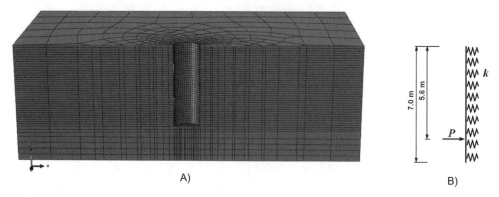

Figure 2. A) Discretisation of 3-D subsoil model; B) outline of 'beam-Winkler' model.

3   DESCRIPTION OF NUMERICAL MODELS

A working shaft made of precast caissons with outer diameter $D_s = 2.3$ m, wall thickness 0.15 m and total height 7.0 m., was subjected to the analysis. The following values of elasticity parameters of the shaft model were assumed $E_s = 30$ GPa, $v_s = 0.2$. The task of the interaction between the shaft and the subsoil was formulated in terms of the finite element method. Three numerical models were considered: two – where subsoil was seen as a 3D body and one – where Winkler's model of subsoil was used. The two former – were solved with the help of ABAQUS system, whereas the third was solved with the authors' program.

In the first model – 'shell-3D', the shaft was mashed with the use of 8-nod shell elements (S8R), integrated in the reduced way. In the second model – 'solid-3D', the shaft was discretised with the use of 8-nod spatial elements (C3D8). In the third model – 'beam-Winkler', beam elements on elastic Winkler's subsoil were used. In this case, the system was divided into only 35 finite elements.

In models "shell-3D" and "solid-3D", the mass of the subsoil 30 m long and 10 m thick was analysed. The layout of layers and the parameters were assumed as in table 1. In both models the mass was subjected to an identical discretisation with the use of C3D8 elements. More than 20 thousand finite elements were applied. The mash of elements was fined in the central part of the mass, in the region shaft contact. On the outer side surfaces as well as on the bottom surface, boundary conditions, restricting displacements in the directions normal to each of the mentioned surfaces, were assumed.

Between the shaft structure and the subsoil, unilateral contact conditions were assumed. Master-slave algorithm was applied. On the surface of the contact, Coulomb friction law was assumed. The friction coefficient was assumed 0.3.

The basic load of the system was a horizontal force of the value 700 kN, acting at the level of 5.6 m. The load acts statically.

Figure 2 presents discretisation of the subsoil for models "shell-3D" and "solid-3D" as well as the outline of the task and the loads for the "beam-Winkler" model.

Model "shell-3D" of the system, is regarded as the basic one. It allows a direct determination of internal forces in the shaft structure, which are used in designing, as well as displacements, strains and stresses in arbitrary elements of the system. Model "solid-3D" is a subsidiary model. It enables the determination of strains and displacements. Its aim is to control the performance of the "shell-3D" model. The "beam-Winkler" model is an additional model which enables only the estimation of the shaft displacements.

4   RESULTS

The carried out analyses allowed the determination of displacements, strains and internal forces in the system elements. Figure 3 presents the outline of the shaft together with the auxiliary lines and systems of coordinates, making the analysis of the results simpler. Figure 3a demonstrates a

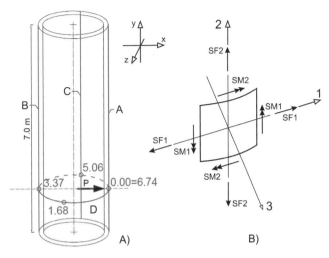

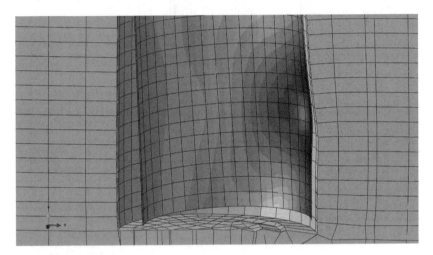

Figure 3. A) Auxiliary lines A, B, C and D; B) local coordinate system and internal forces sign convention.

Figure 4. Contours of displacements and shaft-soil separation for 'solid-3D' model.

global system of coordinates x-y-z-, in which the displacements values will be presented. In the outline of the shaft, vertical auxiliary lines A, B, and C are added, as well as a circumferential line D, along which internal forces will be presented. Since the charts of the internal forces in the horizontal cross-sections will be shown in a developed view, along line D, the length ordinates of the arch have been marked, starting with the point where the force is imposed. The perimeter of the analysed shaft is 6.74 m long.

Figure 3b presents a local system of coordinates 1-2-3, as well as the sign convention of internal forces. Description of section forces symbols is as follows:

- SF1 – direct membrane force per unite width in local 1-direction,
- SF2 – direct membrane force per unite width in local 2-direction,
- SM1 – bending moment per unite width in about 2-axis,
- SM2 – bending moment per unite width in about 1-axis.

The further part of the paper considers only the distribution of internal forces SF1 and SM2 obtained in the analysis of the "shell-3D" model.

Figure 4 presents the displacements of the foot of the shaft as well as the displacements of the soil surrounding the foot. The results refer to the "shell-3D" model. The figure demonstrates the

275

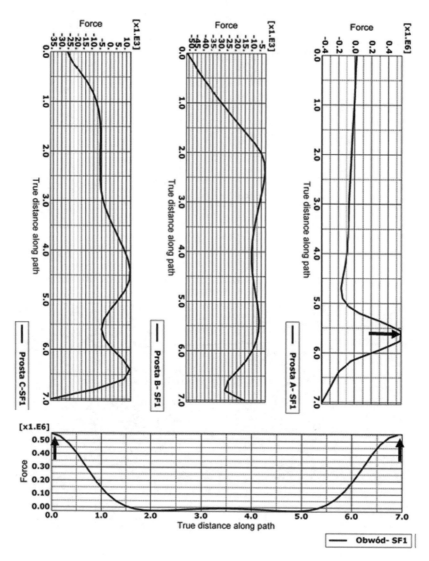

Figure 5. Distribution of membrane force SF1 [N/m] on vertical sections C, B, A, and on the horizontal circumferential section D. Notice different scale factor multiplier on particular sections.

cross-section x-y, i.e. in the plane of the acting load. On the right wall of the shaft, deformations caused by a horizontal force are visible. In the neighbourhood of the left wall of the shaft, soil-shaft separations are distinctly visible. In order to better demonstrate the phenomenon, the scale of displacements is one hundred times bigger than the scale of the structure.

Figure 5 presents the distribution of horizontal membrane forces SF1 generated by the load. The results refer to "shell-3D" model. The charts were performed for four cross-sections: three vertical cross-sections – along lines C, B and A, and one in a horizontal cross-section D (for notations of cross-section reference lines see fig. 3a). The values of the membrane forces in the sections were presented in various scales. The multiplier of the scale factor for cross-sections C and B equalled 1000, whereas for cross-sections A and D – 1000,000. Analysing the charts in cross-sections, it may be concluded that the greatest values of membrane forces SF1 occur in the neighbourhood of the load and they are tension forces. Because of the anisotropy of concrete, of which the shaft is made, the values of the forces determine the amount of circumferential steel reinforcement. In the lower part of cross-section A, compression occurs, which is advantageous from the point of view

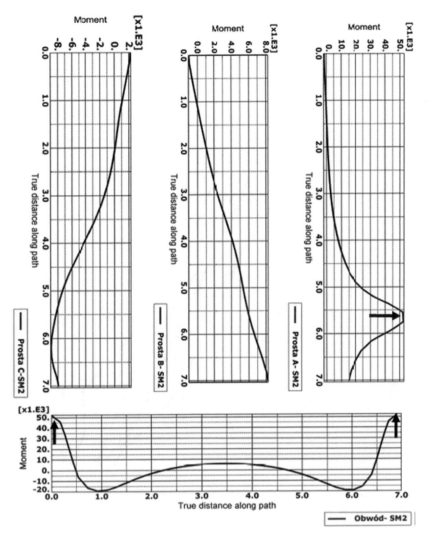

Figure 6. Distribution of bending moments SM2 [Nm/m] on vertical sections C, B, A, and on the horizontal circumferential section D.

of the reinforced concrete mechanics. In cross-section B, on the wall opposite to the point of load, membrane forces SF1 generate compression of the material. In cross-section C, along the height of the shaft, the compression and tension of the material occurs alternatively.

Figure 6 presents the distribution of bending moments SM2, acting along to the horizontal, local axis 1. The layout of this figure is identical to Figure 5. The considerable values of bending moments occur only in the neighborhood of the imposed load (cross-sections A and D). Cross-section B demonstrates almost linear changeability of values of moments along the height of the shaft.

Figure 7 presents the comparison of the horizontal displacements of the shaft, obtained for all the three considered models. The charts were performed for cross-section x-y, i.e. in the plane of the acting load. The solid lines refer to the displacements of the front wall – A, the dashed lines refer to the back wall – B. The line with triangles refers to displacements of the shaft axis in Winkler's model.

The comparison of the charts suggests that for all the three models, the values of displacements are very close. In models 3-D, the operating horizontal load generates characteristic deformations

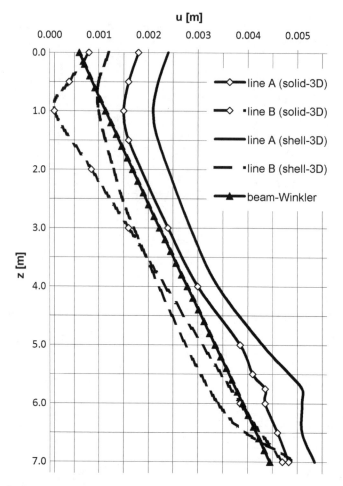

Figure 7. Distribution of horizontal displacements u for particular shaft models.

of the front wall, in the bottom part of the shaft. The load makes the shaft turns left. As a result of the turn, the pressure of the soil on the back wall causes deformations of the upper part of the shaft.

Surprisingly good results were obtained for "beam-Winkler" model. The displacements of the beam model nearly overlap the displacements of the axis of the shaft, obtained for the spatial models. Despite the fact that the inertia moment of the cross-section in the beam model was calculated as for a ring of dimensions responding to the cross-section of the shaft, the system seems to be much stiffer and does not suffers any local deformations caused by bending in the neighbourhood of the acting force.

5 CONCLUSIONS

Working shafts are essential elements in microtunnelling technology. Both from the point of view of the realisation of the microtunnel (laser measurement of the direction of the work) and from the point of view of the design of the shaft structure, it is essential to estimate properly the displacements generated by horizontal reaction of the jacking force. Designing a shaft structure requires additionally the determination of the values of internal forces in the structure. To determine the internal forces and the displacements of the system, the task of the shaft-subsoil interaction needs to be formulated and solved. The primary elements of such task are: a model of the subsoil, a model of the shaft structure, and a model of contact conditions. Because of the degree of the problem complexity, the task of the interaction between the structure and the subsoil has to be solved with the use of numerical models.

A very important element of the process of the development of the subsoil model is the determination of strength and stiffness parameters. Because of the strength anisotropy, the direction of the loading path during the soil tests should be as close as possible to the direction of the loading generated by the structure. To estimate the stiffness parameters properly, it is essential for the values of the strains generated during the tests to be as close as possible to the values of the average strains generated by the structure. Dilatometer tests fulfil all the above requirements. They allow realistic estimation of stiffness and strength of soil.

The application of the finite element method to the modelling of the subsoil involves discretisation of a considerable volume of subsoil, to avoid any interference caused by boundary conditions in the zone of direct influence of the structure. It introduces a considerable number of unknown nodal parameters to the model. Hence, any models limiting the number are valuable and preferable.

The subsequent important element in the task of the interaction between the shaft and the subsoil is modelling the contact conditions. Unilateral constrains on the shaft-subsoil contact area prevent the transition of the tensile stresses onto the ground. It results in a realistic representation of the interactions between the elements of the system. However, it complicates the task in terms of calculations – the model becomes nonlinear.

The most useful way to determine the internal forces is to discretise the shaft with the use of shell elements. The considerable values of horizontal jacking loads result in generation of huge membrane forces, which may only be overtaken by steel reinforcement.

In the calculated example, the displacements obtained for a shaft modelled with spatial elements, a shaft modelled by shell elements as well as a shaft modelled by beam elements, demonstrate a considerable consistency. Each of the analysed spatial models had over 80,000 degrees of freedom, whereas the beam model on Winkler foundation had only 72. The consistency should be confirmed with further tests. In the analysed case, Vesić's formula (2), represented the stiffness of the horizontally layered subsoil very well. Unfortunately, the beam model is useless for the determination of internal forces.

REFERENCES

Chapman, D.N., Rogers, C.D.F., Burd, H.J., Norris, P.M. & Milligan, G.W.E. 2007. Research needs for new construction using trenchless technologies. *Tunnelling and Underground Space Technology* 22: 491–502.

Madryas, C., Kolonko, A., Szot A. & Wysocki, L. 2006. *Mikrotunelowanie*. Wrocław: Dolnośląskie Wyd. Edukacyjne. (in polish).

Marchetti, S., Monaco, P., Totani, G. & Calabrese, M. 2001. *The flat dilatometer test (DMT) in soil investigation*. A Report by the ISSMGE Committee TC16, Proc. Int. Conf. on In-Situ Measurement of Soil Properties, Bali, Indonesia.

Mayne, P.W. 2001. *Stress-strain-strength-flow parameters from enhanced in-situ tests*. Proc. Int. Conf. on In-Situ Measurements of Soil Properties. Bali, Indonesia: 27–48.

Potts, D.M. & Zdravković, L. 2001. *Finite element analysis in geotechnical engineering. Application.* London: Thomas Telford.

Vesic, A.S. 1977. *Design of pile foundations.* NCHRP Synthesis of Practice No. 42. Transportation Research Board. Washington DC.

Underground Infrastructure of Urban Areas 2 – Madryas, Nienartowicz & Szot (eds)
© 2012 Taylor & Francis Group, London, ISBN 978-0-415-68394-4

Topping of sewage wells with concrete cones or reinforced lids on reinforced concrete relief rings – myths and reality

G. Śmiertka
ZPB KaczmareK Polska sp. z o. o. S.K.A., Poland

ABSTRACT: The article addresses the subject of design solutions for toppings of the sewer wells manufactured according to PN-EN 1917:2004/AC: 2009 standard. Functional patterns of reinforced concrete cones and reinforced concrete lids arranged on top of reinforced relief rings have been compared. Impact of the well topping type on its load-bearing capacity in the transverse direction (compressive strength) was estimated. Subsequently, concrete precasts with reinforcement inserts were analyzed in terms of strength and durability. As the summary, results of laborator tests of concrete DN1000/600 cones manufactured by UPB Kaczmarek were quoted.

1 INTRODUCTION

Since introduction – by PKN – of harmonized European standards in Poland, a basic reference document quoted when declaring an adequate quality of sewage wells is BS EN 1917 standard: 2004 "Manholes and inspection chambers of non-reinforced concrete, steel fibre reinforced and reinforced concrete" (PN-EN 1917: 2004/AC: 2009). As its wording describes, no matter what precast design is employed (concrete, reinforced concrete or dispersed reinforcement concrete) the same technical and operational requirements are valid at the time of embeddence on sites of identical operating conditions (communication load values). Concerning load ranges, vertical communication load carrying prefabricates should be tested in accordance with Annex B (PN-EN 1917: 2004/AC: 2009), while the elements subjected to horizontal earth pressure in accordance with Annex A to standard (PN-EN 1917: 2004/AC: 2009) and (PN-EN 476: 2001). Concrete in all elements is subject to compressive strength testing, performed on the test core in accordance with the standard procedure EN 13791 "Assessment of concrete compressive strength in structures and precast concrete elements" (PN-EN 13791) and (PN-EN 1917: 2004/AC: 2009).

While comparing the production range of both reinforced and non-reinforced elements, it should be noted that it is genuinely possible to design and subsequently build the complete non-reinforced sewage wells.

2 REINFORCED CONCRETE

In accordance with PN-B-03264 standard: 2002 "Concrete, reinforced and prestressed concrete structures. Static calculations and design" (PN-B-03264: 2002), we refer to "structures of concrete, reinforced with slender steel bars in such a manner that the stiffness and load capacity of the structure is determined by the steel and concrete cooperation" as reinforced concrete prefabricates, and the range of reinforcement is greater than the minimum, dependant upon the grade of steel and concrete. The use of reinforcement, transferring tensile stresses in load-bearing cross-sections, resulted in reduction of wall thicknesses and thus precast weight by 30% to 50% compared to non-reinforced elements. In the case of designing the reinforced elements, the width of cracks appearing under load when the concrete exceeds its tensile strength shall be checked. The cracks facilitate the penetration of harmful agents into the structure, which in return accelerates its degradation and thus significantly shortens its life. Durability of reinforced structure depends on the durability of cover protecting the reinforcement against corrosion. Immature concrete has a pH of approximately 13. During the natural life of reinforced concrete structures, CO_2 from the air causes a slow decrease of the pH value of concrete. From pH of approx. 9.5, the protection of steel provided by the cement alkalinity vanishes completely and the process of steel corrosion proceeds rapidly (Madryas et al. 2002). Sections designed as reinforced concrete ones are not able to transfer the external loads through the concrete core section, which in turn leads to failure of the structure. Therefore, it should be noted that, in terms of durability, age works against the precast elements.

3 NON-REINFORCED CONCRETE

To embrace the above issues, a return to the use of precast concrete has been observed in the sewage systems in recent years. Assumption of the lack of structural steel forced the designers to increase the wall thickness, which resulted in greater weight of the components. It should however be noted that due to the substantial weights of reinforced concrete precasts heavy equipment maintains the choice for network works utilizing such elements. In addition, the larger mass of assembled components improves buoyancy properties of the well. The secondary and key positive aspect of utilizing the design without rebars is its durability. Concrete and reinforced concrete structures embedded under similar climatic conditions, are subject to the same process of carbonatization. In the case of structure without reinforcement, the progressive neutralization advancing in the surface direction does not significantly affect the strength properties of the structure. Therefore, it may be stated that in terms of durability, the age of precast elements is a positive factor.

4 SEWER WELLS

The standard (PN-EN 1917: 2004/AC: 2009) presents the possible design solutions for the sewage wells. All components may utilize alternative designs, i.e. as reinforced concrete ones. Due to the horizontal components being subjected to large concentrated forces, prefabricated elements such as lid, reducing lid, covering component, have to be mandatory reinforced because of the large tensile stresses. However, there is possibility to make fully concrete wells consisting of bases, rings and cones, that are resistant to the negative process of carbonization of concrete. The nature of rings operation involving the axial transfer of longitudinal compressive forces and normal circumferential loads generating a quasi-membrane state in the structural core, results in the components carrying the normative loads in unobruptive manner. Similarly, the bottoms, thanks to appropriate stiffness of their bottoms transfer the loads onto the subsoil. Up to now, the major problems were caused by the concrete components used for closing the wells from the top or changing the well diameter at its height, the so-called cone.

5 CONES

Only few years ago, the level of vibration press engineering prevented the mechanical creation of the cones, concrete compound of which as well as the components alone would meet the requirements of standards (PN-EN 1917: 2004/AC: 2009, PN-EN 206-1: 2002/AC: 2005). This was due to problematic charging and then the difficulties with the proper per vibration of concrete on the sloping wall of the precast. Attempts to manually fill and compact the wet compound in the mould gave no result, as it resulted in the cones still not meeting the normative requirements. All these problems contributed to the infamous opinion of the cones, as components suitable only for incorporation in the greens of low communication loads. Manufacture of suitable vibration press machines as well as improvement of the engineering process enabled production of the cones that meet the normative requirements (PN-EN 1917: 2004/AC: 2009) primarily concerning the concrete compressive strength as well as the component capacity to withstand the vertical load.

6 RELIEF RINGS

The Polish market for the sewage wells has been flown with the so-called "relief rings". Among designers, contractors and supervisors many votes for and against the use of these elements can be

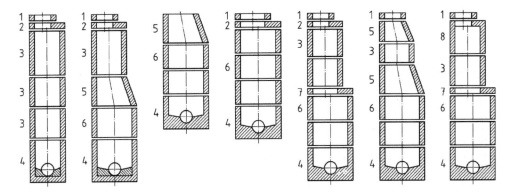

Figure 1.   Concrete well components in a accordance with BS-EN 1917.

heard. Their working principle involves relieving the wells completely from vertical communication loads, towards transferring these onto the soil placed along the outer circumference of the well. Basic requirements for concrete wells are contained in (PN-EN 1917: 2004/AC: 2009). The following are among its components: 1. grade ring, 2. covering lid 3 and 6. ring, 4. base (bottom), 5. cone, 7. reducing lid, 8. covering element (Figure 1). Authors of the original English version did not consider the "relief" ring design, because its use is contrary to the rational modeling of the load direction onto the concrete wells.

Breaking force values that well components should be tested against, apart from (PN-EN 1917: 2004/AC: 2009) can be found in (PN-85/S-10030). The document states, that rigid circular elements should have a crushing strength of at least 25 kN/m. In both standards, the cones, covering and reducing lids to be used under roadways accessible for all types of road vehicles, unpaved road shoulders and within car park boundary should be able to withstand the ultimate load of at least 300 kN. It can be therefore concluded, that the normative requirements with regard to the load-carrying capacity of the "well" concerning the vertical force are more than 10 times larger than the load-carrying requirement for the crushing force.

7   LOADS IMPOSED ON THE WELL

In order to compare the standard-oriented advantages and disadvantages of wells with cones and with the "relief" rings, impact of communication loads on both designs has been examined. Among the load-specific standards, the greatest loads imposed by vehicles wheels are defined in (PN-85/S-10030). Characteristic communication load of class A is generated by the vehicle K of a mass equal to 800 kN. This load is distributed onto eight wheels, providing 100 kN each. Multiplying the individual load by a load factor of 1.5 and a dynamic factor of 1.325, the maximum analytical single wheel load of the vehicle K is derived: $100 \, kN \times 1.5 \times 1.325 = 200 \, kN$.

8   POTENTIAL HAZARDS CONSIDERING THE USE OF TRADITIONAL CONCRETE WELLS

Among the potential threats to the concrete wells, the danger of destroying their structure by the vertical load is being raised. In accordance with the requirements (PN-EN 1917: 2004/AC: 2009) of the precast manufacture, concrete specified by the manufacturer holding the compressive strength of 40 MPa shall be used. After calculating the compressed concrete surface in the wells, it can be said, that the DN500 street inlets are characterized by at least 10-fold margin of load-carrying capacity, while the manholes an average 65-fold safety margin.

The second, often posed argument for, is the insufficient bearing capacity of the subsoil Q_{fNB} under the well bases. Based on (PN-B-03264: 2002), the average load capacity of the subsoil beneath the manhole wells (DN1000 and DN1200) has been calculated for foundation in the irrigated, medium-compacted non-cohesive soils. Depending on the depth of foundation (D_{min}), these amount to: m – 650 kN, 2.0 m – 1200 kN, 3.0 m – 1750 kN, 4.0 m – 2300 kN, 5.0 m – 2850 kN, 6.0 m – 3400 kN. For DN500 street inlets, subsoil load capacity will be approximately: 1.0 m – 300 kN, 2.0 m – 550 kN, 3.0 m – 850 kN, 4.0 m – 1100 kN, 5.0 m – 1400 kN, 6.0 m – 1650 kN. Results of calculations confirm the fulfillment of Q_{fNB} requirements (PN-81/B-03020) for each of the wells inspected.

The third argument of the supporters in favour of "relieving" the well is to reduce their potential subsidence. Assuming a maximum load transferred onto the subsoil of 300 kN (200 kN of potential communication load acc. to (PN-85/S-10030) and the estimated weight of the well of 100 kN), passive earth pressure underneath the manhole wells, will approximately amount for 2.0 kg/cm^2 (200 kPa), while underneath the street inlets it shall be approx. 10 kg/cm^2. It should be noted that the area of the relief ring used in the street inlets would be close to the area of DN500 base, so that the average subsidence in both cases is similar. Load capacity of the subsoil in the range of 2.0 kg/cm^2 is displayed by most of the load-bearing soils without displaying excessive subsidence while not providing any additional reinforcing treatment. For DN500 inlets, a prerequisite for proper operation of the designed and built structure is a proper preparation of the subsoil by providing subcrust of loose soil underneath the bottom base and compacting the layer to an adequate compaction factor of I_S. In the case of the relief rings, precise compaction shall cover a far greater volume of soil along the side surfaces of the well, which poses a far greater danger of excessive subsidence through the accumulation of loads on the backfill soil.

Thus, it can be unequivocally stated, that potential threats for normative wells resulting from vehicular traffic is a myth.

Another very important aspect of the heterogeneous nature of the relief rings operation is the way they transfer the traffic loads onto the subsoil. At the time of the wheel track axial placement on the lid placed in the soil a balanced passive pressure generating an evenly balanced horizontal thrust on the perimeter of the well shaft occurs. When the car wheel enters the lid, the relief ring – which according to (PN-81/B-03020) should be regarded as the foundation – is being ripped off the subsoil because of the resultant traffic loads being shifted outside the core of the ring base. The lids or cones founded directly on the rings transfer the loads through holding down the concrete surfaces, which almost entirely eliminates subsidence of subsoil around the well. In the case analyzed, the soil at the loaded side of the ring will undergo a vertical downward displacement, which in turn may lead to its opposite side being ripped upwards, consequently resulting in damage to the pavement.

9   VALUES OF LOADS IMPOSED ON THE STRUCTURE UTILIZING THE RELIEF RINGS

Communication loads (as they dominate the structures in question) are distributed on the road surface by the wheel track (in (PN-85/S-10030) it is 0.20 m × 0.60 m). Assuming the propagation of stresses across the roadway structural layers onto the larger area, a lower unitary load value being applied to the soil substructure of the road results. In the case of the well utilizing the relief rings, the loads is transferred directly from the cast-iron cover, through the lid and further the "relief" ring onto the roadway soil substructure. Base area of such a ring in case of a street inlet is approximately 0.40 m^2. This means that the force applied in the middle of a cast iron cover may generate stresses of at least 200 kN/0.40 m^2 = 500 kPa. Travel of the wheel throughout the grating surface causes further accumulation of stresses underneath its load-imposed corner or edge, which in turn leads to further increase of stress values in relation to the calculated value of 500 kPa.

In order to convert the imposed vertical load values into the accompanying horizontal forces crushing elements of the well operating chamber, based on (PN-83/B-03010) a stationary thrust factor of $K_0 = 0.40$, and then lateral thrusts directly below the "relief" ring of 200 kPa have been calculated. Additionally, because of the possible inclination of the thrust direction according to (PN-83/B-03010), the value calculated was further reduced by 10%. Local horizontal crushing thrust of 180 kPa has been obtained. Of course, this value vanishes along the depth and because of the thrust graph shape (PN-85/S-10030, PN-83/B-03010), the components most prone to damage are the upper rings up to a datum of approximately – 1.00~1.50 m.

Therefore, it can without abuse be concluded that overstressing of the well structure with the "relief" rings is a reality.

Table 1. Recipe of concrete mix for the production of cones in the EXACT system.

Raw material	Content per 1 m^3
Sand 0–2	1200 kg
Gravel 2–16	720 kg
Blast furnace cement CEM III/A 42.5 N – HSR/NA, incl. additive	450 kg
Plasticizer	2.0 kg
Water	115 kg

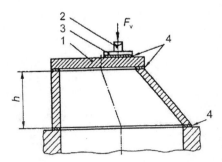

Figure 2. Outline of the cone-testing stand.

10 LABORATORY TESTING

In order to meet the growing demands of the domestic market for the sewage wells, in 2007 ZPB Kaczmarek Company introduced concrete DN1000 and DN1200 cones into production (ZPB KaczmareK sp. z o. o. advertising brochure), for which a concrete mixture of damp consistency and the recipe according to table 1 has been developed.

6 DN1000/600 cones were randomly selected from production for laboratory testing, representing consecutive days of cones production. Items were on average sampled every 2 weeks, collecting one item each in the period from May to August 2010. The testing was held for 6 days, testing one item at a time following a 28 day curing period under laboratory conditions counting from the day of production. The testing post was assembled from the cones and the base intended for laboratory testing (additionally reinforced), providing the precasts with appropriate support. From the top, the cones were topped according to (PN-EN 1917: 2004/AC: 2009), with a 30 mm thick circular metal sheet (1) with an additional sheet measuring 300 m × 300 mm × 30 mm (3). The load of the press was transferred through a 20 mm elastomer spacer with 65° Shore hardness (4) (Figure 2).

Ball joint (2) was mounted on a steel beam transferring the load from the press. Testing was performed on PK-2 Kurzetnik testing machine able to generate maximum force of 700 kN – calibration up to 600 kN. The machine is normally designed for testing of the rings for crushing strength and mates with CL 361 digital meter and CL1 pressure sensor. Due to the oblique shape of its lower section, the test was performed on a steel-supporting stool, enabling preparation of a level, horizontal surface. Next, 3 HEB 120 steel sections were placed on the stand construction to provide a uniform support for 30 mm thick plate, being the foundation for the concrete base (Figure 3).

On such a prepared post the tests commenced, the strength gain adjusted to 1 kN/s. The cones received the loads resulting in destruction of the concrete elements or occurrence of damages making their proper operation impossible. The results obtained are presented in Table 2 (Results of internal tests developed by ZBP Kaczmarek laboratory).

Materials of analogous research by Laboratory of the Civil Engineering Institute of Wroclaw University of Technology have been used as comparative studies, performed using the press providing a maximum load of 500 kN (Test report 2007). Assembled 2.00 m high DN500 concrete street inlet (Figure 4) (Results of internal testing of the street inlets for the vertical force. 2010) has been subjected to similar strength testing in the Laboratory of ZPB Kaczmarek. During all

Figure 3.   View of the testing stand for DN 1200/600 cone.

Table 2.   Results of vertical load testing of DN1000/600 concrete cones.

No.	Raw material	Load capacity [kN]
1.	DN1000/600 cone	520 kN
2.	DN1000/600 cone	608 kN
3.	DN1000/600 cone	598 kN
4.	DN1000/600 cone	579 kN
5.	DN1000/600 cone	566 kN
6.	DN1000/600 cone	691 kN

tests, the loading procedure was interrupted after reaching nominal maximum pressure values of the presses. The testing did not show any damage to the concrete elements being tested, which confirms at least their twofold load capacity safety margin.

11   CONCLUSIONS

Based on laboratory testing, it was found that according to (PN-EN 1917: 2004/AC: 2009) DN1000/600 cones by ZBP Kaczmarek, intended for "embeddence in the areas of roads for all types of wheeled vehicles", meet the requirement with regard to a minimum vertical load. The average breaking force reached 593.7 kN, while the minimal value reached 520 kN. That is almost twice the load capacity reserve in relation to the minimum normative requirements (PN-EN 1917: 2004/AC: 2009) at 300 kN. As the concrete elements, these are resistant (in terms of load capacity) to the carbonization process, shortening the service life of inter alia precast concrete lids. Additionally, by the use of sulphur-resistant blast furnace cement CEM III/A 42.5 N – HSR/NA, the cones are resistant to highly aggressive XA3 environment according to (PN-EN 206-1: 2002/AC: 2005).

Figure 4.   View of the testing stand for DN 500 street inlet.

In economic terms, it should be added that they are far less expensive than their reinforced concrete counterparts are, since they do not contain costly reinforcement (material and labour costs).

Load capacity for the crushing force of assembled wells or their components may be seven times lower than the "potential temporary" overload of their structural stems with the local thrust of 180 kPa, up to a datum of approximately −1.5 m below the roadway surface; introduction of a gap between the lid and the well body results in precipitation entering the inside surface of the well underneath the relief rings. This leads to loosening of the backfilling material underneath the well, which in turn generates additional subsidence in the well area. Use of the "relief" rings results in decrease of the vertical load imposed on the well, thereby resulting in subsequent increase of horizontal crushing load, which stands in complete contradiction with rational modeling of the load direction on the concrete wells.

REFERENCES

Madryas C., Kolonko A., Wysocki L. *Sewer pipe designs*. Wroclaw University of Technology Publishing House, Wroclaw 2002.
PN-81/B-03020 *Construction soils. Direct foundation of the structure. Static calculations and design.*
PN-83/B-03010 *Retention walls. Static calculations and dimensioning.*
PN-85/S-10030 *Bridge facilities. Loads.*
PN-B-03264: 2002 *Concrete, reinforced and prestressed concrete structures. Static calculations and design.*
PN-EN 13791 *Assessment of concrete compressive strength in structures and precast concrete products.*
PN-EN 1917: 2004/AC: 2009 *Manholes and inspection chambers of non-reinforced concrete, reinforced and steel fibre reinforced concrete.*
PN-EN 206-1: 2002/AC: 2005 *Concrete. Part 1: Requirements, properties, production and conformity.*
PN-EN 476: 2001 *General requirements for components used in gravity operated sewage systems.*
Results of internal testing of the street inlets for the vertical force – II 2010.
Results of internal tests developed by ZBP Kaczmarek laboratory.
Test report no. 018/2007/PKB – X 2007.
ZPB KaczmareK sp. z o. o. advertising brochure.

Construction records of major shield tunneling projects with advanced technologies in Japan

Y. Tada
Kajima Corporation, Tokyo, Japan

ABSTRACT: The shield tunneling method, which promoted a first cry in the first half of the 19th century in the U.K., has rapidly become popular in Japan as a versatile tunneling method in urban areas for building infrastructure such as railways, roads, power supply, communication, water supply and sewerage lines. The needs from various clients and the society have always been of a very wide variety and have been continuously changing with the time. They include not only traditional needs such as increasing tunnel diameter, depth, distance and construction speed, but also requirements by modern society, such as branching and widening tunnels responding to non-uniform cross section, reducing environmental load, reducing labor, improving safety and rationalization. Kajima has developed a number of construction methods to meet these needs of client, and has put them into practical use. This paper introduces a few recent construction records of shield tunneling projects with advanced technologies for building infrastructures in Japan.

1 INTRODUCTION

The trend of demand for shield tunnels is mainly three items. The first is large-diameter shield tunnel for large-scale infrastructures such as expressway. The second is long-distance shield tunnel in order to reduce construction costs by skipping the middle shaft for arriving and launching. The last is non-circular shield tunnel to optimize the shape to be more efficiently excavated and used. The following chapters introduce typical project carried out by Kajima as each typed shield tunnel, which are above mentioned large-diameter, long-distance, and non-circular, together with advanced technologies to resolve the concerning on each project.

2 LARGE-DIAMETER SHIELD TUNNEL WITH ADVANCED TECHNOLOGIES

2.1 *The construction records of large-section shield tunnel in Japan*

The demand for shield tunnels having large cross sections is increasing. Kajima has developed the world's top-class large-section shield with diameter of 14 m. By implementing a number of large-diameter shield constructions, Kajima has accumulated rich experience and outstanding result (Fig. 1).

2.2 *Shinagawa line, Metropolitan Expressway Central Circular Route*

2.2.1 *Project outline*
The Metropolitan Expressway being located at the center of Tokyo is a transport network with an overall length of 301 km. It is used each day by an averaged of 1.12 million vehicles and around 2 million passengers, thus serving as a very important support for the daily life and economic activity of the capital region. The Metropolitan Expressway Public Corporation is aiming to improve the service by expanding the network. The society strongly desires the early commencement of service for improving the function of the ring road, which includes the Central Circular Route being closest to the city center. The Central Circular Route has a total length of approximately 47 km, and forms a ring road with an approximately 8 km radius from the city center. The aims of the

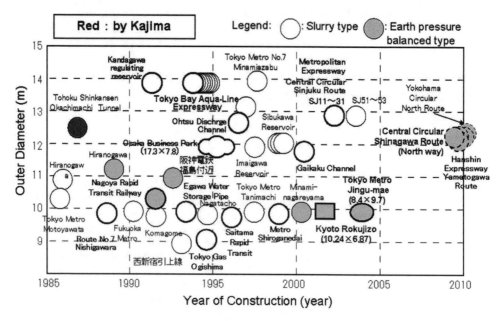

Figure 1.   The Construction Records of Large-diameter Shield Tunnel.

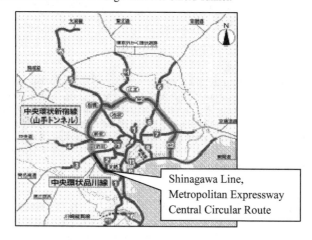

Figure 2.   Location of the Metropolitan Expressway Central Circular Route Shinagawa Line.

Central Circular Route are to mitigate traffic congestion by diverting and detouring through traffic that concentrates in the city center to form a well balanced traffic network of ring and radial configuration. Furthermore, that improved effect of the city centre environment derived from the reduction of exhaust fumes due to the mitigation of traffic congestion is also expected, and early completion of the road is hoped for. Approximately 80% of whole length, about 38 km exept Shinagawa Line, has already been completed and now open (Fig. 2).

The Shinagawa Line is the southwestern part of the Central Circular Route and consists of 9.4 km-long twin tunnels planned under the existing surface highway; namely, No.6 circular road of Tokyo Metropolitan Government and under the Meguro River. The Shinagawa Line is a very complicated large-scale underground structure, composed of 4 ventilation shafts, 2 junctions, and 2 on/off ramp tunnels (Fig. 3). Kajima is responsible for construction of the northbound shield tunnel toward Ohashi Junction under the control of Metropolitan Expressway Co., Ltd. and the other tunnel toward to Oi Junction that is called the southbound tunnel is constructed under the control of the Tokyo Metropolitan Government (Fig. 4).

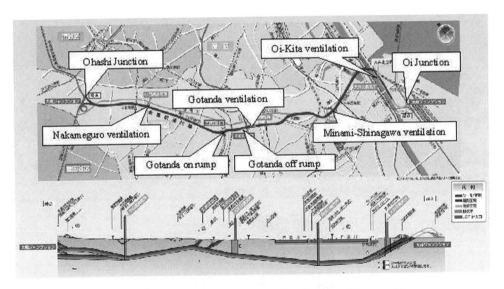

Figure 3. Outline of the Metropolitan Expressway Central Circular Route Shinagawa Line.

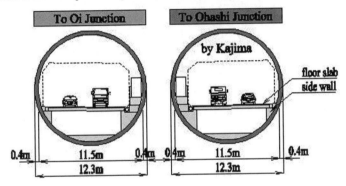

Figure 4. Standard cross-sectional view of the tunnels.

In this construction, the single shield machine passes through a vertical shaft installed in the Oi region where the tunnel is connected with the Bay shore Route, and arrives at the Ohashi region of Route 246. The entire tunnel sits in stable diluvia ground consisting mainly of mudstone layer of the rigid Kazusa layer Group. Maximum and minimum earth coverings are 46 m in the Minami-Shinagawa region and 14 m at the Gotanda entrance/exit, respectively. Minimum radius of curvature is 230 m and maximum slope is 3.0%. The shield machine is just driving at Gotanda rump region, which is around 4.5 km distance from the Oi launching shaft. The maximum monthly advance has recorded more than 500 m/month recently.

2.2.2 Technical characteristics
1) Shield machine and segment lining for long distance excavating

An earth-pressure balanced shield machine with 12.55 m outer diameter is employed to construct the major part of the tunnel of 8 km. In order to shorten construction time, the side wall and floor slabs for the road are to be built almost simultaneously with tunneling. The outer diameter of the tunnel is 12.3 m, lining thickness is 0.4 m and the distance to the other tunnel is about 3.0 m only (Fig. 4, Fig. 5). Reinforced concrete segment for lining has one touch typed joint-system to shorten assembly time and an ability of fire protection itself against tunnel fire without secondary lining.
2) Relay bit method

One category in which new technology is necessary for lengthening shield driving distance is that of techniques for changing cutter bits, which become worn down during excavation. The method of

 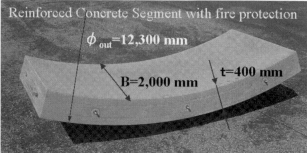

Figure 5.   Shield Machine and Reinforced Concrete Segment.

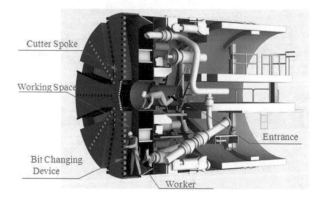

Figure 6.   Schema of Relay Bit Method.

changing bits that has been employed up to now has been either to construct an intermediate shaft for changing bits or to perform ground improvement ahead of the face and have workers going out in front of the cutter disk to change bits.

As a first solution, Spoke Rotation Bit Changing was developed as a mechanical method of changing cutter bits that does not require ground treatment. Following this, the "Relay Bit Method" was developed and put into practical use as a method of changing cutter bits without limitation on the number of changes (Fig. 6). This new technology was applied to the Shinagawa Line project due to shield driving of 8.0 km length.

A special characteristic of this new method is the fact that bits can be changed any number of times without performing ground treatment and without having workers enter space ahead of cutter disk. Another is the fact that bits can be changed at any time and location without any need for restriction. The Relay Bit Method is a simple method in which space that is sufficient for workers to enter is provided within the spokes of the cutter disk and workers enter this space from within the shield machine to change the bits one by one. The bits are enclosed in cylindrical cases that are fitted with special ball valves that are turned to maintain tightness against water so that the bits can be changed. Not only can the state of wear and damage to the bits be directly confirmed by sight, but bits can be changed as needed to suit the type of soil to be excavated. By mounting Relay Bits, the cost of shield machine is somewhat increased (approximately 15% to 20%, although this varies depending on factors such as machine diameter), but trial calculations show that total construction cost is reduced in comparison with conventional methods.

2.3   The construction records of long-distance shield

With the aim of realizing reduction in construction cost and shortening of construction schedule, which are the needs of the time, Kajima has been developing and put various techniques to practical use in order to meet longer distance and higher speed construction of shield tunnels (Fig. 7).

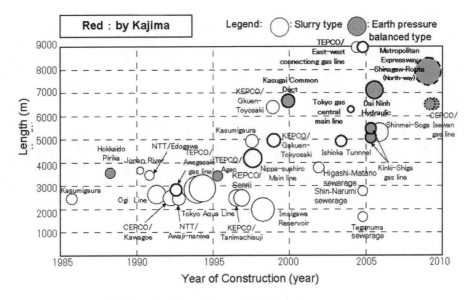

Figure 7. The construction Records of Long-distance Shield Tunnel.

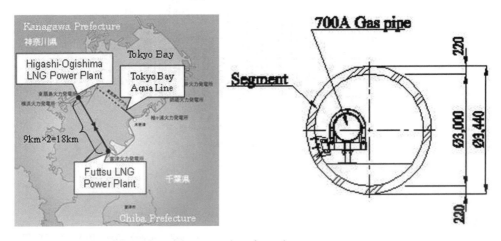

Figure 8. Key plan of the project and cross-section of tunnel.

2.4 *The east-west connecting gas line under Tokyo Bay*

2.4.1 *Project outline*

In order to provide an efficient fuel supply to each existing LNG power plant around Tokyo Bay, Tokyo Electric Power Company (TEPCO) constructed a gas pipeline which stretched 20 km connecting the Higashi-Ohgishima LNG base in Kawasaki City to the Futtsu LNG base in Chiba Prefecture. Of this, the part crossing Tokyo Bay will built as a shield tunnel with an internal diameter 3.0 m and length 18 km, and inside which a single 700 mm diameter gas line was installed inside. Two slurry shield machines working under maximum water pressure of 0.66 MPa each excavated 9.0 km and meet up and connect underneath Tokyo Bay. The project schedule was 37 months to completion, from April 2003 to April 2006 (Fig. 8). Kajima was involved in this project as a contractor of 9.0 km of the Futtsu section.

293

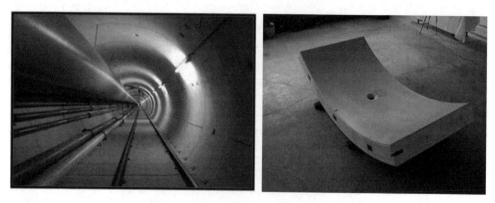

Figure 9.   Inside of tunnel with one touch type segments.

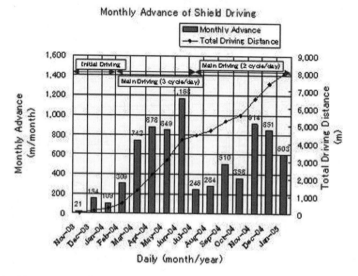

Figure 10.   Monthly advance records.

2.4.2   *Technical characteristics*

1) Long distance excavating and high-speed construction

In this project, a distance of 9 km was driven without replacing the bits. Wear of the preceding and main bits was significantly smaller than the estimated amount. This result was achieved by the use of grade E3 tips excellent in both wear and shock resistances, and arrangement of the preceding and main bits in the same path, and by improved efficiency of soil intake with a larger cutter head opening ratio. In order to complete the tunnel within the designated construction period, it was necessary to excavate the tunnel rapidly at a speed of more than 500 m/month. This was achieved by using one touch type segments with short assembly time, providing the equipment with sufficient capacity to overcome expected problems, and closely monitoring construction (Fig. 9). As a result, major problems that could cause prolongation of construction were avoided and a stable and rapid tunneling speed averaging more than 500 m/month was maintained. The monthly advance was 665 m on the average and 1,168 m at maximum (Fig. 10).

2) Simultaneous segment erection/shield driving system

This project tried to adopt a new technology, "the double jack simultaneous excavation shield method", which allows for the simultaneous difficult erection of segments together with tunnel excavation, speeding up the construction schedule and thus making it possible to shorten the construction schedule, compared to conventional shield machines. As a result, this system was verified to be useful to speed up shield tunneling schedule in trial section of this project.

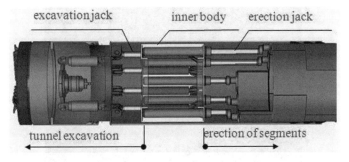

excavation jack   inner body   erection jack

tunnel excavation   erection of segments

Figure 11. Simultaneous Segment Erection/Shield Driving System.

The machine developed for this system is equipped with a drum that can slide inside the machine. The drum has two types of jack; one is specially used for excavation and is fitted on the front side of the drum, and connects the drum to the main body of the machine. The other is mainly used for segment erection and is fitted on the rear side of the drum. Equipped with these two types of jacks, excavation work and segment erection can be performed concurrently, thus making it possible to shorten the construction period.

For the sequence of excavation and erection, by pushing forward with the excavation jack, the whole shield machine is moved forward and as excavation is carried out, the erection jack is extended and segments are assembled into place (Fig. 11).

The most important technical subject for the realization of this method is the dealing with eccentric moments developed caused by the extending and withdrawing of the shield jack during segment erection, and the overall positional stability of the shield machine. For the double jack method simultaneous excavation shield machine, the sliding inner body has been designed to be structurally able to resist these eccentric moments. Then, the unbalanced reaction forces inside the erection jack are counterbalanced by naturally occurring differences in pressure, thus making it able to preserve the forward driving stability of the shield machine.

3) Successful application of mechanical underground docking method

The direct docking tunnel method, which can reduce the work period by approximately a month and a half as compared with the freezing method, was used to connect the shield machine at the extreme water pressure of 0.6 MPa under the seabed of Tokyo Bay. In order to achieve millimeter precision in the mechanical underground docking, precision tunnel surveying and horizontal bore probing from as close as 50 m were utilized. The results of the probing were reflected when implementing modified tunneling, and the machines were successfully aligned with an error of only few millimeters.

The mechanism of the mechanical underground docking method is as follows: the receiver pulls in the inner shell, while the penetrator pushes it out, thus joining the shield machines, and water is kept out by inflating rubber components to block water and sealing the joints with them.

In order to guarantee water cut off, a brush installed at the extension spoke of the penetrator was used to clean the surface of the water cut-off sealing plate. Furthermore, a scope camera was installed inside the chamber to closely monitor the work progress and the cleaning status. As a result, leakage from the joints was kept to a minimum (0.1 liter/minute), the shield machines were dismantled as planned, and the underground docking operations were concluded within approximately two months (Fig. 12).

3 NON-CIRCULAR SHIELD TUNNEL WITH ADVANCED TECHNOLOGIES

3.1 Non-circular shield

There are cases where a shield tunnel can be more efficiently excavated through the use of a shield with a cross-sectional shape other than circular, unlike the ones normally used. Kajima is capable of dealing with all kinds of cross sectional shapes such as horizontal double multi-face. horizontal triple multi-face, vertical double multi-face, rectangle, ellipse, etc., to achieve reduction of excavated soil, low earth covering, and flat structures such as a station.

Figure 12.   Shield machine.

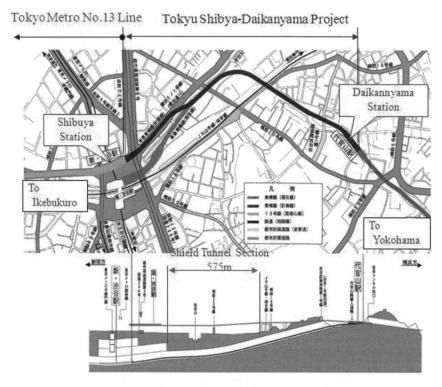

Figure 13.   Outline of the shield tunnel between Sibuya and Daikanyama.

3.2   *Railway tunnel between Shibuya and Daikanyama on the Tokyu Toyoko Line*

3.2.1   *Project outline*

Tokyu Corporation is proceeding with a huge improvement project spanning about 1.5 km from Shibuya metro station to Daikanyama station of Tokyu Toyoko Line. Of the project length, a section 575 m is driven by a shield machine. This shield tunneling is performed just below an existing railway viaduct, arriving at a shaft immediately beneath the viaduct. The geology of the site is composed of cohesive soil stratum, gravel and sand strata. A viaduct protection was placed before excavation just below the viaduct. Since the overburden is shallow, the "APORO Cutter (All Potential Rotary Cutter)" was selected as shield cutter head, which has a low-height double rectangular section capable of boring a special sectional geometry. A shield machine already arrived safely at an arriving shaft at Daikanyama Station in January 2010 (Fig. 13).

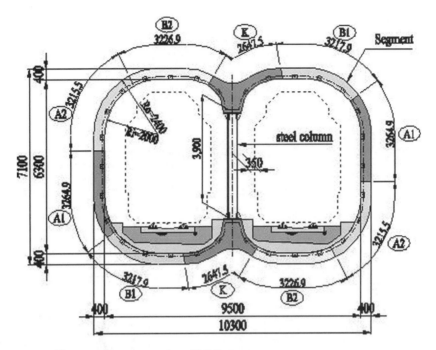

Figure 14. Double rectangular cross-section with RC segments.

Figure 15. Shield machine with APORO cutter and RC segment lining.

3.2.2 *Technical characteristics*
1) Double rectangular segments

This project uses segments of reinforced concrete and steel framed reinforced concrete 0.4 m thick and 1.1 m wide each. The tunnel 10.3 m wide and 7.1 m height is divided into ten parts per ring. The section is composed of a couple of rectangular sections with a composite steel rectangular column at the center (Fig. 14).

2) APORO cutter

The tunneling is being performed with a high-density slurry shield. The overburden rages from 4.5 m to 15.4 m, the minimum curvature radius is 160 m, and the maximum gradient is 3.5%. The shield machine with the new technology, APORO cutter, is an articulated type. A rotating cutter head is mounted, via a wagging frame, on the main drum making orbital revolutions. While the cutter head and main drum rotate and the wagging frame swivels, the shield machine excavates the planned section. The cutter head turn round at 4.7 rpm, and main drum turn around in 4 minutes. The special characteristic of this method is control of movement of wagging angle of frame and revolution of drum should be adjustable for excavation of non-circular cross section (Fig. 15). This

297

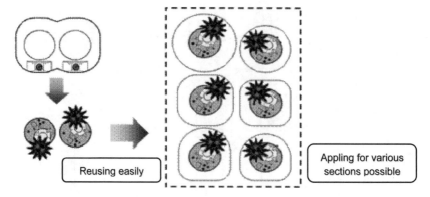

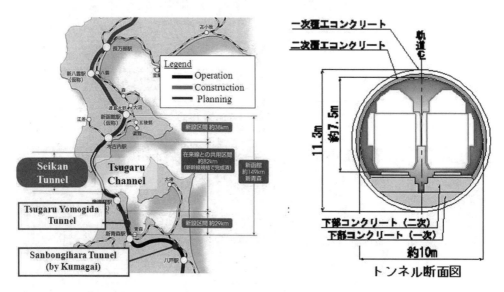

Figure 16. Characteristics of APORO Cutter Method.

Figure 17. Outline of the Hokkaido Shinkansen and cross-section of Tsugaru Yomogida Tunnel.

system is capable of cutting section of various sectional shapes by using the same cutter head and is easy to reuse cutter head including wagging frame and main drum, and is also applicable to hard ground condition (Fig. 16).

4 THE OTHER SHIELD TUNNEL WITH ADVANCED TECHNOLOGIES

4.1 *Hokkaido Shinkansen Tsugaru Yomogida Tunnel*

4.1.1 *Project outline*

The network of high-speed railway lines in Japan known as "Shinkansen" is expanding through construction of lines in various regions with the objective of connecting the major cities in the country. The Tohoku Shinkansen, which runs north from Tokyo, was extended by approximately 80 km, and it reached Aomori City, the capital of Aomori Prefecture, on the northern edge of the Honshu Island on December in 2010 (Fig. 17).

The Hokkaido Shinkansen is an extension, currently under construction, of the Tohoku Shinkansen from Aomori across the Tsugaru Channel into Hokkaido. It includes the longest undersea tunnel in the world, the Seikan Tunnel, which has a total length of 54 km (Fig. 17).

Figure 18.   Shield machine just launching and Primary tunnel lining.

The length of the structure for exclusive use by Shinkansen trains that is constructed from Shin-Aomori Station on the Honshu side to the section for shared use in approximately 29 km. According to the construction plan, it includes eleven tunnels with a total length of approximately 7 km. All these tunnels are below the groundwater level and geology of the area where they will be constructed consists mainly of unconsolidated and semi-consolidated medium- to coarse-grained sandstone. When the tunnel in the section for shared use was excavated in the 1980's, face collapses induced by sediment discharges occurred often, so it was expected that the construction work in this project also would be quite difficult. Furthermore, the portals of all eleven tunnels are located in a mountainous area several kilometers away from any arterial roads, so high costs for building of roads for construction purposes and auxiliary construction methods were anticipated in the budget. That is why, the longitudinal slopes in six tunnels were modified, and the plan was changed to make this section into a new single tunnel, "Tsugaru Yomogida Tunnel", with a length of approximately 6 km. Also, it was decided to utilize mechanized construction by "SENS method" that had been successfully used in similar water bearing unconsolidated grounds in the construction of Sanbongihara Tunnel of the Tohoku Shinkansen and that is characterized by outstanding safety, construction properties and economic efficiency (Fig.17). Kajima is involved in the construction of the Tsugaru Yomogida Tunnel project under the control of Japan Railway Construction, Transport and Technology Agency.

4.1.2   *Technical characteristics*
1) SENS method
   The SENS is a shield tunneling technique to perform excavation while maintaining the face, and to achieve early invert closure with cast-in-place concrete primary lining (ECL). In the SENS practice, secondary lining which carries no load is placed according to the concept of the NATM.

S : Shield Tunneling Method
E : Extruded Concrete Lining (ECL)
N : New Austrian Tunneling Method (NATM)
S : System

The maximum monthly advance recorded with SENS in the construction of Sanbongihara Tunnel of the Tohoku Shinkansen was 173 km/month. In this project, various improvements were implemented in order to enable even faster excavation. These improvements include increase of the number of concrete pumps and widening of the inner formwork. Furthermore, in order to build lining in the conditions with maximum water pressure of 0.4 MPa, a concrete mixture with good workability and anti-washout underwater properties was developed. Also, the structure of the inner formwork to be used repeatedly was improved to avoid the occurrence of shear force in the joints between pieces, and thus to make assembly and dismantlement work easier and safer. Based on these improvements, excavation work was launched in October 2009 (Fig. 18) and will arrive at the intermediate shaft located around 3 km from the launching entrance in June 2011.

a) Construction system

A closed-face type earth and mud pressure shield machine was selected with consideration of the site geology, which is mainly consists of unconsolidated and semi-consolidated medium- to coarse-grained sandstone with high water permeability. Primary concrete is placed under pressure at the tail along with tunneling advance, and jacked stop forms were used to counter slurry pressure. Concrete is placed at 12 points with 12 pumps. The inner tubular form as steel segment is equipped with 16 rings. After having excavated a length of 1.5 m that corresponds to the width of one ring, advance is stopped temporarily, and the form at the rear end is disassembled and assembled again at the top in the tail. Then advance and concrete placement are restarted.

b) Quality of cast-in place concrete

A high performance specialty thickener was mixed with the cast-in-place concrete, because of its excellent properties in term of 1) fluidity, 2) retaining capability of freshness, 3) water sealing capacity, 4) exhibition of early strength and 5) resistance to material segregation. A slump flow of 650 ± 50 mm is maintained even four hours after mixing. The early strength is excellent, that is, unconfined compressive strength of 15 N/m^2 at the age of one day.

c) Tunneling

The planned advance is averaged 12 rings per day and 200 m per month. The maximum monthly advance recorded around 300 m for now which shall be comparable with shield tunneling method. In this project the primary lining concrete is assumed to function as the primary support and the secondary lining is placed last. In fact, tunnel lining has insignificant amount of water leakage and crack which would pose no structural problems. Since no prefabricated segment but cast-in-place concrete is used, the SENS method is possible to be more economical than shield tunneling method.

When the performance of the SENS has been validated, and reasonable methods for construction and quality management have been established, it will find extensive application.

5 CONCLUSION

In this paper, we have introduced some recent construction records of shield tunneling projects with advanced technologies. Shield tunneling has become an indispensable method for the construction of underground spaces at the city center. Kajima shall continue its Research and Development (R&D) efforts with aim of responding to requests from various sectors of the society. Based on rich experience and flexible ideas, Kajima will continue to challenge the evolution of technologies, in its desire to help establish a new area.

REFERENCES

Dobashi, H. 2004. Naka Ochiai Sheild Tunnel on Metropolitan Expressway Central Circular Route. *Tunneling Activities in Japan* 2004: 13–14.

Hasegawa, M. 2010. Plans for the Hokkaido Shinkansen and Utilization of SENS – Tsugaru Yomogida Tunnel – . *Tunneling Activities in Japan* 2010: 2.

Kakuta, T. 2010. Rectangular-Section Shield Tunneling Using the Apollo Cutter – Railway Tunnel between Shibuya and Daikanyama on the Tokyu Toyoko Line – . *Tunneling Activities in Japan* 2010: 8.

Kurosaki, S. 2006. Longest Shield Driven Pipeline Tunnel under Tokyo Bay. *Tunneling Activities in Japan* 2006: 20–21.

Sasaki, M. 2004. Casting Support Tunneling using Shield Machine – Sanbongihara Tunnel of Tohoku Shinkansen (Superexpress Railwau) – . *Tunneling Activities in Japan* 2004: 2.

Shirai, S. 2004. Construction of Long-Distance Undersea Shield Tunnel. *Tunneling Activities in Japan* 2004: 23.

Takeuchi, T. 2008. Mechanical Docking of the Long Shield Tunneling under Tokyo Bay. *Tunneling Activities in Japan* 2008: 22.

Underground Infrastructure of Urban Areas 2 – Madryas, Nienartowicz & Szot (eds)
© 2012 Taylor & Francis Group, London, ISBN 978-0-415-68394-4

Technical and economic effectiveness of using trenchless technologies in constructing roads and railway infrastructure

A. Wysokowski

Head of Road and Bridge Department, University of Zielona Góra, Poland

ABSTRACT: The paper contains information on present development of transport infrastructure in Poland in terms of needs for construction of engineering facilities. Due to the fact that large number of such objects can be made in trenchless technology, the paper contains a short overview of practicable trenchless technologies that can be used for constructing such facilities. The main purpose of this paper is to pay attention to frequently underestimated social costs that can be reduced by applying discussed trenchless construction methods instead of traditional trench-based methods. Economic aspect of this issue is shown in enclosed examples.

1   INTRODUCTION

The present paper concerns issues associated with technical and economic effectiveness of using trenchless technologies in constructing roads and railway infrastructure.

In case of trenchless technologies we focus most frequently on issues associated with technical development of these technologies and scientific basis of their development. Significantly less attention is paid to broadly defined economic aspects.

This is also of great importance when constructing and modernizing motor roads and railway lines. Therefore the author of this paper has undertaken to analyse these issues. He has done it by taking into consideration social costs, which, despite their importance, are often omitted.

2   DEVELOPMENT OF TRANSPORT INFRASTRUCTURE IN POLAND

In the first decade of the XXI century in Poland there has been a very intensive development of transport infrastructure. Particularly great number of infrastructural projects has been carried out following Poland's accession to the European Union, i.e. after 2004.

One should distinctly note that Poland, due to its central location in Europe, is thereby situated on transport route from western to eastern Europe (Portugal, Spain – Belarus, Russia) and from the northern to southern European countries (Norway, Sweden – Austria, Italy).

Therefore, in this context, the transport routes: Berlin (D) – Moscow (RUS), Dresden (D) – Kiev (UA), Stockholm (S) – Prague (CZ), Helsinki (FIN) – Vienna (A) gain particular significance. This applies both to motor roads and railways. It is obvious, that intensification of air links and, associated with it, development of airport infrastructure also take place.

The planed system of motor roads and railway lines in Poland is shown in Figure 1.

Taking into consideration historical determinants, the present density of networks in Poland significantly differs from average densities in other countries of European Union.

At present the total length of public roads in Poland exceeds 380,000 km. These include over 30,000 bridges and tunnels with total length exceeding 550 km. The network of national roads administered by the General Directorate for National Roads and Motorways comprises 18,000 km. These include 4,500 of bridges with a total length of 190 km. Moreover within the network of these main roads, including motorways, there are over 16,000 culverts (3,000 culverts greater than 1.5 m in the clear).

At present the length of motorways in Poland is only 850 km and 560 km of expressways.

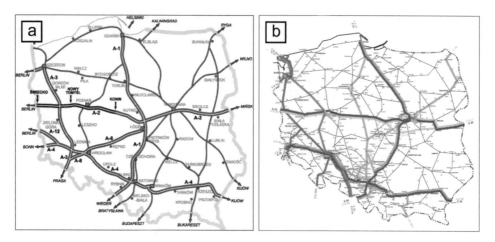

Figure 1.   Planed transport infrastructure in Poland: a) motor roads, b) railway lines.

The network of railway lines administered by the PKP Polish Railways P.L.C. and made available to carriers exceeds 19,000 km of railway lines. Lines important for international railway transits cross Poland and the lines are comprised by international agreements: AGC and AGTC.

AGC – European Agreement on Main International Railway Lines.

AGTC – European Agreement on Important International Combined Transport Lines and Related Installations.

The PKP Polish Railways maintains over 26,000 engineering facilities and installations, including almost 7,000 bridges and viaducts.

As can be seen from information presented above, it is a great need for modernizing, expanding and constructing new, modern transport routes. This applies both to the network of expressways and motorways as well as railway lines including high-speed railways.

The programme of motorway construction in Poland was developed in late 1990s. Despite of many modifications it is still consistently being carried out. Majority of changes made to the original version of the programme applies to the deadlines for completion of individual sections (and also deletion of A3 motorway from the original programme). The present version of the programme comes from the Regulation of the Prime Minister issued on 15 May 2004 (European Union Energy & Transport in Figures). The government plan of 2004 defined the target network of motorways with a total length of approx. 2,000 km. At the end of August 2009 a total 812 km of motorways were given to use. At present subsequent 337 km are in various stages of construction. Till the beginning of the European Football Championship Euro 2012 a total of 1,633 motorways are planned to be given to use (including completion of entire southern A4 motorway, A1 motorway and a section of A2 from the German border to Warsaw). The network of motorways is currently under construction with the aid of EU subsidies. Development of road infrastructure includes also expressroads, the total length of which has been planned for approximately 5,300 km. At present in various stages of construction there are about 600 km of expressroads and ring roads around cities.

The progress of constructing motorway network and high-speed railway lines in Poland is illustrated in Figure 2.

Similarly in the case of railway lines especially the lines comprised by international agreements are successively modernized so as to achieve target standards and ensure interoperability of railway network within the territory of the Republic of Poland in the trans-european traditional railway system and in trans-european high-speed railway system.

In 2005, after many years of discussions in Poland a realistic programme for constructing high-speed railways has been worked out, at least for the period up to 2030. It assumes the construction of a new line from Warsaw via Łódź and further with ramification to Wrocław and Poznań, with inclusion of existing CMK line, the profiles of which enable its adaptation to speeds of at least 250 km/h. Simultaneously a new line from Cracow to the area of Tymbark is to be built with high

302

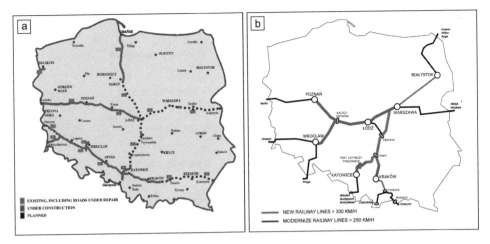

Figure 2.   Advancement state of the transport infrastructure in Poland: a) motorway network, b) high speed railways.

technical parameters, effectively connecting south-eastern ends of Poland with the national network of high-speed railway connections. Thus, along with other modernized lines, in Poland a modern railway network will be created, worthy of the XXI century, which will connect almost all regional centres.

In terms of investment size, the construction of new lines envisaged in the plan and modernization of existing lines is no match for the motorway construction programme, however its effects in relation to the resources invested will be much higher. Even properly developed motorway network may ensure average travel speed between centres of cities not greater than 80–90 km/h. A high-speed railway system consisting, even in large part, of only modernized lines gives the possibility of achieving average speed from 150 to 250 km/h. For example, completion of high-speed lines shall result in travel time from the capital city of Poland (Warsaw) to the capital city of Germany (Berlin) within 3 hours.

Development of transport infrastructure presented above requires and shall require the use of many technologies including also the trenchless technology due to reasons mentioned above.

This technology is and will be used in infrastructural structures, such as:

– sewage pipelines, water pipelines and other network elements,
– tunnels and subways,
– drainage elements of motor roads and railway lines,
– hydrotechnic culverts,
– environment-friendly culverts including passages for animals,
– refurbishments of existing infrastructure elements.

3   TRENCHLESS TECHNOLOGIES USED IN CONSTRUCTING MOTOR ROADS AND RAILWAY LINES

3.1   *Materials for constructing engineering structures made by trenchless technologies*

Terms for completion and operation of underground system of passage pipes under road or railway embankment require specially high quality and strength.

Pipes installed by trenchless technology must meet a series of stringent requirements associated both with their strength- and durability-related features.

Material used in production must have very high compression strength against stresses caused by jacking forces, must be resistant to corrosion both inside and outside the pipe as well as corrosion caused by various fluids and guarantee defect-free operation of culvert structure.

303

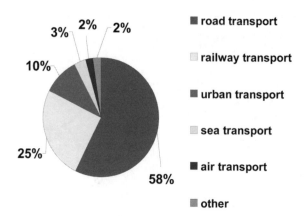

Figure 3. Graph showing EU's financial contribution to the development of individual transport branches in Poland.

Table 1. Scope of infrastructural investments and EU's financial contribution.

Type of transport infrastructure	EU contribution in mln Euro	Investment scope
Roads	11,104.40	– construction of motorway sections, – construction of expressway sections, – conversion of national roads with load capacity 115 kN/axis, – construction or conversion of national roads sections in towns with county rights
Railway	4,863.00	– modernization of railway lines, – interoperability of railway lines, – preparatory work for construction of high-speed lines, – purchase and modernization of rolling stock.
Airports	403.5	– construction and modernization of passenger terminals, – construction or modernization of airport infrastructure: runways, taxiways, apron area, stoppage stands, – support for activities aimed at preparing the construction of the second central airport – development of navigational infrastructure, – development of security systems and airport safety.

Among a wide range of materials for production of pass-through pipes the following can be listed:

– cast iron pipes,
– stoneware pipes,
– polymer-concrete pipes,
– pipes made of polymers reinforced with GRP glass-fibre.

Materials listed above are used in trenchless technologies where significant jacking forces are used by means of hydraulic cylinders.

Additionally in culverts made by trenchless technology by means of soil displacement method where protective steel pipes are used, as a target structure of an engineering object flexible pipes made e.g. of plastics (PP, PE etc.) or corrugated sheet can be used.

As previously mentioned, a very important issue in this technology is the production quality of jacking pipes. As opposed to technologies based on placing pipes in an open trench, outer wall of jacking pipes must also have possibly smooth surface. Smooth outer surface of pipes significantly

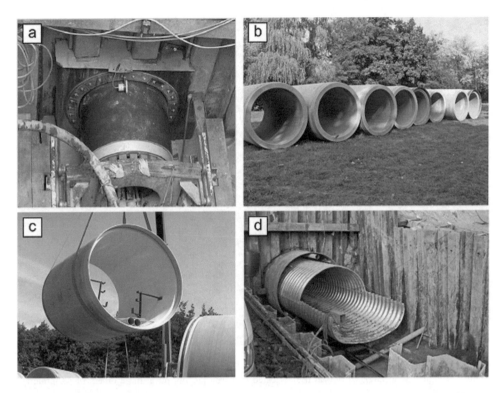

Figure 4. Example materials of jacking pipes used in trenchless technologies: a) stoneware pipes (Keramo Steinzeug), b) concrete reinforced pipes (Haba-Beton), c) CC-GRP pipes (Hobas), d) buried steel pipes (ViaCon).

reduces friction forces on the side surface, and thus reduces costs associated with application of additional lubricants, e.g. bentonite.

Jacking pipes are connected by integrated connector systems, ensuring required stiffness, continuous strength of a pipeline consisting of individual elements, as well as leaktightness. Modern types of connectors also allow pipelines to be placed on an elbow.

Example pipe materials used in trenchless technology are shown in Figure 4.

3.2 Structures of jacking and receiver pits

Jacking and receiver pits, also called shafts, are generally made in the form of sheet pile wall, most often from steel piling sections. Dimensions of the pits should be adapted each time to the specifics of a given investment task. In case of using trenchless technologies on large urbanized areas, in constructions of pipelines and trunk sewers the dimensions of pits should ensure free operation of hydraulic devices, ensure unhindered transport of materials indispensable for constructing the pipeline. In this case, due to considerable lengths of trunk sewer sections, intermediate chambers are used, in which inspection chambers are built in when the work is completed. In case of receiver pit its dimensions should ensure trouble-free transport of drilling equipment onto the surface.

In the jacking pit, in order to ensure stability of the base under the drilling equipment, hardening of the bottom is made by means of ferro-concrete slab with a suitable thickness. Then the equipment is assembled, including pumping station, shell pump, elements of control systems. During assembling it is very important to precisely determine the drilling axis which is a decisive element for the success of entire undertaking. Preparations are finalized by mounting a boring head on supports of the jacking frame and connecting power supply and control cables as well as pipelines of drilling fluid.

Figure 5. Typical jacking pits used in trenchless technology: a) typical pit with application of steel piling sections (Keramo Steinzeug), b) typical prefabricated cast iron jacking pit at the moment of its mounting (Haba-Beton).

Location of pits in investments made by trenchless technologies is a basic task which often conditions the success of the whole investment. Then all aspects of tunnelling technology should be taken into consideration (collisions with territorial development, water and ground conditions, etc.) Also equally important is the selection of the place due to the conditions of the installation and possible recycling of materials.

Figure 5 shows typical jacking pits used in trenchless technology.

3.3 *Equipment used in trenchless technologies*

As a result of fast development of trenchless technologies, the systems and indispensable equipment used for micro-tunnelling work have been substantially developed.

Thanks to micro-tunnelling equipment used at present it is possible to significantly reduce costs associated with construction of pipelines. Moreover, modern micro-tunnelling equipment is much more environment-friendly. This is a result of, among other things, limited usage of combustion engines replaced by hydraulic units.

Thanks to efficient hydraulically operated disks and crushers it is possible to make micro-tunnels practically in each type of soil, including also hard rocks. Cutting wheels mounted on the disk are usually replaceable which ensures quick and efficient adaptation to water and ground conditions.

In modern solutions control sets and peripheral equipment are installed outside jacking pit in mobile containers, compatible with international transport regulations.

Standard hydraulic set is driven usually by an electric motor, powered from generator or external network.

The core of modern micro-tunnelling machines is the control system, by means of which there is a possibility of supervising all parameters of drilling being carried out. All pressures, loads, flows, displacements, rotational speeds are shown in graphical form as well as in numerical values. This ensures appropriate operating accuracy at various lengths of constructed sections including also sections built on elbows. Thanks to monitoring of all parameters on control panel, operator is able to adapt their values to actual soil conditions.

Figure 6. Modern disks of micro-tunnelling heads: a) MTS disk, machining elements of the head visible in detailed drawing (photo by author), b) Herrenknecht AVN disk (Rak & Gajewski2009).

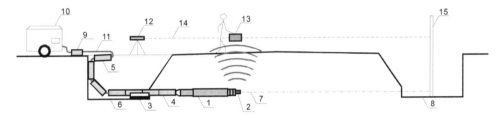

Figure 7. Schema of impact-moling: 1 – pneumatic mole, 2 – steering head, 3 – launching platform, 4 – installed pipe, 5 – elements of pipeline to be connected, 6 – launching excavation, 7 – centre line of pipeline, 8 – reception excavation, 9 – lubricator, 10 – compressor, 11 – air line and torquer, 12 – refractor, 13 – hand-held receiver, 14 – direction of pipeline, 15 – stake (Surmacz & Popielski 2007).

Figure 6 shows moderns disks of micro-tunnelling heads.

3.4 Trenchless methods of constructing engineering structures

Trenchless methods undergo continuous improvements along with the development of technologies associated with other fields of civil engineering. Intensive development of transport infrastructure and progressive urbanization requires substantially greater use of trenchless technologies. Below three most commonly used trenchless technologies are presented, applied in constructing or redeveloping road or railway infrastructure.

3.4.1 Impact molings

Impact molings belong to one of the oldest and, at present, the cheapest trenchless methods. The main device used in this technology is a longitudinal pneumatic mole, commonly known as a "mole". This method is used for installing conduits with small diameters.

Figure 7 shows the idea of impact moling.

3.4.2 Pipe jackings

This technology consists in driving pipes into the soil by means of hydraulic cylinders installed in initial excavation (jacking pit). Arrangement and structure of jacking pits and receiver pits result from the designed route of the conduit, parameters of equipment used, as well as soil and water conditions.

Figure 8 shows the technology of constructing pipelines by pipe jacking method.

3.4.3 Microtunnelling

Microtunnelling technology was developed in the 1970s as modernization of pipe jacking method. Similarly as in pipe jacking method, here similar elements of the system are used. Differences

307

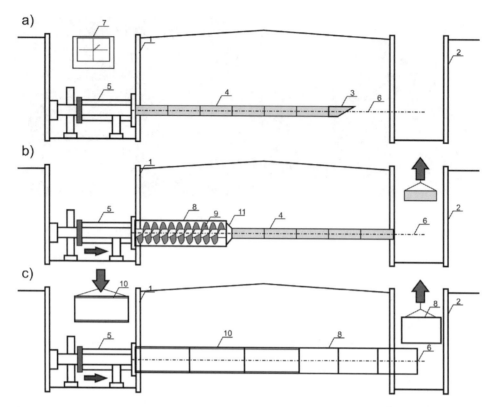

Figure 8.   Schema of three stage pipe jacking: a) pilot bore, b) expanding of borehole and installing sleeve pipes, c) pipe jacking of product pipe; 1 – launching excavation, 2 – reception excavation, 3 – drill head, 4 – drilling rods, 5 – main jacking station, 6 – centre line of pipeline, 7 – view from target disc, 8 – steel sleeve pipe, 9 – screw conveyor, 10 – product pipe, 11 – expander (Surmacz & Popielski 2007).

consist in considerable increase in automation of pipe driving to the soil. In this technology conduits are constructed by means of mechanical head equipped with machining disk appropriately selected for soil conditions. Undeniable advantages of this technology include, among others:

– high precision,
– short completion of tasks,
– possibility of constructing conduits in difficult soil and water conditions.

The most significant difference between technologies based on pipe jacking and microtunnelling is the method of mining and transporting the soil. The soil is bored by rotating disc which loosens the soil and initially grinds down the rocks.

Then the material mined by the cutting disc is grinded in the crusher and mixed with water or betonite flushing. The flushing system operates in closed circuit. In the separator the drillings are separated from the drilling fluid that is reused. Flushing system can be supported by high-pressure water nozzles which substantially improve the productivity of the entire mining transport system. Figure 9 depicts the technology of constructing pipelines by microtunnelling technology.

3.5   *Examples of using trenchless technologies in roads and railways in Poland*

For illustration the author has compiled examples of using trenchless technologies in constructing and modernizing national road infrastructure and railway lines.

3.5.1   *A culvert under the A4 motorway in Kraków – Tarnów section completed in 2007*
The discussed example comprised the construction of a culvert made of corrugated sheets in an embankment of A4 motorway in Kraków – Tarnów section. The culvert structure was made by

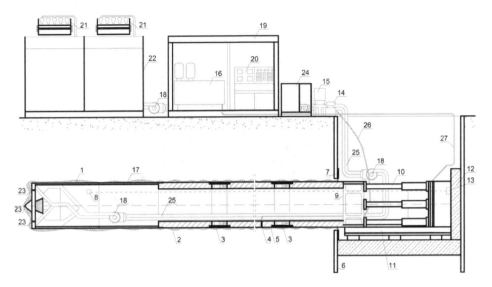

Figure 9.   Schema of microtunnelling: 1 – articulated shield, 2 – product pipe, 3 – intermediate jacking station, 4 – lubricant jet, 5 – space between product pipe and firm ground filled with lubricant, 6 – tunnel eye, 7 – launching shaft, 8 – target camera, 9 – thrust ring, 10 – main jacks, 11 – main jacking station, 12 – reaction block, 13 – laser thedolite, 14 – lubricant pump, 15 – lubricant mixer, 16 – aggregate and hydraulic pump for main jacking station, 17 – machine can, 18 – bentonite pump, 19 – control container, 20 – control computer, 21 – bentonite recycling plant, 22 – bentonite tanks, 23 – jet, 24 – jet's pump, 25 – bentonite pipelines, 26 – lubricant pipelines, 27 – hydraulic lines (Surmacz & Popielski 2007).

Figure 10.   Overall view of construction site of the culvert under motorway A4 in Kraków-Tarnów section (Gwioździk & Sosna 2008).

pipe jacking method. Protective pipes were made of steel pipes, diameter of 2,820 mm, connected by welding in jacking pit made by using steel piling sections. Next, into the interior of such a steel protective pipe – an inner, final, structural pipe was inserted, made of steel corrugated sheets connected by high-strength screws. Total length of completed jacking was 65.0 m. Figure 10 shows general view of the culvert construction site.

3.5.2   *A culvert under railway line in Ostrów Wielkopolski*
The culvert with outer diameter DA 2047 mm was made by microtunnelling technology using CC-GRP pipes. The microtunnelling was carried out under two parallel railway embankments of the line no. 272 on total length of 127 m. The structure was built in 2004.

Figure 11. Overall view of jacking pit during constructing a culvert by microtunnelling technology under railway line in Ostrów Wielkopolski (Hobas System Polska).

Figure 11 shows overall view of construction site during microtunnelling work.

4 ECONOMIC EFFECTIVENESS OF TRENCHLESS TECHNOLOGIES IN CONVERTING AND CONSTRUCTING MOTOR ROADS AND RAILWAY LINES

4.1 *Introduction*

Economic effectiveness reflects the relation between the amount of incurred outlays, materials used and value of effects obtained as a result of these outlays. Trenchless technologies used in redevelopments of motor roads and railway lines at increased production expenditures give imponderably greater effects than traditional open excavations.

With increasing density of underground infrastructure, the costs of constructing new structures and costs associated with damages to the existing ones become increasingly higher. Closing main traffic routes caused by work in open excavations also generates high losses, these are mainly social and environemental costs. Therefore new solutions are still searched for, which from technical and economic point of view can produce best results and simultaneously generate lowest costs.

The analysis presented below is aimed at presenting trenchless technologies as economically justified alternative for works in open excavations.

4.2 *Ways to carry out the analyses*

One of the main ways to carry out economic and technical analyses of investment undertakings is the Social Cost-Benefit Analysis (SCBA).

This method is widely used for evaluating investment projects. It treats equally all obtained profits, independently of in which social and economic groups they grow. Thus, the projects and programmes may be subject to the analysis of social costs and benefits so as to determine their influence on various social and economic groups. All essential costs and benefits must be identified and calculated (weighed). Additionally, the calculation system must take into consideration the project influence on various groups and the analysis must achieve appropriate discount rate.

This method uses mainly multi-criterial investment assessment based on a weighing system referred to selected assessment criteria. The analysis takes into consideration qualitative, quantitative, pecuniary and non-pecuniary effects.

Multi-criterial models enable indepth information on partial assessments to be collected and integrated in a manner that does not limit and deform their contents.

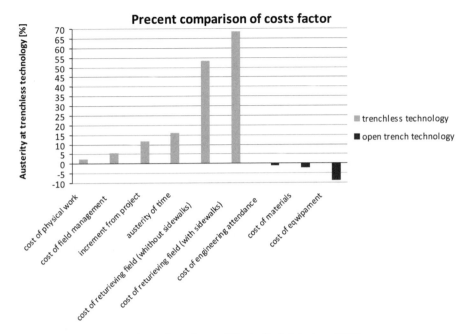

Figure 12.   Percentage comparison of cost factors (Woodroffe & Ariartnam 2007).

4.2.1 *Diagram of analysis procedure*
Multi-criterial assessment can be carried out in two systems:

– zero-one system – consisting in accepting or negating an undertaking in relation to adopted
  criterion
– weighing system – consisting in assigning certain weight to individual criteria; a weight reflects
  enterprise priorities; the sum of weights for all criteria should total 1.

The zero-one system is usually insufficient and if several undertakings get successfully through
the preliminary stages, they are subject to assessment with the application of weighing system.

4.2.2 *Parameters considered in the analysis*
In order to carry out economic analysis of a project, the costs associated with it are taken into
consideration. The costs to be considered are classified as follows:

- Direct structural costs
  – material costs,
  – labour costs,
  – reconstruction costs.
- Indirect structural costs
  – expenditures on materials, machines and work not associated directly with construction
    process.
- Social costs
  – vehicle operational costs – associated with traffic reorganization,
  – costs of delays in public and private transport,
  – costs incurred by people living within the project area.
- Environment – ecology related costs.

Direct costs of constructing a culvert in a road embankment include costs of materials, i.e. pipes
used for making the culvert, costs of labour, including also costs of equipment used and costs
associated with reconstruction of work site.
Figure 12 shows costs and savings that can be obtained by using trenchless technologies.

311

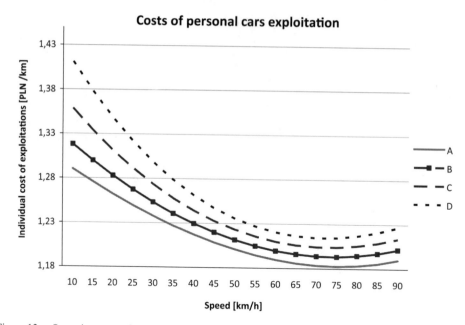

Figure 13.   Operating costs of passenger cars (*Instructions for evaluating economic* ... 2008).

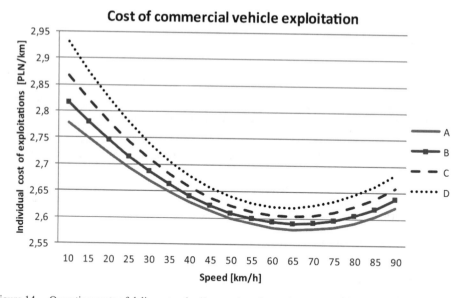

Figure 14.   Operating costs of delivery trucks (*Instructions for evaluating economic* ... 2008).

Indirect costs – include expenditures on materials, machines and labour not associated directly with construction process. The following costs are considered in the analysis:

1. costs resulting from destruction of private property,
2. costs resulting from destruction of adjacent infrastructure,
3. costs resulting from destruction and use of road surface,
4. costs caused by increased expenditures on road maintenance,
5. costs caused by the necessity of relocating existing infrastructure colliding with construction work.

Cost of bus exploitation

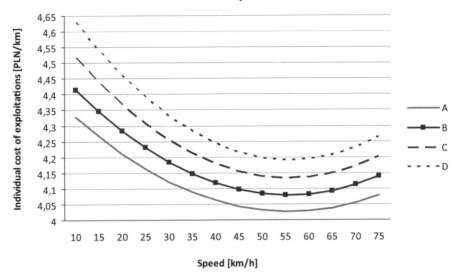

Figure 15.   Operating costs of buses (*Instructions for evaluating economic* . . . 2008).

Cost of losses of time in personal and commertial transport

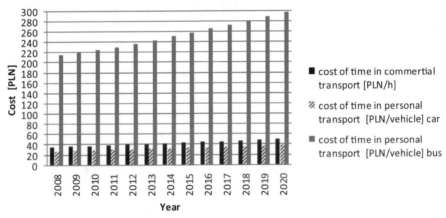

Figure 16.   Costs of time loss in passenger and goods transport (*Instructions for evaluating economic* . . . 2008).

Social costs – include all losses incurred by road users and local community residing in the neighborhood of investment project under construction. Social costs include the following: Vehicle operating costs – costs resulting from increased fuel consumption and deprecitation of a car stuck in a traffic jam caused by road section narrowing or closure, on which work is being carried out.

Figures 13–15 show operating costs of passenger cars, delivery trucks and buses depending on travel speed and class of road technical condition (where: A – good condition, B – satisfactory condition, C – disappointing condition, D – bad condition). Stoppages and traffic jams generate much higher operating costs than driving with constant speed.

Costs of delays in public and private transport.

Operating costs are closely connected with transport costs, and more precisely costs resulting from delays in transport both in passenger and goods transport.

Daily costs of stoppages caused by traffic hold-ups in selected towns/places [thous. PLN]

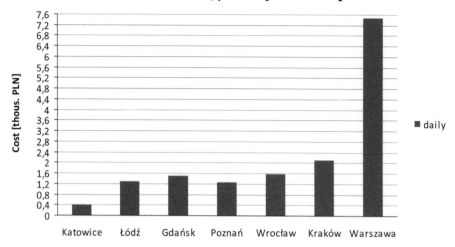

Figure 17. Daily costs of stoppages caused by traffic hold-ups in selected towns/places (*Report on traffic jams in 7 largest cities of Poland* ... 2010).

Downtime cost caused roadtraffic jams for econom [thous. PLN]

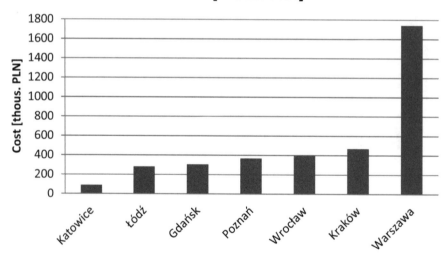

Figure 18. Yearly costs for the State budget due to stoppages caused by traffic hold-ups in selected towns/places (*Report on traffic jams in 7 largest cities of Poland* ... 2010).

An average loss caused by one-hour stoppage of a delivery truck in 2010 was estimated at approx. 40 PLN (10 €), whereas one-hour stoppage of a bus generates cost of 220 PLN (55 €).

Figure 16 shows costs in passenger and goods transport in recent years and admissible increase in upcoming years.

Costs caused by road closures are perfectly depicted on graphs shown on Figures 17–18. The Report on traffic jams in 7 largest Polish cities drawn up in 2010 shows daily losses incurred by household budgets on travel to work and back. These costs would have been much lower if there was no necessity of closing the roads.

Cost of fuel toxic components emmited by personal cars

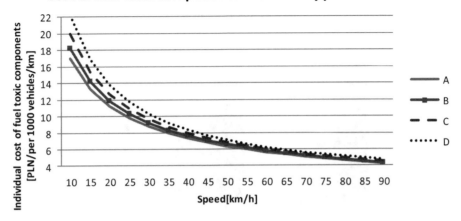

Figure 19. Costs caused by emission of toxic components of fuels from passenger cars (*Instructions for evaluating economic* ... 2008).

Cost of fuel toxic components emmited by commertial vehicles

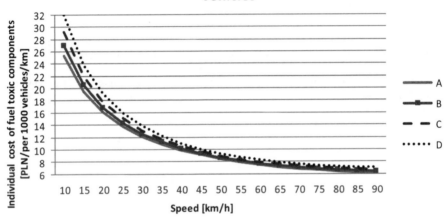

Figure 20. Costs caused by emission of toxic components of fuels from delivery trucks (*Instructions for evaluating economic* ... 2008).

Costs incurred by people residing within the area of investment project.

Environmental (ecology-related) costs – environmental costs include costs connected with natural environment pollution by generated noise, vibrations and air pollution, mainly through waste gas emmisions to the environment. Figures 19 and 20 show environmental emission costs caused by toxic gases from fuels of passenger cars and delivery trucks depending on their speed.

5   EXAMPLE ANALYSES OF ECONOMIC EFFECTIVENESS FOR SELECTED INVESTMENT PROJECTS CARRIED OUT BY TRENCHLESS TECHNOLOGIES WITH INCLUSION OF ASPECTS ASSOCIATED WITH SOCIAL COSTS

5.1   *Examples take to the considerations*

In order to illustrate the necessity of considering social costs in making decisions on selecting a technology for constructing an engineering object (culvert, passage for animals etc.) appropriate

Comparison of trenchless technology investment with trench technology

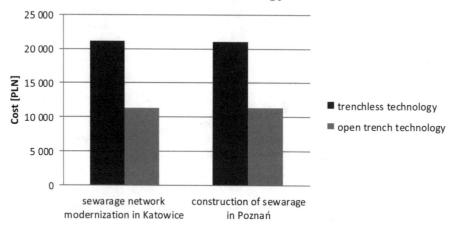

Figure 21.   Cost comparison between trenchless project and open excavation.

examples are presented below:

1. Modernization of sewage network in the centre of Katowice
   – completion of 8 sections consisting of jacking pipes made from polyester resins reinforced with glass fibre, in trenchless technology with a total length of 1054.0 m (DN 2000 – 573.0 m; DN 1600 – 299.0 m; DN 1200 – 182.0 m),
   – contractor: consortium of Hydrobudowa 6 S.A., Bilfinger Berger AG, Ludwig Pfeiffer GmbH & Co.KG,
   – customer: Katowice City,
   – contract value: 31 781 183 PLN.
   By calculating the cost of task completion per one running meter of a DN 2000 tunnel we receive the value of approx. 21 130 PLN.
2. Construction of the Right-bank II sewer trunk in Poznań
   – the trunk sewers will be made in microtunnelling technology (3.9 km of new sewage trunk and 1 km of rain collecting pipe) with diameters up to 2400 mm with total length of 4900.0 m,
   – contractor: consortium Pol-Aqua (leader) and Sonntag Baugesellschaft and Pharmgas,
   – customer: Aquanet SA.,
   – contract value: 103 080 447 PLN.
   By calculating the cost of task completion per one running meter of a DN 2000 mm tunnel we receive the value of approx. 21 036 PLN.
3. Construction of pipe culvert by open excavation method
   The task comprised constructing a culvert under the voivodship road by open excavation method (engineering object in Lubuskie voivodship n. Będów).

The cost of construction made by open excavation method per one meter of a culvert with a diametr of DN2200 made from polyester resins reinforced with glass fibre amounts to approx. 11 364 PLN.

5.2   *Comparison of costs for investments made in trenchless technology and in open excavation*

Comparison of investment costs without considering social costs for projects carried out by trenchless method and in open excavation is shown in figure no. 21.

In the analyses of above examples only direct costs are inlcuded, i.e. materials, labour and equipment.

In case of considering direct costs as well as indirect costs, social and environmental costs of completed investments, their value looks definitely different.

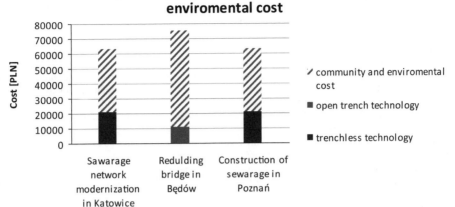

Comparison of trechless technology investment with trench technology taking into account community enviromental cost

Figure 22.   Cost comparison between trenchless project and open excavation taking into consideration social and environmental costs.

Figure 21 shows comparison of a/m investments with inclusion of all above specified costs including also social costs.

Data in figure 22 clearly show that influence of social costs on economic effectiveness of completed investment projects is very high. Moreover it should be emphasized that it even exceeds other costs considered in engineering practice.

Thus, in the author's opinion, further analyses based on other examples are necessary both for investments already completed and under preparation. Such comparisons shall definitely contribute to making proper economic decisions in the discussed scope.

6   CONCLUSIONS

The status of development and implementations of trenchless technologies in Poland is high enough to promote this technology in constructing and redeveloping motor roads and railway lines.

When making economic analyses in selecting technology for engineering objects on the routes of motor roads and railway lines (trench-based or trenchless technology) it is absolutely necessary to consider social costs.

In the paper the author has shown that these costs often dominate in total investment cost, which is not always taken into consideration.

Analyses presented in the paper should be carried out on many other examples, so as to prepare and unify a method of economic assessment for trenchless technologies taking into consideration all components.

Appropriate instructions aimed at achieving economic effectiveness should be worked out and published for investments consisting in constructing engineering structures of roads and railway lines by using trenchless and trench-based technologies.

The methods and instructions worked out should be widely promoted through a cycle of trainings and seminars on this subject.

Development of infrastructure in Poland and the needs in this scope show that subject discussed in this paper may produce measurable social and economic benefits.

REFERENCES

Boterro M., Peila D. 2006. Microtunnelling or trench excavation – comparison of the two alternatives, *Trenchless Engineering*, January–March.
Czerska J. 2002. *Methods for evaluating the effectiveness of investment projects*, Gdańsk.

Gwioździk D. Sosna M. 2008. 65 m in length, 15 m under the road and almost 3 m in diameter, *Trenchless Engineering* no. 2, p. 102–103.

Kuliczkowski A. & All. 2010. *Trenchless technologies in engineering of environment.* Seidel-Przywecki Publishing Sp. z o.o.

Madryas C., Kolonko A., Szot A., Wysocki L. 2006. *Microtunelling.* Lower-silesian Educational Publishing House, Wrocław, Poland.

Madryas C., Wysokowski A., Gaertig M., Skomorowski L. 2011. "*Innovative tunnelling and microtunnelling technologies of record parameters used in the construction of the sewage transfer system connected to the Czajka sewage treatment plant in Warsaw*", 29th International No-Dig Conference and Exhibition, Berlin.

Rak G., Gajewski M. 2009. Construction of sanitary sewage system by mirotunnelling technology for Chorzów city. *Trenchless Technology* no. 3, p. 72–73.

Surmacz A., Popielski P. 2007. Analysis of trenchless technologies of pipelines and tunnels construction In condition of dense urbanized area. *Technical Magazine of Cracow University of Technology* no. 1-Ś/2007.

Woodroffe N.J.A, Ariartnam S.T. 2007. Economic factors determining the advantage of HDD technique over alternative open excavation, *Trenchless technology*, January–March.

Wysokowski A., Madryas C., Skomorowski L. 2010. *Development of the transport infrastructure in Poland with the application of no-dig methods with CC-GRP materials* 28th International No-Dig International Conference and Exhibition, Singapore.

Zwierzchowska A. 2006. Economic aspects of trenchless construction of underground entworks, *Use your head when building* no. 1/2006.

"*European Union Energy & Transport in Figures 2004*". European Commission Directorate-General for Energy and Transport.

"*Report on traffic jams in 7 largest cities of Poland: Warsaw, Lodz, Wroclaw, Cracow, Katowice, Poznan, Gdansk*", January 2010. Deloitte Company & Targeo.pl.

"*Instructions for evaluating economic effectiveness of road and bridge projects for voivodship roads*", February 2008. Road and Bridge Research Institute, Warsaw,

Technical materials of Hobas System Sp. z o.o. Poland.

Technical materials of Haba-Beton Johann Bartlechner Sp. z.o.o. Poland.

Technical materials of Keramo Steinzeug N.V. Poland.

Author index